National Center for Construction Education and Research

Welding Level Three

With Learning Objectives from AWS Advanced and Expert Level Welder Programs

PEARSON

Prentice Hall

Upper Saddle River, New Jersey
Columbus, Ohio

contren®
Learning Series

National Center for Construction Education and Research

President: Don Whyte
Director of Curriculum Revision and Development: Daniele Dixon
Welding Project Manager: Daniele Dixon
Production Manager: Debie Ness
Quality Assurance Coordinator: Jessica Martin
Editor: Cristina Escobar
Desktop Publishers: Rachel Ivines, Laura Parker

The NCCER would like to acknowledge the contract service provider for this curriculum:
Topaz Publications, Liverpool, New York.

This information is general in nature and intended for training purposes only. Actual performance of activities described in this manual requires compliance with all applicable operating, service, maintenance, and safety procedures under the direction of qualified personnel. References in this manual to patented or proprietary devices do not constitute a recommendation of their use.

10 9 8 7 6
ISBN 0-13-102586-4

Preface

This volume was developed by the National Center for Construction Education and Research (NCCER) in response to the training needs of the construction, maintenance, and pipeline industries. It is one of many in NCCER's *Contren® Learning Series*. The program, covering training for close to 40 construction and maintenance areas, and including skills assessments, safety training, and management education, was developed over a period of years by industry and education specialists.

NCCER also maintains a National Registry that provides transcripts, certificates, and wallet cards to individuals who have successfully completed modules of NCCER's *Contren® Learning Series*, when the training program is delivered by an NCCER Accredited training Sponsor.

The NCCER is a not-for-profit 501(c)(3) education foundation established in 1995 by the world's largest and most progressive construction companies and national construction associations. It was founded to address the severe workforce shortage facing the industry and to develop a standardized training process and curricula. Today, NCCER is supported by hundreds of leading construction and maintenance companies, manufacturers, and national associations, including the following partnering organizations:

PARTNERING ASSOCIATIONS

- American Fire Sprinkler Association
- American Petroleum Institute
- American Society for Training & Development
- American Welding Society
- Associated Builders & Contractors, Inc.
- Association for Career and Technical Education
- Associated General Contractors of America
- Carolinas AGC, Inc.
- Carolinas Electrical Contractors Association
- Citizens Democracy Corps
- Construction Industry Institute
- Construction Users Roundtable

- Design-Build Institute of America
- Merit Contractors Association of Canada
- Metal Building Manufacturers Association
- National Association of Minority Contractors
- National Association of State Supervisors for Trade and Industrial Education
- National Association of Women in Construction
- National Insulation Association
- National Ready Mixed Concrete Association
- National Systems Contractors Association
- National Utility Contractors Association
- National Vocational Technical Honor Society
- North American Crane Bureau
- Painting & Decorating Contractors of America
- Plumbing-Heating-Cooling Contractors National Association
- Portland Cement Association
- SkillsUSA
- Steel Erectors Association of America
- Texas Gulf Coast Chapter ABC
- U.S. Army Corps of Engineers
- University of Florida
- Women Construction Owners & Executives, USA

Some features of NCCER's *Contren® Learning Series* are:

- An industry-proven record of success
- Curricula developed by the industry for the industry
- National standardization providing portability of learned job skills and educational credits
- Credentials for individuals through NCCER's National Registry
- Compliance with Apprenticeship, Training, Employer, and Labor Services (ATELS) requirements for related classroom training (CFR 29:29)
- Well-illustrated, up-to-date, and practical information

FEATURES OF THIS BOOK

Capitalizing on a well-received campaign to redesign our textbooks, NCCER is publishing select textbooks in a two-column format. *Welding Level Three* incorporates the design and layout of our full-color books along with special pedagogical features. The features augment the technical material to maintain the trainees' interest and foster a deeper appreciation of the trade.

Think About It uses "What If?" questions to help trainees apply theory to real-world experiences and put ideas into action.

Hot Tip provides a head start for those entering the trade by highlighting important safety requirements and presenting tricks of the trade from experienced welders.

We're excited to be able to offer you these improvements and hope they lead to a more rewarding learning experience.

As always, your feedback is welcome. Please let us know how we are doing by visiting NCCER at www.nccer.org or e-mail us at info@nccer.org.

UNIQUE ASPECTS OF THE NEW AWS WELDING PROGRAM

In this new edition of the *Contren®* Welding series, you'll note features and icons that relate to the American Welding Society's (AWS) *Guides for the Training and Qualification of Welding Personnel: Expert and Advanced Level Welders.* These features correlate with specific guidelines of AWS S.E.N.S.E. (Schools Excelling through National Skills Education) programs. These features can also be used as helpful tools in delivering and evaluating NCCER performance examinations. The table below shows an alignment matrix between the *Contren®* Welding series and the AWS S.E.N.S.E. programs for expert and advanced level welders.

This unique version of the *Contren®* Welding series allows dual credentialing through both NCCER and AWS, provided all applicable requirements are met by the participating organizations. For more information on AWS's S.E.N.S.E. program, contact AWS at (305) 443-9353 or visit www.aws.org. For information on NCCER's Accreditation and National Registry, contact NCCER Customer Service at 1-888-622-3720 or visit www.nccer.org.

Welding Three Modules	Corresponding AWS Learning Objectives
29301-03 Preheating and Postweld Heat Treatment of Metals	None
29302-03 Physical Characteristics and Mechanical Properties of Metals	AWS EG3.0-96, Paragraph 3.2.1.4, Introduction to Welding Metallurgy: Learning Objectives 1–3
29303-03 GMAW – Pipe	AWS EG3.0-96, Paragraph 3.2.1.6, Unit 2, GMAW: Learning Objectives 1, 3, 4, and 11 AWS EG4.0-96, Paragraph 3.2.1.6, Unit 3, GMAW: Learning Objectives 1, 3–5, and 7
29304-03 FCAW – Pipe	AWS EG3.0-96, Paragraph 3.2.1.6, Unit 3, FCAW: Learning Objectives 1, 3, 4, 7, and 9 AWS EG4.0-96, Paragraph 3.2.1.6, Unit 4, FCAW: Learning Objectives 1, 3–5, and 7
29305-03 GTAW – Carbon Steel Pipe	AWS EG3.0-96, Paragraph 3.2.1.6, Unit 4, GTAW: Learning Objectives 1, 3, 4, and 13 AWS EG4.0-96, Paragraph 3.2.1.6, Unit 5, GTAW: Learning Objectives 1, 3–5, and 10
29306-03 GTAW – Low-Alloy and Stainless Steel Pipe	AWS EG3.0-96, Paragraph 3.2.1.6, Unit 4, GTAW: Learning Objectives 3, 4, and 15 AWS EG4.0-96, Paragraph 3.2.1.6, Unit 5, GTAW: Learning Objective 11
29307-03 GTAW – Aluminum Pipe	AWS EG3.0-96, Paragraph 3.2.1.6, Unit 4, GTAW: Learning Objectives 3, 4, and 14 AWS EG4.0-96, Paragraph 3.2.1.6, Unit 5, GTAW: Learning Objective 12
29308-03 GMAW – Aluminum Plate and Pipe	AWS EG3.0-96, Paragraph 3.2.1.6, Unit 2, GMAW: Learning Objectives 6 and 7 AWS EG4.0-96, Paragraph 3.2.1.6, Unit 3, GMAW: Learning Objective 8

Acknowledgments

This curriculum was revised as a result of the farsightedness and leadership of the following sponsors:

Applied Welding Technology
Exxon Mobile
Flint Hills Resources
Fluor Daniel Inc.
Kawerak
Koch Refining
Lee College
Northern Arizona Vocational
  Institute of Technology
  Northland Pioneer College

Spec-Weld Technologies, Inc.
Texas A & M University
  System—Texas Engineering
  Extension Service
William H. Turner Adult
  Education Center
Zachry Construction
  Corporation

This curriculum would not exist were it not for the dedication and unselfish energy of those volunteers who served on the Authoring Team. A sincere thanks is extended to:

Tom Atkinson
Rex Ball
Curtis Casey
Bill D. Cherry
John Gault
Terry A. Lowe

David McGrath
John D. Murray
Joe D. Sanders
Jerry Trainor
John R. Yochum

Contents

*Please note that Modules 29307-03 and 29308-03 are electives for those progressing
through the *Welding Level Three* program.

Preheating and Postweld Heat Treatment of Metals

COURSE MAP

This course map shows all of the modules in the third level of the Welding curriculum. The suggested training order begins at the bottom and proceeds up. Skill levels increase as you advance on the course map. The local Training Program Sponsor may adjust the training order.

WELDING LEVEL THREE

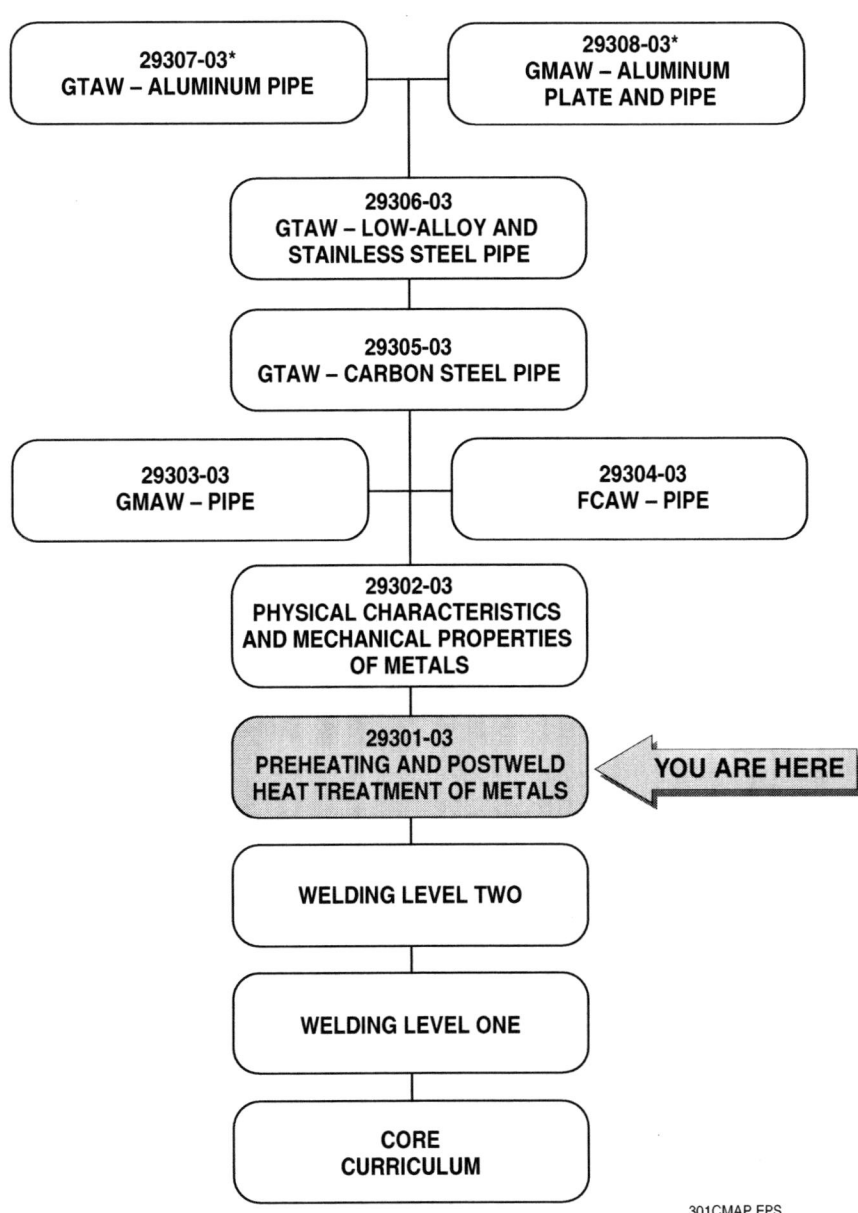

301CMAP.EPS

*Please note that Modules 29307-03 and 29308-03 are electives
for those progressing through the *Welding Level Three* program.

MODULE 29301-03 CONTENTS

Figures

Tables

Preheating and Postweld Heat Treatment of Metals

Objectives

When you have completed this module, you will be able to do the following:

1. Explain how to preheat metals.
2. Describe maintaining interpass temperature.
3. Explain postweld heat treatment of metals.
4. Identify and explain the effects of welding on metals:
   - Heat-affected zone (HAZ)
   - Cracking
   - Face changes/grain structure

Prerequisites

Before you begin this module, it is recommended that you successfully complete the following: Core Curriculum; Welding Levels One and Two.

Required Trainee Materials

1. Pencil and paper
2. Appropriate personal protective equipment

1.0.0 ◆ INTRODUCTION

This module explains preheating, **interpass temperature** control, and postheating procedures that sometimes need to be done to preserve weldment strength, **ductility**, and weld quality. Proper preheating and interpass temperature control can reduce shrinkage, prevent excessive hardening, and prevent **underbead cracking**. Interpass temperature control is also required during certain welding procedures to prevent or reduce localized hardening and shrinkage. If certain metals cool too quickly between weld passes, undesirable changes occur in the base and weld metals, and the heat from successive passes will not reverse these changes. Postheating may be necessary to reduce **residual stresses** and improve corrosion resistance, dimensional stability, fatigue resistance, impact, and low temperature strength.

2.0.0 ◆ PREHEATING AND INTERPASS TEMPERATURE CONTROL

Preheat weld treatment (PHWT) is the controlled heating of the base metal immediately before welding begins. Its main purpose is to keep the weld and the base metal within the **heat-affected zone (HAZ)** along the weld line from cooling too fast. In some cases, the whole structure is preheated. Sometimes, it may not be possible or practical to heat the whole structure. When this is the case, only the section near the weld is preheated.

Interpass temperature control is used to maintain the temperature of the weld zone during welding. It is used when the cooling rate is too high or

Heat Treatments

Different forms of heat treatments are used to alter the properties of metals. Soft metals can be made hard, brittle, or strong using the correct heat treatment, and they can be returned back to their original soft form with heat treatment. Welding specifications often require that welded joints be heat treated before welding or after fabrication. To avoid mistakes, heat treatments made to metals of any kind must be done in a planned and systematic way because the use of temperatures that are too low or too high can result in more harm than any good that may result.

too low to maintain the correct temperature of the weldment between weld passes. The minimum interpass temperature is usually the same as the preheat temperature. If welding is ever interrupted, the temperature of the weldment must be brought back to the interpass temperature before welding is continued. Usually, there is also a maximum interpass temperature, which is especially important in preventing overheating from the heat buildup in smaller weldments.

2.1.0 Temperature and Metal Structure

At room temperature, most metals are solid with a crystalline structure. They become liquids if heated to a high enough temperature. Temperature is a measure of molecular activity, so when the temperature is high enough, the forces that hold the molecules in their crystalline structures break down and the metal becomes a liquid. At even higher temperatures, metals become gases. The temperature at which a metal becomes liquid is its melting point, and this point is different for every metal. In the case of **alloys** (metal mixtures), the various **constituents** melt at different temperatures. Some, like carbon, may not melt at all.

Below the melting point, there is a temperature zone at which crystalline changes take place. Depending on the metal or alloy, these changes can seriously affect the metal's physical structure and mechanical characteristics. Quick cooling (**quenching**) of a metal while it is in this temperature zone can freeze some of the modified crystalline structure. The result is that the mechanical properties and physical traits of the cooled metal will be very different than if the same metal were cooled more slowly. Quenching does not have much of an effect on low-carbon and mild steels because they do not contain much carbon. However, quenching medium- and high-carbon steels produces a hard and brittle crystalline structure, which reduces the metal's toughness. *Table 1* shows the minimum preheat temperatures for various base materials.

When medium-carbon and high-carbon steel are welded, the temperature in the weld zone always reaches the melting point (above the **critical temperature**), and the crystalline structure in the weld zone changes. If the metal is not preheated enough, the large mass of the cool base metal will quench the weld zone and cause localized hardness and brittleness. If the metal is adequately preheated, the weld zone cools much more slowly, allowing the

Table 1 Minimum Preheat Temperatures for Various Base Materials

BASE MATERIALS	WELDING PROCESSES	MATERIAL THICKNESS	MINIMUM PREHEAT TEMPERATURE (°F)
ASTM A36 ASTM A53 GR. B ASTM A106 GR. B API 5L GR. B API 5L GR. X42	SHIELDED METAL ARC WELDING WITH OTHER THAN LOW HYDROGEN ELECTRODES	LESS THAN OR EQUAL TO ¾"	NONE (SEE NOTE)
		OVER ¾" THROUGH 1½"	150
		OVER 1½" THROUGH 2½"	225
		OVER 2½"	300
ASTM A36 ASTM A53 GR. B ASTM A106 GR. B ASTM A572 GR. 42 ASTM A572 GR. 50 ASTM A586 API 5L GR. B API 5L GR. X42	SHIELDED METAL ARC WELDING WITH LOW HYDROGEN ELECTRODES, SUBMERGED ARC WELDING, GAS METAL ARC WELDING, AND FLUX CORED ARC WELDING	LESS THAN OR EQUAL TO ¾"	NONE (SEE NOTE)
		OVER ¾" THROUGH 1½"	50
		OVER 1½" THROUGH 2½"	150
		OVER 2½"	225
ASTM A572 GR. 60 ASTM A572 GR. 65 ASTM 5L GR. X52	SHIELDED METAL ARC WELDING WITH LOW HYDROGEN ELECTRODES, SUBMERGED ARC WELDING, GAS METAL ARC WELDING, AND FLUX CORED ARC WELDING	LESS THAN OR EQUAL TO ¾"	50
		OVER ¾" THROUGH 1½"	150
		OVER 1½" THROUGH 2½"	225
		OVER 2½"	300
ASTM A514 ASTM A517	SHIELDED METAL ARC WELDING WITH LOW HYDROGEN ELECTRODES, SUBMERGED ARC WELDING, GAS METAL ARC WELDING, AND FLUX CORED ARC WELDING	LESS THAN OR EQUAL TO ¾"	50
		OVER ¾" THROUGH 1½"	125
		OVER 1½" THROUGH 2½"	175
		OVER 2½"	275

NOTE: If below 32°F, preheat to 70°F

301T01.EPS

crystalline structure to transform back to the crystalline form of the surrounding base metal.

Preheating may be required to:

- Reduce shrinkage in the weld and adjacent base metal
- Prevent excessive hardening and reduced ductility of weld filler and base metals in the HAZ
- Reduce hydrogen gas in the weld zone
- Ensure compliance with required welding procedures

2.1.1 Reducing Shrinkage

Restrained assemblies are held firmly in position to maintain their shape and can build up residual stresses when they cool from welding. During welding, the heat causes the section that is welded to expand. When it cools, it shrinks, and stresses may occur if the surrounding area does not shrink along with it. With steel weldments, this can cause distortion, warping, and even cracking. With cast-iron weldments, shrinkage usually causes the base metal to crack or the weld to fail. For example, if a broken cast-iron pulley spoke is being repaired, the rim and other spokes must be adequately preheated, or they will be placed under heavy strain by the thermal expansion of the welded spoke. When the weld cools, the residual strains could cause the rim to crack or the welded spoke to crack again. If the rim and spokes near the weld are adequately preheated, they will shrink along with the welded spoke, reducing or eliminating most of the stress on the rim. *Figure 1* shows the areas of strain caused by inadequate preheating of a cast-iron pulley.

2.1.2 Preventing Excessive Hardening and Reduced Ductility

All welding processes use or generate temperatures higher than the melting point of the base metal. If the welded metal is a high-carbon alloy steel or cast iron that is quenched, a hard, brittle metal called martensite will form in the weld zone. This causes

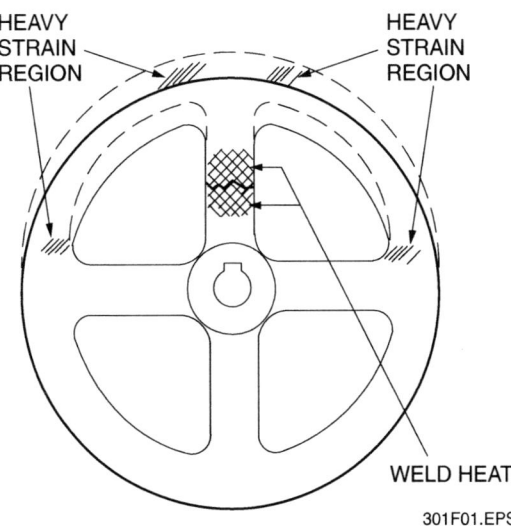

DASHED LINES REPRESENT
THERMAL EXPANSION MOVEMENT

HEAVY STRAIN REGION   HEAVY STRAIN REGION

WELD HEAT

301F01.EPS

Figure 1 ◆ Strain caused by inadequate preheating of a cast-iron pulley.

the entire heat-affected zone (HAZ) parallel to the weld fusion line to become hard and brittle (see *Figure 2*). The large temperature difference also causes differential thermal expansion and creates stresses in the weld region. Heavy plate and castings have a high heat-absorption capacity, which will cause the weld metal and adjacent base metal to quench unless they are adequately preheated.

2.1.3 Reducing Hydrogen Gas in the Weld Zone

The welding arc can split moisture in the weld zone into hydrogen and oxygen. The hydrogen dissolves in the molten base metal, and the oxygen combines with other elements to form slag. As the metal begins to cool, the hydrogen forms gas bubbles in the base metal along the weld boundary. These bubbles create pressure that strains the metal and causes underbead cracking along the weld boundary and at the toe. Preheating helps

Hot Tip

Classification of Carbon Steels

The classification of carbon steels depends mainly on their carbon content, as summarized here.

Common Name	Carbon Content
Low-carbon steel	0.15% maximum
Mild steel	0.15% to 0.30%
Medium-carbon steel	0.30% to 0.50%
High-carbon steel	0.50% to 1.00%

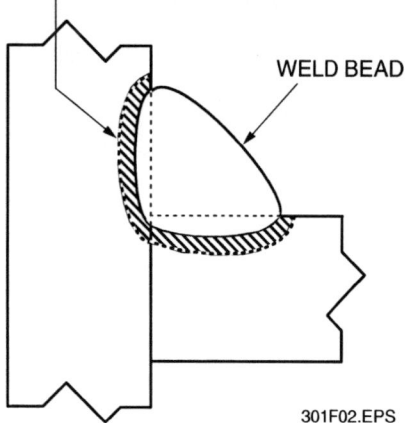

Figure 2 ◆ Heat-affected zone (HAZ).

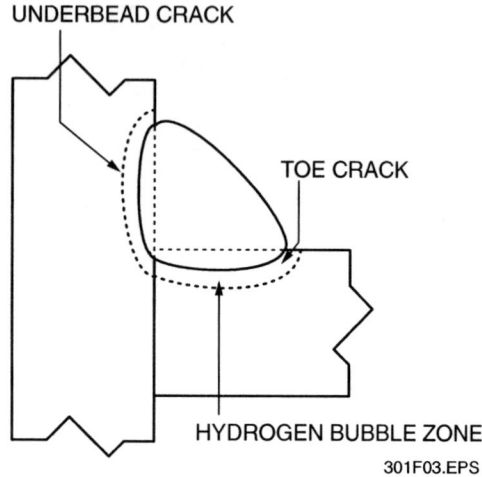

Figure 3 ◆ Underbead cracking caused by hydrogen bubbles.

remove moisture from the weld zone and results in slower cooling, which allows the hydrogen more time to diffuse up through the filler metal and down into the adjacent base metal. The more time the hydrogen has to diffuse, the less pressure it causes and, therefore, the less chance that underbead cracking will occur. *Figure 3* is a simplified diagram showing underbead cracking caused by hydrogen bubbles.

2.1.4 Complying with Required Welding Procedures

If a site has welding procedure specifications (WPS) or quality specifications for particular welding procedures, it is usually mandatory that they be followed exactly. Check with your supervisor if you are not absolutely sure of the specifications that apply to your task.

Underbead Cracking

Always consider preheating and postheating a weld when there is a danger of underbead cracking, also called hydrogen cracking. The chance of underbead cracking is increased when welding carbon and low-alloy steels. Underbead cracking can take place if there is a combined presence of the following:

- Diffusible (atomic) hydrogen in the steel
- A steel with a partly or wholly martensitic microstructure
- Tensile stress at the weldment and heat-affected zone
- A temperature below 300°F

Proper preheat, interpass temperature maintenance, and postweld heat treatment, in conjunction with the proper electrode/shielding gas combination, can eliminate this problem.

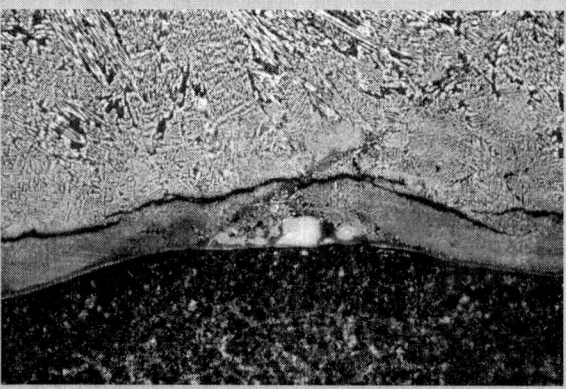
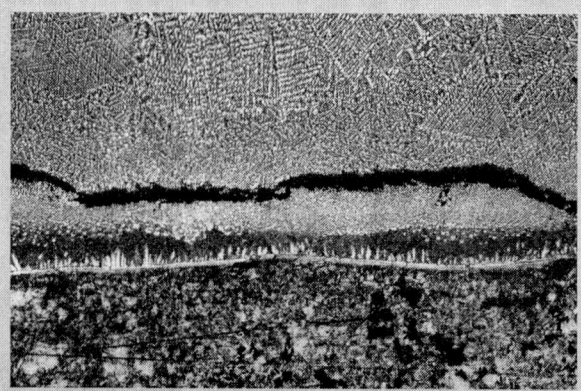

2.2.0 Metals That Require Preheating

It can be very difficult to determine whether metal assemblies and conditions require preheating. Preheating and interpass temperature control depend on the base metal composition, thickness, and degree of restraint. Some base metals only require preheating when they exceed a certain thickness or are too cold. For example, metals exposed to freezing winter temperatures usually require some preheating. Complex-shaped welding assembles and castings usually require preheating to avoid severe stresses or warping. Some alloys, such as high-carbon or alloy steels, require preheating to avoid brittleness and cracking. Not all metals require preheating. Low-carbon steels rarely need preheating. *Table 2* shows the metals that usually require preheating.

2.2.1 Preheating Needs Field Test

A simple test called the preheating needs field test can be used in the field to determine if a calculated preheat temperature is correct or if preheating is required when an alloy is not known.

To perform the test, weld one edge of a thick vertical lug (approximately 2" by 2" by ¾" thick) to the base metal with a slightly convex fillet weld along one side of the lug. Use the same welding current and electrode type and size that will be used on the actual welding job. Allow the weld to cool completely. With a heavy hammer, strike the lug on the side opposite the welded side to break it off the base metal. If the lug bends and then finally breaks through the weld filler metal, the preheating was adequate. If the weld tears out of the base metal or if underbead cracks have occurred, the test is a failure. More preheating is required. *Figure 4* shows the preheat needs field test.

Table 2 Metals That Usually Require Preheating

Metal or Alloy	Conditions or Forms
Aluminum	Large or thick section castings
Copper	All (prevents too-rapid heat loss)
Bronze, copper-based	All (prevents too-rapid heat loss)
Mild carbon steels	Restricted joints, complex shapes, freezing temperatures, and carbon content over 0.30%
Cast irons	All types and all shapes
Cast steels	Higher carbon content or complex shapes
Low-alloy steels	Thicker sections or restricted joints
Low-alloy steels	Heating temperature dependent on carbon or alloy content
Manganese steels	Heating temperature dependent on carbon or alloy content
Martensitic stainless steels	Carbon content over 0.10%
Ferritic stainless steels	Thick plates

2.2.2 Preheating Temperatures

Preheating requirements are affected by the alloy composition and thickness of the base metal, the welding process to be used, and the ambient temperature. The **hardenability** of a steel is directly related to its carbon content and alloying elements. Because different types of steel have different amounts of carbon and alloying elements, they have different preheat requirements. As a general rule, the higher the carbon content, the higher the preheating temperature required. *Table 3* lists preheat requirements for steels.

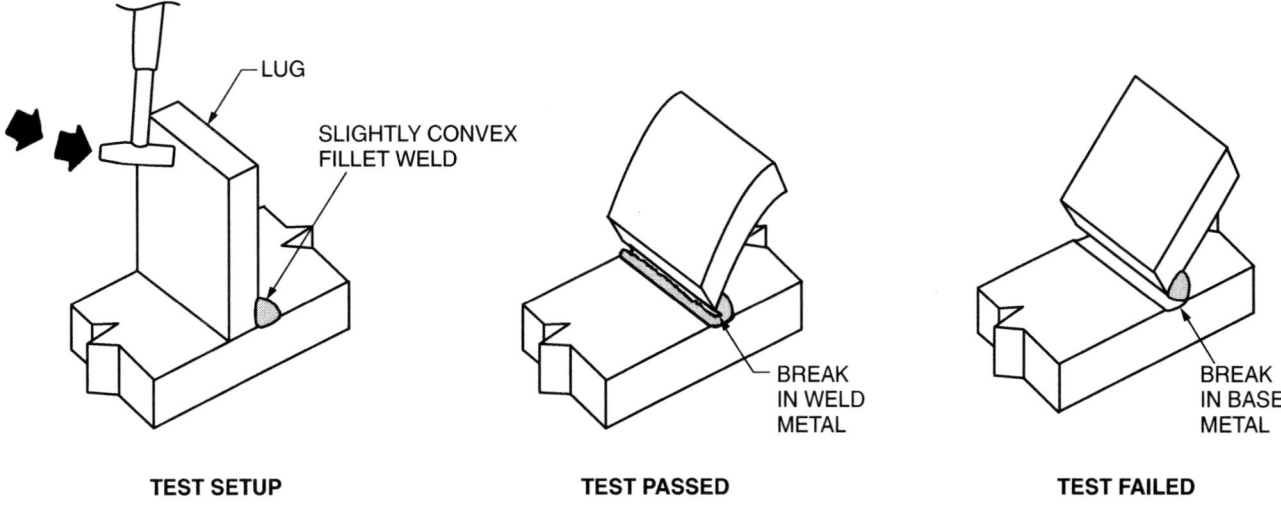

TEST SETUP   TEST PASSED   TEST FAILED

301F04.EPS

Figure 4 ◆ Preheat needs field test.

Table 3 Preheat Requirements for Steels

Steel Type	Shapes or Conditions	Preheat Temperature
Low-carbon steel (0.10%–0.30% C)	Freezing temperatures	Above 70°F
Low-carbon steel (0.10%–0.30% C)	Complex large shapes	100°F–300°F
Medium-carbon steel (0.30%–0.45% C)	Complex large shapes	300°F–500°F
High-carbon steel (0.45%–0.80% C)	All sizes and shapes	500°F–800°F
Low-alloy steel	Lower C or alloy %	100°F–500°F
Low-alloy steel	Higher C or alloy %	500°F–800°F
Manganese steel	Lower C or alloy %	100°F–200°F
Manganese steel	Higher C or alloy %	200°F–500°F
Cast iron	All shapes	200°F–400°F
Martensitic stainless steel	0.10%–0.20% C	500°F
Martensitic stainless steel	0.20%–0.50% C	500°F and postheat
Ferritic stainless steel	¼" and thicker	200°F–400°F and postheat

Temperature-indicating crayons (*Figure 5*) are often used to measure the preheat temperature of the base metal. They are used by marking the workpiece before heating or by stroking the workpiece with the crayon during heating. When the rated temperature has been reached, a distinct melt or smear will become evident. Other methods to measure preheat temperature will be discussed later in this module.

301F05.EPS

Figure 5 ◆ Temperature-indicating crayon.

2.3.0 Preheating Methods

Preheating is most effective when the entire welding assembly is preheated (general heating) in an oven or preheating furnace because the heating is even. However, because of site conditions or the size or shape of the welding assembly, preheating in this manner is not always possible. Therefore, a variety of heating devices can be used for preheating. Localized heating of specific regions can be done with gas torches, blowtorches, electric resistance heaters, radio frequency induction heaters, or partial ovens that are built only around the region to be heated. Parts of the oven or enclosure can be removed so that the welder can have access and still maintain the preheating temperature.

Most welding shops contain one or more types of preheating devices, which may include:

- Oxyfuel torches
- Portable preheating torches
- Open-top preheaters
- Electric resistance heaters
- Induction heaters
- Enclosed furnaces

2.3.1 Oxyfuel Torches

If properly monitored, oxyfuel torches equipped with specialized heating tips can be used for localized preheating.

Hot Tip

Preheating Steel in Low-Temperature or High-Humidity Conditions

Steel should not be welded if the steel temperature is below 32°F. If below 32°F, steel should be heated to at least 70°F prior to welding. Under humid conditions, steel should be heated to a higher preheat temperature in order to dry any moisture from its surface.

Preheating

When preheating metal, make sure to apply heat to the region of the intended weld long enough to sufficiently heat the area. Insufficient heating will allow a relatively large weld to cool too fast. To ensure proper preheating, the temperature of the preheated region should be measured again after waiting about 15 to 20 minutes.

The specialized heating tips are designed to produce blowtorch flame patterns. Heating tips, like the rosebud-style multiflame heating heads, are designed to mount directly to torch handles like welding tips. Heating tips designed to replace the cutting tips on straight cutting torches or combination cutting torch cutting attachments are also available. *Figure 6* shows different types of heating tips.

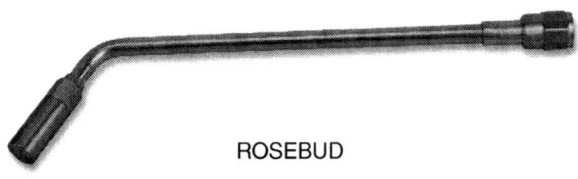

ROSEBUD

HEATING TIP FOR STRAIGHT AND COMBINATION CUTTING TORCHES

301F06.EPS

Figure 6 ◆ Types of heating tips.

2.3.2 Portable Preheating Torches

Portable preheating torches are designed to burn either oil or gas combined with compressed air. Some use just gas and have an air blower to increase the intensity of the flame. *Figure 7* shows a gas and compressed air torch suitable for preheating large weldments.

2.3.3 Open-Top Preheaters

Open-top preheaters come in several basic designs. The simplest is a large, horizontal gas burner with an adjacent or overhead support structure. The object to be heated is mounted on the support structure directly over the gas burners. A variation is the open flat-top preheater, which consists of a horizontal grate or series of bars with gas burners located underneath (see *Figure 8*). The item to be preheated is placed on the grating. This type of preheater works similarly to a kitchen gas cooking range.

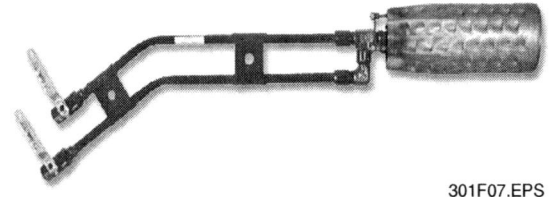

301F07.EPS

Figure 7 ◆ Gas preheating torch.

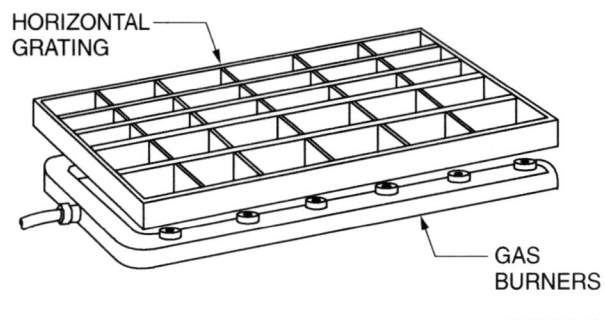

HORIZONTAL GRATING

GAS BURNERS

301F08.EPS

Figure 8 ◆ Open flat-top preheater.

2.3.4 Electric Resistance Heaters

Electric resistance heaters consist of tubular resistance elements that are formed in several shapes or styles to fit various applications (see *Figure 9*).

Oxyfuel Welding of Low- and Mild-Carbon Steels

Oxyfuel welding can be used to weld low-carbon and mild steels, but the process is much slower than with arc welding. The slow heating characteristics of the process result in rather extensive heating of the steel. As a result, the mechanical properties developed in the base metal by earlier heat treatment or cold working may deteriorate.

Air Tube Burner

This is an example of a gas fuel compressed-air tube burner that can be used for the heat treatment of large weldments. The tube section has evenly spaced multiple heating tips along its length. A burner of this type provides more even heat distribution of the weldment over the area encompassed by its length.

301SA02.EPS

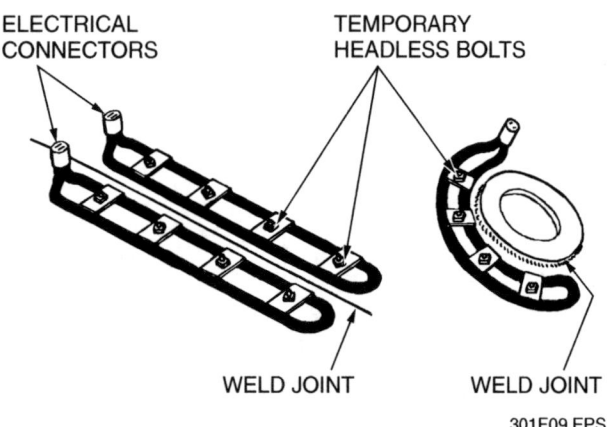

Figure 9 ◆ Resistance heating elements.

They are temporarily fastened against the surface of the metal to be heated. Electricity passing through the elements heats the weldment.

2.3.5 Induction Heaters

In **induction heating**, an alternating magnetic field is generated within the metal body to be heated. A magnetic coil is placed around the metal object, and a strong alternating current is passed through the coil. The resulting alternating magnetic field generated within the base metal creates circulating electrical currents known as eddy currents. The interaction of these currents with the electrical resistance of the base metal produces an electrical heating effect, raising the temperature of the base metal. *Figure 10* shows an air-cooled

301F10.EPS

Figure 10 ◆ Induction heating system.

induction heating system consisting of a power supply, induction blanket, and associated cables. Air-cooled systems are typically used for preheat applications up to 400°F. This type of system may use a controller to monitor and automatically control the temperature. If not equipped with a controller, the use of a temperature indicator is required.

Induction Heaters

A water-cooled induction heater is shown here. Because water cools more efficiently than air, this type of induction heating system can be used for applications requiring higher temperatures such as high-temperature preheating and stress relieving. This type of system is normally equipped with a temperature controller and temperature recorder because they are important components related to stress-relieving applications.

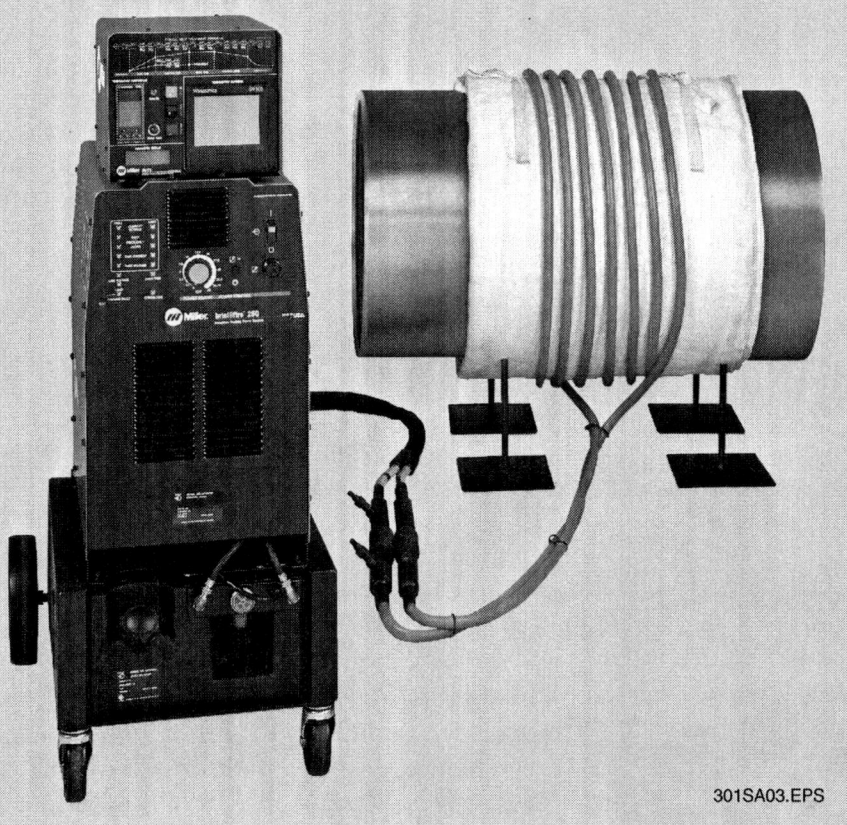

301SA03.EPS

2.3.6 Enclosed Furnaces

Enclosed furnaces (ovens) include both permanent structures and temporary structures or partial enclosures that must be made of firebrick and clay, sheet metal, or some other noncombustible material. The item to be heated may be totally or partially enclosed. With very large items, it may not be practical or possible to enclose the entire item. Furnaces and ovens are heated by gas or oil burners, torches, or electric resistance heaters. *Figure 11* shows an example of a gas-fired box furnace.

3.0.0 ◆ MEASURING TEMPERATURES

Controlling preheating, interpass, and postheating temperatures depends on accurately determining the temperature of the weld zone or assembly. Both underheating and overheating can negatively affect a metal's physical and mechanical characteristics. There are a number of ways to determine the surface temperature of an item

being heated. Some involve complex manufactured products, while others use less sophisticated materials.

Devices and materials that can be used to determine surface temperatures include the following:

- Pyrometers
- Thermocouple devices
- Temperature-sensitive indicators

3.1.0 Pyrometers

A pyrometer is a handheld device that measures temperature from the frequencies of the emitted electromagnetic waves (infrared or visible light). Optical pyrometers are non-contact instruments commonly used to measure temperature. They work on the principle of using the human eye to match the brightness of the hot object to the brightness of a calibrated lamp filament inside the instrument. Optical pyrometers like the one shown in *Figure 12* are easy to operate.

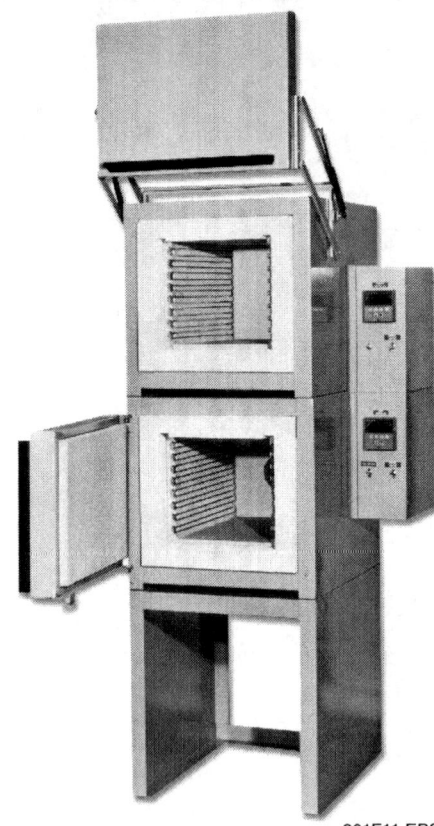

Figure 11 ◆ Furnace.

Figure 12 ◆ Optical pyrometer.

The instrument's optics serve as a telescope to provide a clean, enlarged picture of the area to be measured (target). The operator adjusts a control (rheostat) on the instrument while aiming at and viewing the target through the instrument telescope. By this process, the light radiated from the target is compared to the known brightness of the calibrated lamp inside the instrument. The operator adjusts the control until a color blend is achieved between the radiated target signal and the internal calibrated lamp. The corresponding temperature is then indicated on a direct reading temperature scale. Newer gun-style pyrometer models are available that require the operator only to aim the gun at the target and read the temperature. No adjustments are necessary.

3.2.0 Thermocouple Devices

A thermocouple device is a bimetallic device (usually twisted wires of two dissimilar metals) placed in physical contact with the material whose temperature is to be measured. Thermocouples work on the principle that when two dissimilar metals are joined, a predictable voltage will be generated that relates to the difference in temperature between the measuring junction and the reference junction. Leads from the thermocouple are connected to a device that can measure very small electric voltages and currents. The thermocouple generates a weak electric voltage and current that is in direct proportion to the thermocouple's temperature. The measured voltage or current is fed to a calibrated meter and read directly in degrees of temperature.

3.3.0 Temperature-Sensitive Indicators

Temperature-sensitive indicators are devices that change form or color at specific temperatures or temperature ranges. These include both commercially made materials and devices such as crayon sticks (*Figure 13A*), liquids (*13B*), chalks, powders, pellets, and labels (*13C*) designed to change color or form at precise temperatures, often within a 1% tolerance. An example of a commonly used indicator is the temperature stick, mentioned earlier in this module. Temperature sticks are a series of graduated temperature-indicating crayons that cover the temperature range from 100°F to 1,150°F. These sticks are used to make a mark on the base metal. At a specific temperature, the mark melts or changes color. *Table 4* shows a preheating chart and the temperature-indicating crayons to use for each type of metal.

Table 4 Preheating Chart and Temperature-Indicating Crayons (1 of 2)

METAL GROUP	METAL DESIGNATION	APPROXIMATE COMPOSITION - PERCENTAGE								RECOMMENDED PREHEAT TEMPERATURE	TEMPERATURE - INDICATING CRAYONS (°F)
		C	MN	BI	CR	NI	MO	CU	OTHER		
PLAIN CARBON STEELS	PLAIN CARBON STEEL	BELOW 0.20								UP TO 150°F	150
	PLAIN CARBON STEEL	0.20–0.30								150°F–300°F	150-200-250-300
	PLAIN CARBON STEEL	0.30–0.45								250°F–450°F	250-300-350-400-450
	PLAIN CARBON STEEL	0.45–0.80								450°F–750°F	450-500-600-700-750
PLAIN MOLY STEELS	CARBON MOLY STEEL	0.10–0.20					0.50			150°F–250°F	150-200-250
	CARBON MOLY STEEL	0.20–0.30					0.50			200°F–400°F	200-250-300-350-400
MANGANESE STEELS	SILICON STRUCTURAL STEEL	0.35	0.80	0.25						300°F–500°F	300-400-500
	MEDIUM MANGANESE STEEL	0.20–0.25	1.0-1.75							300°F–500°F	300-400-500
	SAE 1330 STEEL	0.30	1.75							400°F–800°F	400-500-600
	SAE 1340 STEEL	0.40	1.75							500°F–800°F	500-600-700-800
	SAE 1045 STEEL	0.50	1.75							800°F–900°F	600-700-800-900
	12% MANGANESE STEEL	1.25	12.0							USUALLY NOT REQUIRED	
HIGH TENSILE STEELS	MANGANESE MOLY STEEL	0.20	1.65	0.20			0.35			300°F–500°F	300-400-500
	JALTEN STEEL	0.35 MAX	1.50	0.30				0.40		400°F–800°F	400-500-600
	MANTEN STEEL	0.30 MAX	1.35	0.30				0.20		400°F–800°F	400-500-800
	ARMCO HIGH TENSILE STEEL	0.12 MAX				0.50 MIN	0.05 MIN	0.35 MIN		UP TO 200°F	200
	DOUBLE STRENGTH #1 STEEL	0.12 MAX	0.75			0.50–1.25	0.10 MIN	0.50–1.50		300°F–800°F	300-400-500-600
	DOUBLE STRENGTH #1A STEEL	0.30 MAX	0.75			0.50–1.25	0.10 MIN	0.50–1.50		400°F–700°F	400-500-600-700
	MAYARIR STEEL	0.12 MAX	0.75	0.35	0.2-1.0	0.25–0.75		0.80		UP TO 300°F	200-300
	OTISCOLOY STEEL	0.12 MAX	1.25	0.10 MAX	0.10 MAX			0.50 MAX		200°F–400°F	200-300-400
	NAX HIGH TENSILE STEEL	0.15–0.25		0.75	0.60	0.17	0.15 MAX	0.25 MAX	ZR.12	UP TO 300°F	200-300
	CROMANSIL STEEL	0.14 MAX	1.25	0.75	0.50					300°F–400°F	300-400
	A.W. DYN-EL STEEL	0.11–0.14						0.40		UP TO 300°F	200-300
	CORTEN STEEL	0.12 MAX		0.25-1.0	0.5-1.5	0.55 MAX		0.40		200°F–400°F	200-300-400

301T04A.EPS

Table 4 Preheating Chart and Temperature-Indicating Crayons (2 of 2)

METAL GROUP	METAL DESIGNATION	C	MN	BI	CR	NI	MO	CU	OTHER	RECOMMENDED PREHEAT TEMPERATURE	TEMPERATURE-INDICATING CRAYONS (°F)
HIGH TENSILE STEELS (CONT.)	CHROME COPPER NICKEL STEEL	0.12 MAX	0.75		0.75	.75		0.55		200°F-400°F	200-300-400
	CHROME MANGANESE STEEL	0.40	0.90		0.40					400°F-600°F	400-500-600
	YOLOY STEEL	0.05-0.35	0.3-1.0			1.75		1.0		200°F-600°F	200-300-400-500-600
	HI-STEEL	0.12 MAX	0.6	0.3 MAX		0.55		0.9-1.25		200°F-500°F	200-300-400-500
NICKEL STEELS	2¼% NICKEL STEEL	0.25				2.25				200°F-400°F	200-300-400
	3½% NICKEL STEEL	0.23				3.50				200°F-400°F	200-250-300-350-400
MOLY BEARING CHROMIUM AND NICKEL STEELS	SAE 4140 STEEL	0.40			0.90		0.20			600°F-800°F	600-700-800
	SAE 4340 STEEL	0.40			0.80	1.85	0.25			700°F-900°F	700-800-900
	SAE 4615 STEEL	0.15				1.80	0.25			400°F-800°F	400-500-600
	SAE 4820 STEEL	0.20				3.50	0.25			600°F-800°F	600-700-800
LOW CHROME MOLY STEELS	1¼% CR–½% MO STEEL	0.17 MAX			1.25		0.50			250°F-400°F	250-300-350-400
	2¼% CR–1% MO STEEL	0.15 MAX			2.25		1.0			300°F-500°F	300-400-500
MEDIUM CHROME MOLY STEELS	5% CR–½% MO STEEL	0.15 MAX			5.0		0.5			400°F-800°F	400-500-600
	7% CR–½% MO STEEL	0.15 MAX			7.0		0.5			400°F-800°F	400-500-600
	9% CR–1% MO STEEL	0.15 MAX			9.0		1.0			400°F-800°F	400-500-600
PLAIN HIGH CHROMIUM STEELS	11½-13% CR TYPE 410	0.15 MAX			12.0					400°F-800°F	400-500-800
	16-18% CR TYPE 430	0.12 MAX			17.0					300°F-500°F	300-400-500
	23-27% CR TYPE 448	0.20 MAX			25.0					300°F-500°F	300-400-500
CHROME NICKEL STAINLESS STEELS	18% CR 8% NI TYPE 304	0.07			18.0	8.0				USUALLY DO NOT REQUIRE PREHEAT BUT IT MAY BE DESIRABLE TO REMOVE CHILL	200
	25-12 TYPE 309	0.07			25.0	12.0					
	25-20 TYPE 310	0.10			25.0	20.0					
	18-8 CB TYPE 347	0.07			18.0	8.0			CB 10xC		
	18-8 MO TYPE 316	0.07			18.0	8.0	2.5				
	18-8 MO TYPE 317	0.07			18.0	8.0	3.5				
IRONS	CAST IRON									700°F-900°F	700-800-900
	NI RESIST									500°F-1000°F	500-700-900-1000
NON FERROUS	NICKEL, MONEL, INCONEL									PREHEAT NOT USUALLY REQUIRED FOR THIN SECTIONS.	
	ALUMINUM-COPPER									300°F-400°F PREHEAT MAY BE DESIRABLE FOR THICK SECTIONS.	

301T04B.EPS

Furnace Thermocouples

The rate of temperature rise, holding time at temperature, and rate of cooling are critical to proper heat treatment. For this reason, thermocouples used in furnaces must measure the temperature of the metal being heat-treated, not the ambient temperature inside the furnace.

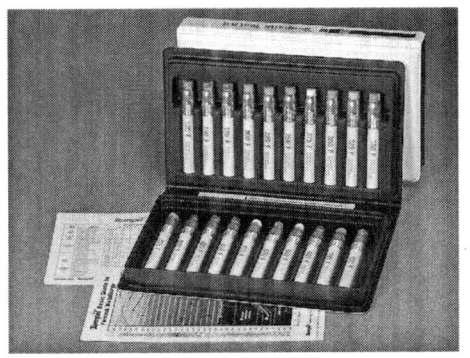

TEMPERATURE STICKS
(A)

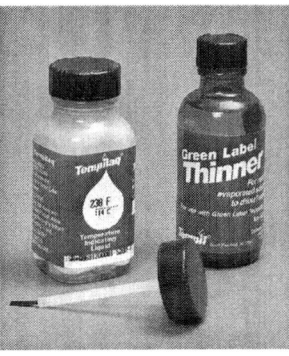

TEMPERATURE LIQUID
(B)

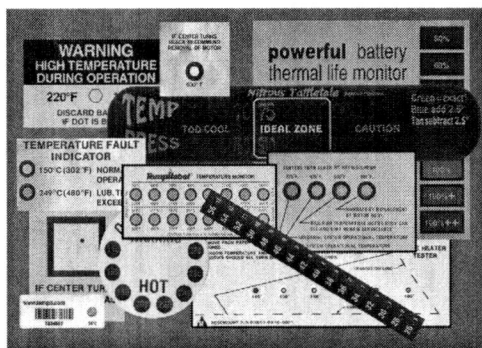

TEMPERATURE LABELS
(C)

301F13.EPS

Figure 13 ♦ Temperature-sensitive indicators.

4.0.0 ♦ INTERPASS TEMPERATURE

When preheating is important, interpass temperature for multiple-pass welds is usually also important. Interpass temperature control is often simply a continuation of preheating. Heating equipment and techniques are the same.

The interpass temperature is usually the same as the preheat temperature. However, some specifications list both a maximum and a minimum interpass temperature. Sometimes, the interpass temperature may be specified as a maximum temperature. This might be the case with small weldments and frequent weld passes, where too much heat could build up.

With large, high-mass weldments and infrequent weld passes, the base metal can cool below the required preheat temperature if additional heat is not supplied. If the base metal cools below the specified minimum preheat temperature, it must be reheated to the preheat temperature before welding continues.

Determining Metal Temperatures by Color

The approximate temperature of a metal heated in a furnace can be determined by the different colors exhibited by the metal, as shown here.

Temperature (°F)	Color
930	Faint red
1,075	Blood red
1,175	Dark cherry
1,275	Medium cherry
1,375	Cherry
1,450	Bright cherry
1,550	Salmon
1,630	Dark orange
1,725	Orange
1,830	Lemon
1,975	Light yellow
2,200	White

5.0.0 ◆ POSTHEATING

Postweld heat treatment (PWHT) is the heating of a weldment or assembly after welding is completed. It can be performed on most types of base metals. Postheat equipment and techniques are the same as those used for preheating.

Postheat treatments include the following:

- Stress relieving
- Annealing
- Normalizing
- **Tempering**

5.1.0 Stress Relieving

Stress relieving is the most commonly used postheat treatment. It is done below the critical (recrystallization) temperature, usually in the 1,050°F to 1,200°F range.

Stress relief treatment is done for the following reasons:

- To reduce residual (shrinkage) stresses in weldments and castings. This is especially important with highly restrained joints.
- To improve resistance to corrosion and caustic embrittlement.
- To improve dimensional stability during machining operations.
- To improve resistance to impact loading and low-temperature failure.

Some codes cover stress relieving. These codes specify the heating rate, holding time, and cooling rate. The heating rate is usually 300°F to 350°F per hour. The holding time is usually one hour for each inch of thickness. The cooling rate is also usually 300°F to 350°F per hour. Stress relieving requires the use of temperature indicators and temperature-control equipment.

5.2.0 Annealing

Annealing relieves stresses much more than a stress relieving postheat treatment can, but it results in a steel of lower strength and higher ductility. Annealing is done at temperatures approximately 100°F above the critical temperature, with a prolonged holding period. The heating is followed by slow cooling in the furnace or by covering, wrapping, or burying the item, although austenitic stainless steel requires rapid cooling. Annealing is used to relieve the residual stresses associated with welds in carbon-molybdenum pipe and also to relieve stresses in welded castings that contain casting strains. Generally, annealing is considered to leave the metal in its softest condition, with good ductility.

5.3.0 Normalizing

Normalizing is the process in which heat is applied to remove strains and reduce grain size. Normalizing is done at temperatures and holding periods comparable to those used in annealing, but the cooling is usually done in still air outside the furnace, and the cooling rate is slightly faster. Normalizing usually results in higher strength and less ductility than annealing. It can be used on mild steel weldments to form a uniform austenite solid solution, to soften the steel, and to make it more ductile. Normalized metal is not as soft and free of stresses as fully annealed metal.

5.4.0 Tempering

Tempering, also called drawing, increases the toughness of quenched steel and helps avoid breakage and failure of heat-treated steel. It reduces both the hardness and the brittleness of hardened steel. Tempering is done at temperatures

---

Importance of Maximum Interpass Temperatures

Austenitic stainless steel typically has a 350°F maximum interpass temperature because it is subject to intergranular migration of alloying elements (sensitization) between 800°F–1600°F. Migration decreases the corrosion resistance of stainless steel.

Stress Relief

As discussed in this module, the most commonly used method for stress relieving weldments is by postweld heat treatment. However, stress relief can also be accomplished mechanically. One method involves attaching a mechanical vibrating device to the weldment during or immediately after welding. When activated, the device vibrates the weldment at a specific resonant frequency. This acts to even the stress distribution within the weldment by means of plastic deformation of the metal's grains. The weight of the weldment generally determines the length of time the weldment is subjected to the vibrations.

Hot Tip

Tempering Temperatures

If carbon steel is heated in an oxidizing atmosphere, a film of oxide forms on a clean surface that changes color as the temperature increases. The colors can be seen with the unaided human eye under low light conditions with each color or shade being related to a specific temperature. The colors listed here for plain carbon steel have been widely used in the trade in the past as a means of determining the correct amount of temper. These colors are affected to some extent by the composition of the metal being heated; therefore, they should be used as a guide only.

Temperature (°F)	Color
430	Very pale yellow
440	Light yellow
450	Pale straw yellow
460	Straw yellow
470	Deep straw yellow
480	Dark yellow
490	Yellow brown
500	Brown yellow
510	Spotted red brown
520	Brown purple
530	Light purple
540	Full purple
550	Dark purple
560	Full blue
570	Dark blue
640	Light blue

below the critical temperature, much lower than those used for annealing, normalizing, or stress relieving. Tempering is commonly done to tool steel.

5.5.0 Time-at-Temperature Considerations

In most heat treatment procedures, temperature and time-at-temperature are both specified. Time is very important because metallurgical changes are sluggish at temperatures below the critical temperatures. The maximum temperature is determined by the metal alloy. The holding time at the maximum temperature is based on the metal's thickness, usually one hour for each inch of thickness. The cooling rate is determined by the specific treatment or any applicable code. Temperature control must also be used for postheating. Postheating temperatures are determined in the same manner as preheating temperatures.

Summary

Preheating, interpass temperature control, and postheating are very important for relieving residual stresses and avoiding weld failures with certain alloy steel weldments and castings. Heating requirements vary with alloy composition, metal thickness, ambient temperature, and weldment configuration or shape. Proper preheating will preserve the weldment's strength, ductility, and weld quality. Postheating relieves residual stresses, improves corrosion resistance, restores ductility, and reduces brittleness. These processes result in a stronger, tougher metal. Temperature control during preheating and postheating is very important because a temperature that is too high can weaken the base metal, and a temperature that is too low will not meet the heating objectives.

Review Questions

1. The temperature at which a metal turns liquid is its _____.
   a. welding temperature
   b. melting point
   c. preheating point
   d. postheating temperature

2. Hydrogen gas in the weld zone will cause _____.
   a. slow melting of the base metal
   b. underbead cracking
   c. slag
   d. hardening

3. The hardenability of steel is directly related to its _____.
   a. carbon content
   b. preheat temperature
   c. thickness
   d. welding temperature

4. Preheating is most effective when the welding assembly is heated _____.
   a. with a torch
   b. with induction heaters
   c. in an oven
   d. rapidly

5. Open-top preheaters work similarly to a(n) _____.
   a. oxyfuel torch
   b. kitchen gas range
   c. oven
   d. charcoal grill

6. Air-cooled induction heating systems are typically used for preheat applications up to _____.
   a. 300°F
   b. 400°F
   c. 600°F
   d. 1,000°F

7. A thermocouple device is a(n) _____ device.
   a. bimetallic
   b. plastic
   c. alloy
   d. noncontact

8. Temperature-indicating crayons can be used to indicate temperatures ranging from 100°F to _____.
   a. 400°F
   b. 980°F
   c. 1,150°F
   d. 1,350°F

9. The interpass temperature is usually the same as the _____ temperature.
   a. welding
   b. preheat
   c. cutting
   d. annealing

10. Annealing is done at temperatures approximately _____ above the critical temperature.
    a. 100°F
    b. 200°F
    c. 300°F
    d. 400°F

Trade Terms Introduced in This Module

Alloys: Substances with metallic properties and composed of two or more chemical elements, at least one of which is a metal.

Constituents: The elements and compounds, such as metal oxides, that make up a mixture or alloy.

Critical temperature (transformation temperature): The temperature at which iron crystals in a ferrous-based metal transform from being face-centered to body-centered. This dramatically changes the strength, hardness, and ductility of the metal.

Ductility: The characteristic of metal that allows it to be stretched, drawn, or hammered without breaking.

Hardenability: A characteristic of a metal that makes it able to become hard, usually through heat treatment.

Heat-affected zone (HAZ): The part of the base metal that has not been melted but has been altered structurally by the weld heat.

Induction heating: The heating of a conducting material by means of circulating electrical currents induced by an externally applied alternating magnetic field.

Interpass temperature: In a multiple-pass weld, the temperature of the base metal at the time the next weld pass is started. It is usually the same temperature as used for preheating.

Postweld heat treatment (PWHT): The process of heat treating the base metal and weld after making the weld.

Preheat weld treatment (PHWT): The process of heating the base metal before making a weld.

Quench: To rapidly cool a hot metal using air, water, or oil.

Residual stress: The strain that occurs in a welded joint after the welding is completed, as a result of mechanical action, thermal action, or both.

Tempering: The process of reheating quench-hardened or normalized steel to a temperature below the transformation range and then cooling it at any rate desired.

Underbead cracking: Subsurface cracking in the base metal under or near the weld.

Additional Resources

This module is intended to present thorough resources for task training. The following reference works are suggested for both instructors and motivated trainees interested in further study. These are optional materials for continued education rather than for task training.

The Procedure Handbook of Arc Welding. Cleveland, OH: The Lincoln Electric Company.

Welding Handbook: Welding Processes. Miami, FL: American Welding Society.

Welding Essentials: Questions and Answers. William Galvery and Frank Marlow. New York, NY: Industrial Press.

Figure Credits

Alloy Steel International	301SA01
Gerald Shannon	301F05
Victor Division of Thermadyne Industries, Inc.	301F06
Belchfire Corp.	301F07, 301SA02
Miller Electric Mfg. Co.	301F10, 301SA03
Lucifer Furnaces, Inc.	301F11
Pyrometer Instrument Co., Inc.	301F12
Tempil Inc.	301F13

The NCCER makes every effort to keep these textbooks up-to-date and free of technical errors. We appreciate your help in this process. If you have an idea for improving this textbook, or if you find an error, a typographical mistake, or an inaccuracy in NCCER's Contren® textbooks, please write us, using this form or a photocopy. Be sure to include the exact module number, page number, a detailed description, and the correction, if applicable. Your input will be brought to the attention of the Technical Review Committee. Thank you for your assistance.

Instructors – If you found that additional materials were necessary in order to teach this module effectively, please let us know so that we may include them in the Equipment/Materials list in the Instructor's Guide.

Write: Curriculum Revision and Development Department
National Center for Construction Education and Research
P.O. Box 141104, Gainesville, FL 32614-1104

Fax: 352-334-0932

E-mail: curriculum@nccer.org

Craft   Module Name

Copyright Date   Module Number   Page Number(s)

Description

(Optional) Correction

(Optional) Your Name and Address

Physical Characteristics and Mechanical Properties of Metals

COURSE MAP

This course map shows all of the modules in the third level of the Welding curriculum. The suggested training order begins at the bottom and proceeds up. Skill levels increase as you advance on the course map. The local Training Program Sponsor may adjust the training order.

WELDING LEVEL THREE

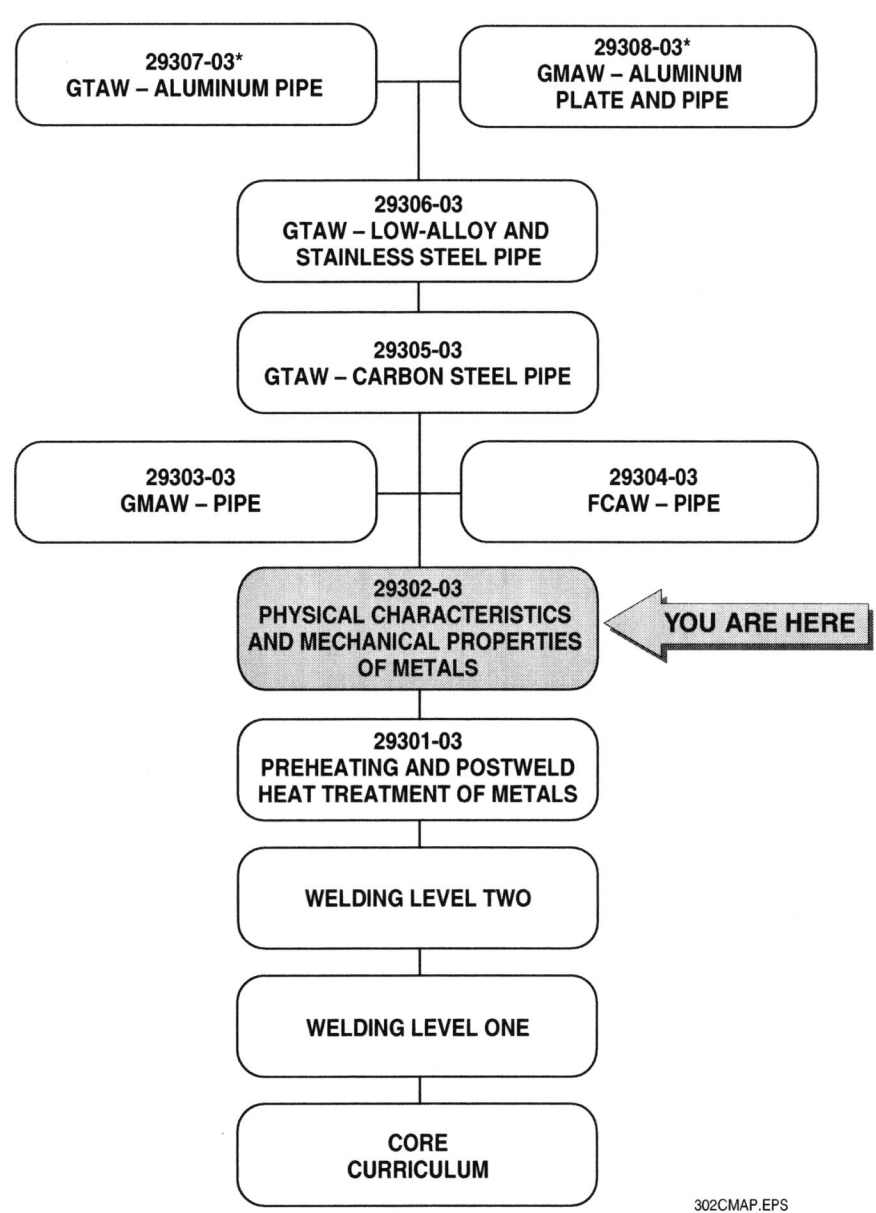

*Please note that Modules 29307-03 and 29308-03 are electives for those progressing through the *Welding Level Three* program.

Figures

Tables

Physical Characteristics and Mechanical Properties of Metals

Objectives

When you have completed this module, you will be able to do the following:

1. Identify and explain the composition and classifications of base metals.
2. Explain and demonstrate field identification methods for base metals.
3. Identify and explain the physical characteristics and mechanical properties of metals.
4. Identify and explain forms and shapes of structural metals.
5. Explain metallurgical considerations for welding metals.

Prerequisites

Before you begin this module, it is recommended that you successfully complete the following: Core Curriculum; Welding Levels One and Two; Welding Level Three, Module 29301-03.

Required Trainee Materials

1. Pencil and paper
2. Appropriate personal protective equipment

1.0.0 ◆ INTRODUCTION

Welders must understand the physical characteristics and mechanical properties of metals to properly identify base metals and select weld filler metals. They must also understand how welding heat affects base material and filler metals.

This module explains physical characteristics, mechanical properties, classification systems, and weldability of common metals and their alloys. It describes the important physical characteristics and mechanical properties that influence the selection of a metal for a specific application. It explains how to identify common base metals and describes the standard commercial structural steel shapes used for fabrication and construction.

2.0.0 ◆ METAL COMPOSITION AND CLASSIFICATIONS

The composition of common welding base metals varies widely, from metals made of essentially one **metallic** element to alloys (mixtures) of metallic and nonmetallic components. Metals are classified into two basic groups: **ferrous** metals, which are composed principally of iron, and nonferrous metals, which contain very little or no iron. Ferrous metals include all the steels, cast irons, **wrought** irons, **malleable** irons, and **ductile** (nodular) irons. Nonferrous metals and their alloys include the light metals (aluminum, magnesium, titanium), the heavy metals (copper, nickel, lead, tin, zinc), and the precious metals (platinum, gold, silver).

2.1.0 Ferrous Metal Composition and Classification Systems

Ferrous metals contain mostly iron and some carbon. The amount of carbon ranges from almost none in commercially pure iron to about 1.4%. Steel contains less than 1.7% carbon. Above 1.4% carbon, steel develops weak and brittle properties similar to those of cast iron. Carbon steels, the largest group of the ferrous-based metals, contain much smaller amounts of carbon than cast irons. As the carbon content increases in carbon steels, the steel becomes stronger and more wear resistant. The carbon content of a particular carbon steel is the largest factor in determining its use. Steel **castings** usually contain a higher percentage of carbon than rolled plate and other rolled shapes. The casting foundry can usually provide the alloy formula.

Steels also contain impurities and trace amounts of elements such as manganese, silicon, phosphorus, copper, and lead. Sometimes additional amounts of these elements are intentionally added to steel to change its characteristics, such as to increase its forgeability (the ability to be heated in a forge and then beaten or hammered into shape) or to decrease a tendency toward blowholes (the vents that form when air or gases escape).

Steels, including the carbon steels, low-alloy carbon steels, alloy steels, and stainless steels, are classified by various systems such as specification number and grade or manufacturer's trade name and number. Steel classification systems include those developed by the **American Iron and Steel Institute (AISI)**, the **American Society for Testing Materials (ASTM)**, and the Unified Numbering System (UNS), as well as manufacturer's trade names and identification numbers. Stainless steels have their own classification systems. They are sometimes referred to by the percentages of their chromium and nickel content, such as 18/8, 25/20, and 18/10, but this system has largely been replaced by the AISI stainless steel classification system. Common classification systems for steels are explained in the following sections.

2.1.1 Carbon Steel Classification

The principal classification system for carbon steels is the AISI Numerical Designation of Standard Carbon and Alloy Steels. It was originally a four-digit system developed by the **Society of Automotive Engineers (SAE)** for carbon steels commonly used in structural shapes, plate, strip, sheet, and welded tubing. Five digits are now used for some alloys. The classification system may be referred to as the AISI, the SAE, or the AISI/SAE system.

The AISI and SAE designations are essentially the same, except that the AISI system sometimes uses a letter prefix to denote the manufacturing process that produced the steel. The following are examples of letter designations:

- *A* – Open-hearth steel
- *B* – Acid-Bessemer carbon steel
- *C* – Basic open-hearth carbon steel
- *D* – Acid open-hearth carbon steel
- *E* – Electric furnace steel

Carbon and Alloy Steels

Steel is an alloy of iron and carbon typically containing less than 1% carbon. All steels also contain varying amounts of other elements such as manganese, silicon, phosphorus, sulfur, and oxygen. In addition, some standard alloy steels can also contain elements such as nickel, chromium, and molybdenum. There are currently about 3,500 different grades of steel with many different properties. Seventy percent of these steels have been developed in the past 20 years. Most of the steels produced in the world are carbon and alloy steels. This is because over 95% of the construction and fabrication metals used worldwide consist of carbon and alloy carbon steels.

302SA01.EPS

Steel Alloying Elements

Some alloying elements used with steel and the reasons they are added include the following:

- *Manganese* – To promote forgeability
- *Silicon* – To prevent blowholes
- *Phosphorus* – To improve machinability
- *Copper* – To inhibit corrosion
- *Lead* – To improve machinability

Unified Numbering System (UNS) of Metals

The UNS numbering system used to identify metals was developed jointly by the ASTM and SAE as a way of relating the different numbering systems for metals and alloys. Its objective is to eliminate the confusion caused when more than one identification number is used to identify the same material or when the same identification number is used to identify two different materials. It is anticipated that this system will eventually replace the AISI and other systems.

Each UNS number consists of a letter prefix followed by five digits. Often the prefix letter relates to the family of metals. For example, UNS numbers A00001 through A99999 pertain to aluminum and aluminum alloys and UNS numbers C00001 through C99999 pertain to copper and copper alloys. When possible, the digits in the UNS sequence contain the same numbering sequences used in the other systems. For example, the UNS number G10200 corresponds to the AISI/SAE number 1020 for carbon steel. Most engineering and materials reference books contain tables that cross-reference AISI and SAE numbers for metals to the corresponding UNS numbers.

The absence of a letter prefix indicates basic open-hearth or Acid-Bessemer carbon steel. The prefixing letter (if any) designates the manufacturing process. The first two numerical digits represent the series (type and class) of steel. The third and fourth (and sometimes fifth) numerical digits specify the approximate percentage of carbon content. *Table 1* shows the AISI/SAE system for designating carbon and alloy steel types and classes.

The following examples explain the AISI/SAE numbers for carbon steels:

- AISI number C1020:

  C = Indicates basic open-hearth carbon steel
  10 = Carbon steel, nonresulfurized
  20 = Contains approximately 0.20% carbon
- AISI Number E2512:

  E = Indicates electric furnace steel
  25 = Designates steel alloyed with approximately 5% nickel
  12 = Designates steel containing approximately 0.12% carbon
- AISI number E52100:

  E = Indicates electric furnace steel
  5 = Contains either approximately 0.50%, 1.00%, or 1.45% chromium, designated by the next digit: 0, 1, or 2, respectively

2 = Designates approximately 1.45% chromium
100 = Designates approximately 1% carbon

The common group classification of carbon steels is primarily based on carbon content. Plain carbon steels (AISI series 10XX through 11XX) are basically iron-carbon alloys. High-strength low-alloy (HSLA) carbon steels (AISI series 13XX through 98XX) have small amounts of alloying elements to improve strength, hardness, and toughness or to increase resistance to corrosion, heat, and environmental damage. The plain carbon steels are commonly grouped as follows:

- *Low-carbon* – 0.10% to 0.15% carbon, 0.25% to 1.50% manganese
- *Mild-carbon* – 0.15% to 0.30% carbon, 0.60% to 0.70% manganese
- *Medium-carbon* – 0.30% to 0.50% carbon, 0.60% to 1.65% manganese
- *High-carbon* – 0.50% to 1.00% carbon, 0.30% to 1.00% manganese

The HSLA carbon steels are alloyed with one or more of the elements manganese, nickel, chromium, or molybdenum to provide higher strength, better toughness, weldability, and, in

Table 1 AISI/SAE System for Designating Carbon and Alloy Steel Types and Classes

AISI/SAE Number	Carbon Steel Types and Classes
1XXX	Carbon steels
10XX	Plain (nonresulfurized) carbon steel grades
11XX	Free machining, resulfurized (screw stock)
12XX	Free machining, resulfurized, rephosphorized
13XX	Manganese 1.75%
15XX	High-manganese (1.00% to 1.65%) carburizing steels
2XXX	Nickel steels
23XX	3.50% nickel
25XX	5.00% nickel
3XXX	Nickel-chromium steels
31XX	1.25% nickel, 0.65% or 0.80% chromium
32XX	1.75% nickel, 1.07% chromium
33XX	3.50% nickel, 1.50% or 1.57% chromium
34XX	3.00% nickel, 0.77% chromium
4XXX	Molybdenum steels
40XX	0.20% or 0.25% molybdenum
41XX	0.50%, 0.80%, or 0.95% chromium, 0.12%, 0.20%, 0.25%, or 0.30% molybdenum
43XX	1.82% nickel, 0.50% or 0.80% chromium, 0.25% molybdenum
46XX	0.85% or 1.82% nickel, 0.20% or 0.25% molybdenum
47XX	1.05% nickel, 0.45% chromium, 0.20% or 0.35% molybdenum
48XX	3.50% nickel, 0.25% molybdenum
5XXX	Chromium steels
50XX	0.27%, 0.40%, 0.50%, or 0.65% chromium
51XX	0.80%, 0.87%, 0.92%, 0.95%, 1.00%, or 1.05% chromium
5XXXX	1.00% carbon, 0.50%, 1.02%, or 1.45% chromium
6XXX	Chromium-vanadium steels
61XX	0.60%, 0.80%, or 0.95% chromium, 0.10% or 0.15% (minimum) vanadium
86XX	0.55% nickel, 0.50% chromium, 0.20% molybdenum
92XX	0.65%, 0.82%, or 0.85% manganese, 1.40% or 2.00% silicon, 0.00% or 0.65% chromium
93XX	3.25% nickel, 1.20% chromium, 0.12% molybdenum
94XX	0.45% nickel, 0.40% chromium, 0.12% molybdenum
97XX	0.55% nickel, 0.20% chromium, 0.20% molybdenum
98XX	1.00% nickel, 0.80% chromium, 0.25% molybdenum
XXBXX	Boron steels
XXLXX	Leaded steels

some cases, greater resistance to corrosion. As the percentage content of these elements increases, changes in the weldability of the low-alloy steels make electrode selection and welding procedures more critical. HSLA steels include the following:

- Low-nickel steels
- Low-nickel chrome steels
- Low-manganese steels
- Low-alloy chromium steels
- Weathering steels

Another class of steel in the AISI series is sulfur steel. It is not an alloy steel but is still included in the series.

2.1.2 ASTM Steel Specification System

The ASTM develops and publishes specifications for use in material production and testing. Weld-ing codes and procedures typically use the ASTM designation to identify metal types because it includes specifications for mechanical and physical characteristics of the material as well as chemical composition. Piping and tubing are the largest group of materials fabricated by welding covered by ASTM specifications.

Prefix letters are a general part of each specification and provide a general idea of the specification content. For example, *A* is used for ferrous metals, *B* for nonferrous metals, and *E* for miscellaneous subjects such as examination and testing.

Sometimes an *S* is added before the prefix letter. This indicates that ASME International, publishers of the *Boiler and Pressure Vessel Code*, has adopted that specification.

With the exception of a few mild steel grades such as SA-36 (mild steel plate) and SA-53 (mild steel pipe), ASTM specifications are three-digit numbers. Most specifications also have various grades, classes, or types. *Table 2* gives examples of some of the ASTM specifications pertaining to alloy steels.

NOTE

The complete collection of ASTM standards consists of 15 sections comprising 65 volumes, with a separate index. The following is provided only to illustrate sample ASTM numbers.

2.2.0 Low-Alloy Steel Groups

Low-alloy steels fall into the following general groups:

- **High-strength low-alloy (HSLA) steels**
- Quench-and-tempered steels
- **Heat-treatable low-alloy (HTLA) steels**
- Chromium-molybdenum steels

2.2.1 High-Strength Low-Alloy Steels

High-strength low-alloy (HSLA) steels are designed to meet specific mechanical requirements per ASTM specifications. The principal difference between plain-carbon steels and HSLA steels is the addition of alloys such as manganese, chromium, nickel, copper, molybdenum, and vanadium. Typically, these alloys total less than 2%. The carbon content of HSLA steels is about the same as that in low-carbon and mild steels (0.06% to 0.28%). Compared to mild steels, HSLA steels are designed to provide a combination of higher strength, better corrosion resistance, and improved notch toughness. The weldability of HSLA steel is similar to that of mild steel.

HSLA steels can be divided into two general groups: Group A and Group B. Steels in Group A are designed for high strength, while steels in Group B are designed for high strength, improved atmospheric corrosion resistance, and notch toughness. Typical steels in Group A are A441, A572, and A633, which are used for structural applications such as pipelines, buildings, bridges, machinery, and railroad equipment. Other steels in Group A are A225 and A737, which are used for pressure vessels. Typical steels in Group B are A242 and A588, which are used for high-strength structural applications that may include unpainted applications. Group B steels specifically intended for low-temperature applications (the nickel content is as high as 9%) are A203 and A353.

CAUTION

The welding recommendations provided are general guidelines only. When welding HSLA steel, the site WPS or site quality standards must be followed. Failure to follow proper standards can result in defective welds.

HSLA steels can be welded using the common welding processes, although successful welding requires consideration of preheating, interpass temperature, and control of hydrogen, including the use of low-hydrogen electrodes when applicable. Typically, only the root pass and possibly the

Table 2 ASTM Specification Standards

ASTM Number		Tensile	Type of Base Metal
A209	Grade T1	55,000	C-0.05 Mo tubes
	Grade T1a	60,000	C-0.05 Mo tubes
	Grade T1b	53,000	C-0.05 Mo tubes
A217	Grade T17	60,000	1.00 Cr-0.15 V tubes
A334	Grade 5	65,000	5.00 Ni-Steel pipe
A335	Grade P1	55,000	C-0.05 Mo pipe
	Grade P2	55,000	0.60 Cr-0.85 Mo pipe
	Grade P3	60,000	1.57 Cr-0.70 Mo pipe
	Grade P3b	60,000	2.00 Cr-0.55 Mo pipe
	Grade P11	60,000	1.25 Cr-0.55 Mo-Si pipe
	Grade P12	60,000	0.95 Cr-0.55 Mo pipe
	Grade P15	60,000	1.40 Si-0.55 Mo pipe
A517	Grade A	115,000	0.05 Cr-0.025 Mo-Si plates
	Grade B	115,000	0.05 Cr-0.02 Mo-V plates
	Grade P	115,000	0.125 Ni-0.10 Cr-0.05 Mo plates

C – Carbon	Cr – Chromium	Mo – Molybdenum	Ni – Nickel	Si – Silicon	V – Vanadium

hot pass are made with the GTAW process on HSLA steels. Fill passes and cover passes are often made with faster processes such as SMAW or GMAW. One advantage of using the GTAW process is that hydrogen is easier to control because fluxes are not used. Minimum preheat and interpass temperatures are determined based on the alloy type and the thickness of the thinner part being welded. Generally, HSLA steels less than ¾ inch thick do not require preheating.

CAUTION

Steels that are wet or below 32°F should always be preheated to at least 70°F and maintained at that temperature to remove and prevent moisture. Failure to preheat metals will result in defective welds.

Table 3 provides minimum preheat and interpass temperatures for typical high-strength low-alloy (HSLA) steels.

Filler metal for welding HSLA steel depends on the alloy and the application. Mild steel filler metal can be used when joint strength is not critical or when alloy steel is being joined to mild steel. Alloy steel should be used when the strength of the joint must equal or approach that of the base metal. High-alloy stainless steel is used for special applications such as welding dissimilar metals or welding high-nickel steel such as A353.

Typically, HSLA steels are not postweld heat-treated; however, this may be required to improve ductility or to maintain the dimensions of the weldment during machining.

2.2.2 Quench-and-Tempered Steels

Quench-and-tempered (Q&T) steels are furnished in the heat-treated condition, which generally consists of **austenitizing**, quenching, and tempering. Some Q&T steels fall into the carbon steel classification, some into the low-alloy steel classification, and others into the alloy classification of the AISI/SAE system. Q&T steels combine high yield and tensile strength with good notch toughness,

Table 3 Minimum Preheat and Interpass Temperatures for Typical High-Strength Low-Alloy Steels

ASTM Steel	Thickness (Inches)	Minimum Temperature (°F)
A242	Up to 0.75	32
A441	0.81 to 1.50	50
A572, Gr 42, 50	1.56 to 2.50	150
A588	Over 2.50	225
A633, Gr A, B, C, D		
A572, Gr 60, 65	Up to 0.75	50
A633, Gr E	0.81 to 1.50	150
	1.56 to 2.50	225
	Over 2.50	300

ductility, corrosion resistance, or weldability, depending on their intended use. Q&T steels have less than 0.25% carbon, with yield strengths of 50,000 to 180,000 psi. Most Q&T steels are furnished as plate, although some are available as seamless pipe.

Q&T steels are used where high strength and minimum weight are essential. Many of the Q&T steels are covered by ASTM classifications. A common group of Q&T steels known as HY steels are covered by military specifications (MIL). The most common HY steel is HY-80, which has a yield stress 80,000 psi. HY-100, HY-130, HY-150, and HY-180 are also available. Types and typical uses of Q&T steels include the following:

- *A514 and A517* – Used for earth-moving equipment, pressure vessels, bridges, TV towers, and ships.
- *A533 Grade B* – Used for nuclear pressure vessels.
- *A543 Type B* – Used in nuclear reactor vessels, ships, and submarines. Classes 1 and 2 steels are similar to HY-80 and HY-100.
- *A553* – Used for cryogenic applications. It contains 8% to 9% nickel for high strength at extremely low temperatures.
- *A678* – Used for high-strength structural applications.

Most Q&T steels can be welded by common welding processes without preheat or postheat.

Hot Tapping

Steel pipes are commonly hot-tapped to install a bypass around a repair area. With hot tapping, taps for the bypass are welded to the pipe while material or liquid is still flowing in the pipe. Low-alloy pipes cannot be hot-tapped because the material flowing through the pipe will cool the pipe rapidly and prevent proper preheat, interpass, and postheat temperature maintenance. This can result in defective welds as well as pipe failure during welding.

However, thick and highly restrained joints may require some preheat to prevent cracking. Because Q&T steels obtain their strength from heat treatment, slow cooling can destroy the benefits of the original heat treatment. Slow cooling is generally caused by the following:

- Excess preheat
- Exceeding the recommended interpass temperature
- High heat input, such as high amperage with slow travel speed and the use of weave beads instead of stringers

On the other hand, cooling too quickly can cause a hard, brittle zone that may crack. Fast cooling is caused by the following:

- Insufficient preheat
- Insufficient interpass temperature
- Insufficient heat input during welding

 **CAUTION**
When welding Q&T steels, pay careful attention to the welding procedures concerning preheat and interpass temperature recommendations to prevent loss of strength or cracking.

The preheat and interpass temperature recommendations for Q&T steels vary depending on the type of steel being welded. *Table 4* shows minimum preheat and interpass temperatures for Q&T steels.

Table 4 Minimum Preheat and Interpass Temperatures for Q&T Steels

Thickness Range	A514/ A517	A533	A537	A543	A678
Up to 0.50	50	50	50	100	50
0.56 to 0.75	50	100	50	125	100
0.81 to 1.00	125	100	50	150	100
1.10 to 1.50	125	200	100	200	150
1.60 to 2.00	175	200	150	200	150
2.10 to 2.50	175	300	150	300	150
Over 2.50	225	300	225	300	—

When welding Q&T steels, always use low-hydrogen techniques and stringer beads to minimize heat input. Typically, the GTAW process is used only on thin sections or for the first one or two passes.

Most Q&T steels are designed to be used in the as-welded condition, with no postheat. Sometimes, thick sections of A533 or A537 steels may require postheat treatment. If postheat treatment is required, use extreme care to follow the procedure exactly to prevent adversely affecting the base metal.

2.2.3 Heat-Treatable Low-Alloy Steels

Heat-treatable low-alloy (HTLA) steels are alloy steels with carbon in the range of about 0.25% to 0.45% and small amounts of chromium, nickel, and/or molybdenum to enhance hardenability. They are generally welded in the annealed condition, and then the entire weldment is heat treated. The relatively high carbon content of HTLA steels allows them to be heat treated to a very high strength and hardness. Because HTLA steels must be heat treated after welding, they are typically used for smaller parts that can be placed in tempering ovens.

HTLA steels are produced to contain a very low level of impurities. During welding, low hydrogen techniques are used, and extreme care must be taken to keep the weld joint and filler metal absolutely clean and free of impurities. For this reason, GTAW is the preferred process for welding HTLA steels. Heat input is also very important, and welding procedures that take into account current, voltage, travel speed, and preheat temperatures must be followed.

HTLA steels generally require higher preheat and interpass temperatures than other types of alloy steels to prevent hardening of the weld zone and hot cracks. They also require postweld heat treatment. *Table 5* shows minimum preheat and interpass temperatures for HTLA steels.

HTLA steels should receive postweld heat treatment immediately after they are welded. The type of postweld heat treatment depends on the type of steel and the preheat and interpass temperature maintained during welding. Allowing the weldment to cool after welding without some type of heat treatment generally results in cracks. Immediately after it is welded, HTLA steel should be stress-relieved following strict guidelines depending on the alloy and the preheat and interpass temperature used. Finally, to achieve the desired mechanical properties, the weldment should be austenitized, quenched, and tempered.

2.2.4 Chromium-Molybdenum Steels

Chromium-molybdenum (Cr-Mo) steels contain 0.5% to 9% chromium and 0.5% to 1% molybdenum. The carbon content is generally less than 0.20%. Cr-Mo steels, which are sometimes called chrome-moly or heat-resisting alloy steels, are widely used in the petroleum industry and in steam-generating plants for elevated-temperature

Table 5 Minimum Preheat and Interpass Temperatures for HTLA Steels

AISI Steel	Thickness Range (Inches)	Minimum Preheat and Interpass Temperature (°F)
4027	Up to 0.50	50
	0.60 to 1.0	150
	1.1 to 2.0	250
4037	Up to 0.50	100
	0.60 to 1.0	200
	1.1 to 2.0	300
4130/5140	Up to 0.50	300
	0.60 to 1.0	400
	1.1 to 2.0	500
4135/4140	Up to 0.50	350
	0.60 to 1.0	450
	1.1 to 2.0	500
4340/8630	Up to 2.0	550
	Up to 0.5	200
	0.6 to 1.0	250
	1.1 to 2.0	300
8640	Up to 0.50	200
	0.60 to 1.0	300
	1.1 to 2.0	350

applications. The chromium provides improved resistance to oxidation and corrosion, and the molybdenum increases strength at elevated temperatures. Cr-Mo steels are also used in the aircraft industry as tubing for highly stressed parts. Cr-Mo steels that are in the AISI/SAE 4100 series are available in forging, castings, plate, pipe, and tubing to various ASTM specifications.

Cr-Mo steels are hardenable, so a welding procedure that includes preheating and sometimes postheating must be used. The preheat temperature depends on the amount of alloy, the carbon content, the weld material, and base metal thickness. Generally, the higher the alloy, carbon content, or thickness, the higher the preheat temperature. *Table 6* shows minimum preheat and interpass temperatures for Cr-Mo steels.

Once the Cr-Mo base metal has been preheated, welding should proceed without interrupting the heating cycle in order to avoid hard spots and cracking. If the base metal contains less than 4% chromium or is less than 1" thick, the heating cycle can sometimes be interrupted if at least 33% or two weld layers are completed.

CAUTION

Check your welding procedure for any special precaution necessary when interrupting a heating cycle. Failure to follow proper welding procedures will result in defective welds.

In addition, when the welding consumables are a potential source of hydrogen, such as with SMAW, it may also be necessary to raise the preheat temperature and hold the weldment at this elevated temperature for a period of time. This allows the hydrogen to escape before the base metal cools, which reduces the risk of cracking. To avoid potential problems that can occur if the heating cycle of Cr-Mo base metal is interrupted, always plan to complete the heating cycle once it has started.

Some Cr-Mo steels containing less than 1.25% Cr and 0.05% Mo can be placed in service as-welded if a high preheat temperature was used

Table 6 Minimum Preheat and Interpass Temperatures for Cr-Mo Steels

ASTM Steel	Type	Up to 0.5	0.5 to 1.0	Over 1.0
A335-P2	1/2 Cr - 1/2 Mo	300°F	300°F	300°F
A335-P12	1 Cr - 1/2 Mo			
A335-P11	11/4 Cr - 1/2 Mo			
A369-FP3b	2 Cr - 1/2 Mo	300°F to 350°F	300°F to 350°F	350°F
A335-P22	21/4 Cr - 1Mo			
A335-P21	3 Cr - 1 Mo	300°F to 350°F	300°F to 350°F	400°F
A335-P5	5 Cr - 1/2 Mo	400°F	400°F	500°F
A335-P7	7 Cr - 1/2 Mo			
A335-P9	9 Cr - 1 Mo			

*Maximum carbon content of 0.15%. For higher carbon content, preheat should be increased 100°F to 200°F.

and the section is relatively thin. Other Cr-Mo steels should either be stress-relieved, annealed, or normalized and tempered.

Stress-relief heat treatment is used to reduce welding stresses and to increase the ductility and toughness of the weld metal and the heat-affected zone. The stress-relief temperature ranges from about 1,150°F to 1,400°F and depends on the alloy being treated. The entire weldment or the weld zone is generally held at the stress-relief temperature for one hour per inch of thickness.

The procedures for annealing or normalizing and tempering are generally the same as that used for HTLA steels discussed earlier in this module.

2.3.0 Common Grade Stainless Steel Classifications

Stainless steels are typically classified as being common grade or specialty grade. This section covers common grade stainless steels. The specialty grades are covered later in this module. Common grade stainless steels (corrosion-resistant steels) are iron-based alloys that normally contain at least 11% chromium. Other alloying elements, including nickel, carbon, manganese, and silicon, may be present in varying quantities depending on the specific type of stainless steel to enhance its physical and mechanical properties. The principal physical characteristics of all stainless steels are their resistance to corrosion and heat. Some have good low- and high-temperature mechanical properties. When compared with mild steels, stainless steels have the following characteristics:

- Lower **coefficients** of thermal conductivity that increase the chances of distortion
- Higher coefficients of thermal expansion that increase the chances of distortion
- Higher electrical resistances that increase the tendency to build up heat from welding current

Common grades of stainless steels are classified by their grain structure. The type of grain structure is determined by the specific alloy content of the stainless steel and its heat treatment during manufacture. Based on their microcrystalline structures, the classification of common grade stainless steels is divided into three groups: austenitic, ferritic, and martensitic.

2.3.1 Austenitic Stainless Steels

Austenitic stainless steels are nonmagnetic in the annealed condition and nonhardenable by heat treatment. However, they can be hardened significantly by cold working. Austenitic stainless steels combine excellent corrosion and heat resistance with good mechanical properties over a broad temperature range. Austenitic stainless steels make up the largest of the three common grade stainless steel groups. For this reason, they are the type of stainless steel encountered most often by welders. They include the AISI 200 and 300 series stainless steels (*Table 7*) that all contain significant amounts of both chromium and nickel. Austenitic steels are sometimes further subdivided in two classifications based on their compositions: chromium-nickel (AISI 300 series) and chromium-nickel-magnesium (AISI 200 series). The range of applications for austenitic stainless steels includes housewares, containers, industrial piping and vessels, and architectural facades.

Austenitic stainless steels are also available in low-carbon (L-grade) and high-carbon (H-grade) types. The letter L after a stainless steel type, such as 304L, indicates a low-carbon content of 0.03% or under. Similarly, the letter H after a stainless steel type indicates a high-carbon content, ranging between 0.04% and 0.10%. L-grades are typically used where annealing after welding is impractical, such as in the field where pipe and fittings are welded. H-grades are used when the steel will be subjected to extreme temperatures because the higher carbon content helps it retain strength.

2.3.2 Ferritic Stainless Steels

Ferritic stainless steels are always magnetic. They are hardened to some extent by cold working, not heat treatment. Ferritic stainless steels combine corrosion and heat resistance with fair mechanical properties over a narrower temperature range than austenitic steels. Ferritic stainless steels are straight chromium stainless steels containing 11.5% to 27% chromium, about 1% manganese, and little or no nickel. Carbon content is 0.20% or less. Examples of ferritic stainless steels include AISI types 405, 409, 430, 442, 444, and 446. They typically are used for decorative trim, sinks, and automotive applications, particularly exhaust systems.

2.3.3 Martensitic Stainless Steels

Martensitic stainless steels are magnetic and can be hardened by quenching and tempering. They are excellent for use in mild environments such as the atmosphere, freshwater, steam, and weak acids. However, they are not resistant to severely corrosive solutions. Martensitic stainless steels comprise two groups: chromium-martensitic and martensitic. The chromium-martensitic stainless steels contain from 11.5% to 18% chromium, about

Table 7 AISI Classification Numbers and Composition for Selected Stainless Steels

AISI NO.	C%	MN%	SI%	CR%	NI%	OTHER ELEMENTS
CHROMIUM-NICKEL-MAGNESIUM-AUSTENITIC (NONHARDENABLE)						
201	0.15 Max	5.5–7.5	1.0	16.0–18.0	3.0–5.5	N 0.25 Max
202	0.15 Max	7.5–10	1.0	17.0–19.0	4.0–6.0	N 0.25 Max
CHROMIUM-NICKEL-AUSTENITIC (NONHARDENABLE)						
301	0.15 Max	2.0	1.0	16.0–18.0	6.0–8.0	—
302	0.15 Max	2.0	1.0	17.0–19.0	6.0–10.0	—
303	0.15 Max	2.0	1.0	17.0–19.0	8.0–10.0	Mo 0.0–0.6
304	0.08 Max	2.0	1.0	18.0–20.0	8.0–10.5	—
304L	0.03 Max	2.0	1.0	18.0–20.0	8.0–12.0	—
308	0.08 Max	2.0	1.0	19.0–21.0	10.0–12.0	—
309	0.20 Max	2.0	1.0	22.0–24.0	12.0–15.0	—
310	0.25 Max	2.0	1.5	24.0–26.0	19.0–22.0	—
316	0.08 Max	2.0	1.0	16.0–18.0	10.0–14.0	Mo 2.0–3.0
316L	0.03 Max	2.0	1.0	16.0–18.0	10.0–14.0	Mo 2.0–3.0
321	0.08 Max	2.0	1.0	17.0–19.0	9.0–12.0	Ti, 5xC Min
347	0.08 Max	2.0	1.0	17.0–19.0	9.0–13.0	Cb+Ta10xC Min
348	0.08 Max	2.0	1.0	17.0–19.0	9.0–12.0	Ti 0.10 Max
CHROMIUM-MARTENSITIC (HARDENABLE)						
403	0.15 Max	1.0	0.5	11.5–13.0	—	—
410	0.15 Max	1.0	1.0	11.5–13.5	—	—
414	0.15 Max	1.0	1.0	11.5–13.5	1.25–2.5	—
416	0.15 Max	1.0	1.0	12.0–14.0	—	S 0.15 Min
420	0.15 Min	1.0	1.0	12.0–14.0	—	—
422	0.20–0.25	1.0	0.75	11.0–13.0	0.5–10	Mo 0.75–1.25 W 0.75–1.25 V 0.15–0.3
440A	0.60–0.75	1.0	1.0	16.0–18.0	—	Mo 0.75 Max
440B	0.75–0.95	1.0	1.0	16.0–18.0	—	Mo 0.75 Max
440C	0.95–1.2	1.0	1.0	16.0–18.0	—	Mo 0.75 Max
CHROMIUM-FERRITIC (NONHARDENABLE)						
405	0.08 Max	1.0	1.0	11.5–14.5	—	Al 0.10–0.30
409	0.08 Max	1.0	1.0	10.5–11.75	—	Ti, 6xC Min
430	0.12 Max	1.0	1.0	16.0–18.0	—	—
442	0.20 Max	1.0	1.0	18.0–23.0	—	Ti, 0.20 + 4 (%C+%N)
444	0.025 Max	1.0	1.0	17.5–19.5	1.0	Mo 1.75–2.5, N 0.035 Max (Nb+T2), 0.2 + 4 (%C+%N) Min
446	0.20 Max	1.50	1.0	23.0–27.0	1.0	N 0.25 Max
MARTENSITIC						
501	0.11 Max	1.0	1.0	4.0–6.0	—	Mo 0.4–0.65
502	0.10 Max	1.0	1.0	4.0–6.0	—	Mo 0.4–0.65

302T07.EPS

1% manganese, and in some cases 0% to 2.5% nickel. Examples of chromium-martensitic stainless steels are AISI types 403, 410, 414, 420, 422, 431, and 440. Martensitic stainless steels contain only 4% to 6% chromium and no nickel. For this reason, they are not considered true stainless steels, although their corrosion resistance is much greater than mild carbon steels, even at elevated temperatures. Examples of martensitic stainless steels are AISI types 501 and 502.

Why Stainless Steel Doesn't Rust

When steel comes in contact with oxygen and water vapor in the atmosphere, a chemical reaction occurs and the steel begins to change to its original form of iron oxide. The chromium added to steel prevents it from rusting or staining, thus the name stainless steel. The chromium reacts with the oxygen and water in the atmosphere to form a microscopic and transparent layer of chrome-oxide, called a passive film. The sizes of chromium atoms and their oxides are about the same, so they join to form a tight, stable layer only a few atoms thick on the surface of the steel. If the stainless steel surface is cut or scratched, disrupting the passive film, it is self-healing because more oxide immediately forms and covers the exposed surface, protecting it from rust again. This is why after repeated use and sharpening, stainless steel cutlery, like these knives, remains bright and shiny. However, if the surface of stainless steel is contaminated by a ferrous metal, the ferrous metal will rust.

302SA02.EPS

2.4.0 Specialty Grade Stainless Steel Classifications

In addition to the common grades of stainless steel described earlier, four specialty grades of stainless steels are also available: precipitation-hardening, superaustenitic, superferritic, and duplex. Each of these types is briefly described here.

2.4.1 Precipitation-Hardening Stainless Steels

Precipitation-hardening (PH) stainless steels contain alloying additions such as aluminum that enable them to be hardened by a solution and aging heat treatment. Precipitation-hardening stainless steels were developed after World War Two because the design and manufacture of jet aircraft required the use of stainless steels with a better weight-to-strength ratio. Precipitation-hardening stainless steels have designations such as 17-4PH, PH13-8MO, and AM350.

2.4.2 Superaustenitic Stainless Steels

Superaustenitic stainless steels have compositions that fall between 300 series stainless steels and nickel-alloy steels. They are highly resistant to corrosion and stress-corrosion cracking caused by chloride-containing solutions. Many of these alloys contain deliberate amounts of copper and/or nitrogen to increase their strength and corrosion resistance. These stainless steel types are widely used in process industry applications in place of common austenitic stainless steels that had a tendency to fail because of pitting, crevice corrosion, or chloride-induced stress corrosion cracking.

2.4.3 Duplex Stainless Steels

Duplex stainless steels are supplied with approximately equal amounts of ferrite and austenite, thus the name *duplex*. They also contain about 24% chromium and 5% nickel. Developed around 1970, duplex steels are among the most recently developed stainless steels and offer higher yield strength and superior resistance to pitting and stress cracking than 300 series stainless steels. An example of duplex stainless steel is Type 2205, which accounts for more than 80% of duplex use. Duplex stainless steels are used for heat exchangers, tubes, and pipes for applications involving the production and handling of gas and oils and for desalination.

Highly alloyed, high-performance duplex stainless steels, called *super-duplex stainless steels*, are also made. These contain higher levels of chromium, molybdenum, and nitrogen. They were developed to provide outstanding resistance to acids, acid chlorides, and other caustic solutions typical of those encountered in the chemical, petrochemical, and pulp and paper industries.

2.4.4 Superferritic Stainless Steels

Superferritic stainless steels differ from common grade ferritic stainless steels by their lower carbon and nitrogen content and higher molybdenum and/or chromium content. This combination provides much higher corrosion resistance in oxidizing and chloride chemical environments and excellent resistance to corrosion in salt water. Examples of superferritic stainless steel designations include E-Brite®, Sea-Cure®, and AL 29-4-2®.

2.5.0 Nonferrous Metals Classification Systems and Groups

Nonferrous metals include all metals except iron. Commonly used nonferrous metals include aluminum, magnesium, titanium, copper, nickel, zinc, tin, lead, and the precious metals. Nonferrous metals with densities lower than that of steel are considered light metals. They include, in order of increasing density, magnesium, beryllium, aluminum, and titanium. The heavier common nonferrous metals include copper, nickel, lead, tin, and zinc, as well as the precious metals platinum, gold, silver, and their alloys. Each of the nonferrous metals and alloys is classified by a different and unrelated classification system.

2.5.1 Aluminum Alloys

Pure aluminum is a soft, lightweight metal that can be used in its nearly pure form. Because aluminum is very soft and ductile and has a low tensile strength, it is often alloyed with elements such as copper, manganese, silicon, magnesium, and zinc to make different alloys with specific properties for very different purposes. When aluminum is exposed to the atmosphere, a microscopic and transparent aluminum-oxide layer forms immediately and protects the metal from further oxidation. This characteristic gives aluminum its high resistance to corrosion. Aluminum is lightweight, having one-third the weight of steel, and can be made very strong, ductile, and malleable. It is a good conductor of heat and electricity and is highly resistive to weathering and corrosion from many acids. However, aluminum can be corroded easily by alkalis because they attack the oxide layer. It is also noncombustible and nonmagnetic, making it widely used for electrical shielding and near inflammable or explosive substances. It is decorative, easily formed, cast, and machined. Polished aluminum has the highest reflectivity of any material, including mirror glass.

Aluminum alloys are broadly classified as either casting alloys or wrought alloys for rolling, forging, or extruding. The Aluminum Association, Inc. (AA) uses a three-digit plus one decimal system to designate each type of casting alloy (*Table 8*). The first digit (1 through 9) shows the main alloying element or elements. For the 1XX.X series this is essentially pure aluminum (99.00% aluminum). With the exception of the 1XX.X series, the second and third digits identify the specific alloys. For the 1XX.X series only, the second and third digits give the degree of aluminum purity above 99.00%. For example, if the second and third digits are 60, the alloy contains a minimum of 99.60% aluminum. The decimal indicates if the alloy composition is for the final casting (.0) or for ingot (.1 or .2, depending on impurity limits). Modifications to the alloys are indicated by the use of prefix letters such as A, B, C.

Recycled Steel

Steel is the world's most recycled material. According to the International Iron and Steel Institute, more than 435 million tons of steel were recycled in 2001.

Recycling Aluminum

Aluminum can be recycled repeatedly with no loss of quality or properties.

Table 8 AA System Code Designations For Basic Cast Aluminum Alloys

Designation	Major Alloying Element
1XX.X	Near pure aluminum
2XX.X	Copper
3XX.X	Silicon plus copper and/or magnesium
4XX.X	Silicon
5XX.X	Magnesium
6XX.X	Unused series
7XX.X	Zinc
8XX.X	Tin
9XX.X	Other element

Table 9 AA System Code Designations for the Basic Wrought Alloy Groups

Designation	Major Alloying Element
1XXX	Non-aluminum elements are 1% or less
2XXX	Copper
3XXX	Manganese
4XXX	Silicon
5XXX	Magnesium
6XXX	Magnesium and silicon
7XXX	Zinc
8XXX	Other elements
9XXX	Unused serie

Similarly, the AA uses a four-digit numbering system for wrought aluminum alloy types (*Table 9*) that is not related to the three-digit casting alloy designation system. In this wrought aluminum system, the first digit identifies the alloy group by the major alloying element. For the 1XXX group, the second digit designates modifications and impurity limits. A 0 indicates no special control on individual impurities. A second digit of 1 through 9 indicates special control over one or more impurities and is assigned consecutively. For groups 2XXX through 8XXX, the second digit indicates alloy modifications and is consecutively assigned. A second digit of 0 indicates the original alloy. The last two digits indicate the minimum percentage of aluminum.

Why Aluminum Must Be Cleaned Before Welding

The aluminum-oxide coating protecting aluminum must be thoroughly cleaned from the metal before welding. This is because, while pure aluminum melts at about 1,200°F, the aluminum oxide coating that protects the metal does not melt until it reaches the higher temperature of 3,700°F.

302SA03.EPS

The following examples explain AA wrought aluminum numbers:

- AA Number 1075:

  1075 = Indicates aluminum content of 99.0% or greater

  1075 = Indicates no special control on the impurities

  10**75** = Indicates an aluminum content of 99.75%

- AA Number 1180:

  1180 = Indicates aluminum content of 99.0% or greater

  1**1**80 = Indicates impurities are limited

  11**80** = Indicates an aluminum content of 99.80%

In general, aluminum and its alloys can be welded by oxyfuel or airfuel, electric arc, electrical resistance methods, and brazing.

2.5.2 Magnesium Alloys

Magnesium is mainly used in alloy form. It is the lightest of the structural metals, having less than one-quarter the density of steel, and is mainly used in the aerospace and automotive industries. Magnesium is the eighth most common element in the world and the sixth most abundant metal, comprising about 2.5% of the earth's surface. Because seawater contains about 0.13% magnesium, some production facilities process seawater to obtain the magnesium. This is done after the precipitation of other salts from the seawater, which leaves a magnesium-rich brine. The uses of magnesium center on three properties of the metal: its ability to form intermetallic compounds with other metals, its high chemical reactivity, and its low density. It can be extruded, forged, rolled, and machined.

Magnesium alloys are broadly classified as casting alloys or wrought (rolled, forged, or extruded) alloys. Both of these classifications use the ASTM recommendation B275, *Codification of Light Metals and Alloys*. Designations consist of one or two letters representing the alloying elements, followed by their respective percentages, and rounded off to the nearest whole number. The letters representing the alloying elements are as follows:

- *A* – Aluminum
- *B* – Bismuth
- *C* – Copper
- *D* – Cadmium
- *E* – Rare Earths
- *F* – Iron
- *G* – Magnesium
- *H* – Thorium
- *K* – Zirconium
- *L* – Beryllium
- *M* – Manganese
- *N* – Nickel
- *P* – Lead
- *Q* – Silver

Magnesium

Magnesium is widely used in specialty racing cars and some commercial cars as a lightweight material to minimize weight and fuel consumption because of the 20% to 25% weight saving over aluminum. It typically is used in the construction of transmission casings, intake manifolds, and cylinder head covers. Magnesium pistons as well as some other engine parts are also being used in some race cars.

302SA04.EPS

- *R* – Chromium
- *S* – Silicon
- *T* – Tin
- *Y* – Antimony
- *Z* – Zinc

Magnesium alloy designations also contain information indicating the temper condition. This part of the designation, separated by a hyphen from the main designation letters and numbers, consists of a letter or a letter and number. The letters and numbers that indicate the temper are as follows:

- *F* – As fabricated
- *O* – Annealed
- *H10, H11* – Strain hardened
- *H23, H24, H26* – Strain hardened and annealed
- *T4* – Solution heat treated
- *T5* – Artificially aged
- *T6* – Solution heat treated and artificially aged
- *T8* – Solution heat treated, cold worked, and artificially aged

For example, in the alloy AZ92A-T6, the first A is aluminum and the Z is zinc. The 9 indicates that the aluminum percentage is between 8.6 and 9.4. The 2 indicates that the zinc percentage is between 1.5 and 2.5. The final A denotes that this is the first alloy to receive the AZ92 designation. The letter and number T6 indicate solution heat treated and artificially aged tempering.

Magnesium and its alloys are best welded by the gas tungsten arc welding (GTAW) process for thin sections and the gas metal arc welding (GMAW) process for thick sections. The composition of the filler metals should be compatible with the base metal. However, zinc above 1% increases hot shortness and can result in cracking. *Table 10* shows examples of magnesium alloys with their principal uses.

2.5.3 Titanium Alloys

Titanium is mainly used in alloy form. Its alloys are high-strength, lightweight (about one-half the density of steel), corrosion-resistant structural metals

Table 10 Examples of Magnesium Alloys

ASTM Number	NOMINAL ALLOY COMPOSITION (By %)						
	Al	Mn	Zn	Zr	Rare Earths	Th	Mg
SAND AND PERMANENT MOLD CASTINGS							
AZ92A-T6	9.0	0.10	2.0				Balance
AZ63A-T6	6.0	0.15	3.0				Balance
AZ81A-T4	7.6	0.13 Min	0.7				Balance
AZ91C-T6	8.7	0.13	0.7				Balance
EZ33A-T5			2.6	0.8	3.3		Balance
HK31A-T6			0.3	0.7		3.3	Balance
HZ32A-T6			2.1	0.8		3.3	Balance
DIE CASTINGS							
AZ91A-F	9.0	0.13	0.7				Balance
AM60A-F	6.0	0.13					Balance
AZ91D-F	9.0	0.13	0.7				Balance
EXTRUSIONS							
AZ61A-F	6.5	0.15	1.0				Balance
MIA-F		1.20					Balance
AZ80A-T5	8.5	0.12	0.5				Balance
AZ10A-F	1.3	0.20	0.4				Balance
AZ31B-F	3.0	0.20	1.0				Balance
ZK60A-F			5.5	0.45			Balance
SHEET AND PLATE							
AZ31B-H24	3.0	0.20	1.0				Balance
HK31A-H24				0.7		3.3	Balance

302T10.EPS

that retain their strength at elevated temperatures. It is used in aircraft engines and structures, chemical processing, surgical implants, sporting goods, marine, and other corrosion-resisting applications.

Titanium is a reactive metal. At high temperatures, it readily combines with oxygen, hydrogen, and nitrogen to form stable compounds. Similar to aluminum and magnesium, titanium forms a tight protective microscopic oxide film on clean surfaces at room temperature and in contact with oxygen. This makes it highly resistant to corrosion at low temperatures. Titanium alloys are classified into three groups according to the crystal structures that are stable at room temperature.

- *Alpha alloys* – These are the usual crystalline structures of pure titanium. Aluminum is added to stabilize the crystalline structure. These alloys have excellent strength and are oxidation-resistant from 600°F to 1,100°F. They have good ductility and weldability and can be hardened by working but not by heat treatment.

- *Beta alloys* – These alloys can contain vanadium, chromium, and aluminum. Iron and chromium are added to make heat-treatable, high-strength alloys. They have fair weldability.

- *Alpha-beta alloys* – These alloys contain one or more of the elements manganese, molybdenum, tin, vanadium, iron, and aluminum. They have a wide range of mechanical characteristics but have poor weldability.

Titanium is considered difficult to weld because the weld must be kept in an inert (nonreactive) gas atmosphere until it cools to below the reactive temperature. Small amounts of hydrogen or large amounts of oxygen or nitrogen can cause embrittlement in the HAZ, which means hardness that leads to easy breakability. Titanium is usually welded by the GTAW process.

About 34 different titanium alloys are currently being produced. *Table 11* gives examples of titanium grades and alloys.

Table 11 Examples of Titanium Grades and Alloys

Designation	NOMINAL ALLOY COMPOSITION (By %)				
	Al	Sn	Zr	Mo	Other
UNALLOYED GRADES					
ASTM Grade 1	—	—	—	—	—
ASTM Grade 7	—	—	—	—	0.2 Pd
ASTM Grade 11	—	—	—	—	0.12–0.25Pd
ASTM Grade 12	—	—	—	0.3	0.6–0.9Ni
ALPHA AND NEAR ALPHA ALLOYS					
Ti-5Al-2.5Sn	5.0	2.5	—	—	—
Ti-5Al-2.5Sn-ELI	5.0	2.5	—	—	—
Ti-6Al-2Sn-4Zr-2Mo	6.0	2.0	4.0	2.0	—
Ti-6Al-2Nb01Ta-0.8Mo	6.0	—	—	1.0	2 Nb, 1 Ta
ALPHA-BETA ALLOYS					
Ti-6Al-4V	6.0	—	—	—	4.0 V
Ti-6Al-4V-ELI	6.0	—	—	—	4.0 V
Ti-6Al-6V-2Sn	6.0	2.0	—	—	0.75 Cu, 6.0V
Ti-8Mn	—	—	—	—	8.0 Mn
Ti-7Al-4Mo	7.0	—	—	4.0	—
Ti-6Al-2Sn-4Z-6Mo	6.0	2.0	4.0	6.0	—
BETA ALLOYS					
Ti-13V-11Cr-3Al	3.0	—	—	—	11.0 Cr, 13.0 V
Ti-8Mo-8V-2Fe-3Al	3.0	—	—	8.0	8.0
Ti-11.5Mo-6Zr-4.5Sn	—	4.5	6.0	11.5	—

302T11.EPS

Titanium

Commercial production of titanium did not begin until the 1950s. At that time, titanium was recognized for its importance and developed as unique lightweight, high-strength alloys for use in engine and airframe components of high-performance jet aircraft. Today, titanium alloys are widely used in many applications that previously used metals such as stainless and specialty steels, copper alloys, and nickel-based alloys.

2.5.4 Beryllium Alloys

Beryllium (Be) is a high-strength, lightweight metal. It is only about two-thirds as heavy as aluminum, but six times stiffer than steel. Beryllium's heat capacity is five times that of copper, meaning that one pound will absorb as much heat as five pounds of copper. For this reason, it is widely used in aerospace and nuclear reactor applications. It is a reactive metal and, at high temperatures, readily combines with oxygen, hydrogen, and nitrogen to form stable compounds. At low temperatures, it is highly resistant to corrosion. Small amounts of nitrogen and hydrogen or large amounts of oxygen cause embrittlement. *Table 12* lists beryllium alloys and some mechanical properties.

Beryllium is considered a difficult metal to weld because it must be kept in an inert gas atmosphere while being welded.

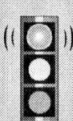 **WARNING!**
Dusts, fumes, vapors, and fine machining particles of beryllium are cancer-causing agents. If inhaled, they can cause chronic beryllium disease. An independent air supply is required when working with beryllium.

Beryllium

Beryllium allows the transmission of X-rays better than glass or other metals. Because of this property, along with its high melting point, it is widely used as the window portion of high-intensity X-ray tubes such as those used in high-resolution X-ray machines and imaging equipment.

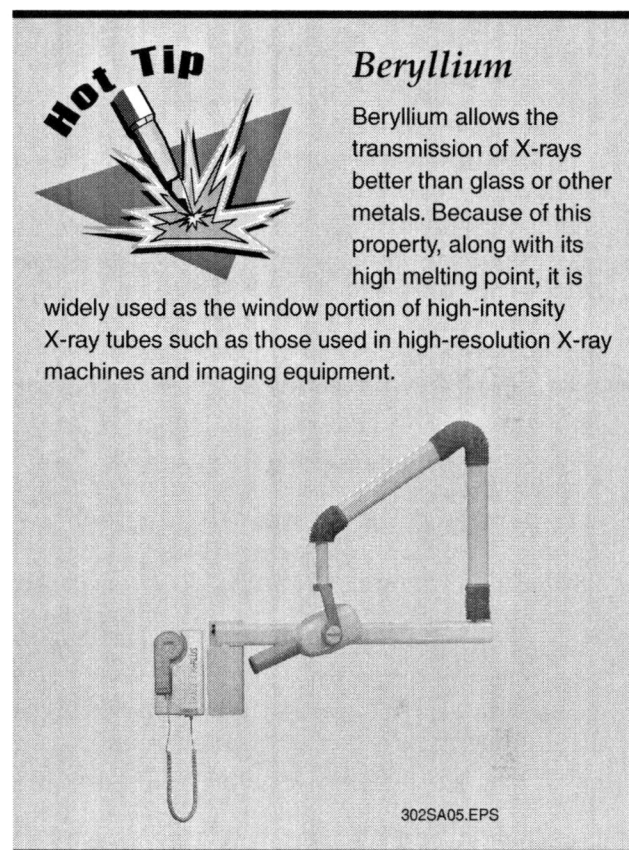

302SA05.EPS

Table 12 Examples of Beryllium Alloys and Mechanical Properties

Designation and % Be	Tensile Strength (psi)	Yield Strength (psi)	Commercial Form	Description or Properties
HP-20 (98.0%)	48,000	35,000	Block	Pressed powder
HP-8 (99.0%)	37,000	27,000	Block	Pressed powder
HP-12 (98.5%)	38,000	30,000	Block	Pressed powder
HP-40 (92.0%)	65,000	60,000	Block	Pressed powder
C-10 (99.0%)	20,000	—	Block	Pressed powder
PR-2 (98.0%)	77,000	55,000	Sheet/plate	Some form at 1,350°F
IS-2 (99.0%)	45,000	31,000	Sheet/plate	High form at 700°F
XT-20 (98.0%)	77,000	40,000	Extrusion	Good mechanical properties
XT-40 (92.0%)	90,000	50,000	Extrusion	High strength
Lockalloy	60,000	50,000	Extrusion	Service at 600°F to 800°F

2.5.5 Copper and Copper Alloys

Copper, like aluminum, is often used in nearly pure form for electrical and heat conductors. It is alloyed with zinc to make brass and with tin to make bronze. Other bronzes are formed by alloying copper with aluminum, silicon, or beryllium. Many variations of alloys exist within each of the brass and bronze groups, made by varying the percentages of the alloying metals and by adding other metallic and nonmetallic elements. These include zinc, tin, nickel, silicon, aluminum, cadmium, and beryllium.

Copper and copper alloys are classified into eight major groups:

- *Coppers* – Coppers are metals that have a designated minimum copper content of 99.3% or higher.
- *High copper alloys* – For the wrought-type products, these are alloys with copper contents less than 99.3% but more than 96% that do not fall into any other copper alloy group. Cast-type high copper alloys have copper contents in excess of 94%, to which silver may be added for special properties.
- *Brasses* – These alloys contain zinc as the principal alloying element with or without other alloying elements. The wrought-type alloys are grouped into three categories of brasses: copper-zinc alloys, copper-zinc-lead alloys (leaded brasses), and copper-zinc-tin alloys (tin brasses). Cast alloys are grouped into five categories of brasses: copper-tin-zinc alloys (red, semi-red and yellow brasses), manganese bronze alloys (high-strength yellow brasses), leaded manganese bronze alloys (leaded high-strength yellow brasses), copper-zinc-silicon alloys (silicon brasses and bronzes), and cast copper-bismuth and copper-bismuth-selenium alloys.
- *Bronzes* – Bronzes are copper alloys in which the major alloying element is not zinc or nickel. For wrought-type alloys, there are four groups of bronzes: copper-tin-phosphorus alloys (phosphor bronzes), copper-tin-lead-phosphorus alloys (leaded phosphor bronzes), copper-aluminum alloys (aluminum bronzes), and copper-silicon alloys (silicon bronzes). Cast-type alloys are divided into four groups of bronzes: copper-tin alloys (tin bronzes), copper-tin-lead alloys (leaded and high leaded tin bronzes), copper-tin-nickel alloys (nickel-tin bronzes), and copper-aluminum alloys (aluminum bronzes). Alloys known as manganese bronzes, in which zinc is the major alloying element, are included in the brasses.

- *Copper-nickels* – Copper-nickel alloys have nickel as the principal alloying element, with or without other alloying elements.
- *Copper-nickel-zinc alloys* – These alloys contain zinc and nickel as the principal and secondary alloying elements, with or without other elements. They are commonly called nickel silvers.
- *Leaded coppers* – Leaded coppers comprise a series of cast alloys of copper with 20% or more lead, sometimes with a small amount of silver but without tin or zinc.
- *Special alloys* – These are alloys whose chemical compositions do not fall into any of the above categories.

Two classification systems for copper alloys are the ASTM designation system (B224) (*Table 13*) and the Copper Development Association, Inc. designation system (*Table 14*). The Copper Development Association system has been updated to fit the Unified Numbering System (UNS) and includes all commercially available metals and alloys. The format for the UNS designation is CNNN00, in which NNN designates numbers. As shown in *Table 14*, the alloys are divided into wrought and cast alloy categories. For this reason, an alloy made in both a wrought and cast form can have different UNS numbers, depending on the method of manufacture.

Tough-pitch copper is oxygen bearing, and most of the impurities have been oxidized during refining to increase conductivity. Modern continuous casting methods have allowed the oxygen content to be reduced to 0.03% or less to achieve the desired results. During fire refining, oxygen from air injected to cause oxidation of impurities also combines with hydrogen to form steam, which creates gas pockets (porosity) in the metal during solidification. (In the past, high porosity was a characteristic of tough-pitch copper.) When tough-pitch copper is reheated by welding or annealing operations, the hydrogen trapped in the copper is diffused into the metal and reacts with cuprous oxide to form insoluble water vapor. The expansion associated with this reaction forces the grains apart, which causes embrittlement of the metal. For applications where copper must be resistant to gassing, the oxygen is sometimes eliminated by adding a deoxidizer, such as phosphorus, to the molten metal before casting. Such copper is specified as deoxidized copper.

For welding copper alloys, GMAW and GTAW are most commonly used.

Table 13 ASTM Classification System for Copper Alloys

ASTM Designation	Alloy Type	Typical Uses
TOUGH-PITCH COPPERS		
ETP	Electrolytic, tough-pitch	Bus bars, brazing rods, wire anodes, forgings
FRHC	Fire-refined, high-conductivity, tough-pitch	Mechanical applications
FRTP	Fire-refined, tough-pitch	Sheets, strips, plate
ATP	Arsenical, tough-pitch	Roofing, radiator cores
STP	Silver-bearing, tough-pitch	Pans, printing rolls, fasteners
SATP	Silver-bearing, arsenical, tough-pitch	Pans, printing rolls, fasteners
OXYGEN-FREE COPPERS		
OFHC	Oxygen-free (without residual deoxidants)	Tubing wave guides, starting anodes, wire
OFP	Oxygen-free, phosphorus-bearing	Welding rods, forging
OFTPE	Oxygen-free, tellurium- and phosphorus- bearing	Welding rods, forging
OFS	Oxygen-free, silver-bearing	Plate, sheets, rods, forging
OFTE	Oxygen-free, tellurium-bearing	Bars, free-machining
DEOXIDIZED COPPERS		
DHP	Phosphorized, high-residual phosphorus	Tubes, pipes, anodes, projectile rotating bands
DLP	Phosphorized, low-residual phosphorus	Tubes, wave guides, general use
DPS	Phosphorized, silver-bearing	Heat exchangers, steam lines,
DPA	Phosphorized, arsenical	condenser tubes, tubes for general use
DPTE	Phosphorized, tellurium-bearing	Free-machining

302T13.EPS

Table 14 UNS Designations for Classifying Copper Alloys

Copper Designation	Composition
WROUGHT ALLOY GROUPS	
C11X00	99.95% oxygen-free, high-conductivity copper
C12X00	99.88+% copper (tough-pitch copper)
C19X00	96+% copper (high-copper alloys)
C2XX00	Copper-zinc alloys (brasses)
C3XX00	Copper-zinc-lead alloys (leaded brasses)
C4XX00	Copper-zinc-tin alloys (tin brasses)
C51X00	Copper-tin alloys (phosphor bronzes)
C54X00	Copper-tin-lead alloys (leaded phosphor bronzes)
C62X00	Copper-aluminum alloys (aluminum bronzes)
C65X00	Copper-silicon alloys (silicon bronzes)
C70X00	Copper-nickel alloys
CAST ALLOY GROUPS	
C80X00	99+% copper, copper alloys
C81X00	High-copper alloys (beryllium copper)
C83X00	Red brasses and leaded red brasses
C84X00	Yellow brasses and leaded yellow brasses
C86X00	Manganese and leaded manganese-bronze alloys
C87X00	Silicon bronzes and brasses
C90X00	Tin bronzes

302T14.EPS

Statue of Liberty

The Statue of Liberty contains 179,000 pounds of copper. When copper oxidizes, it forms a blue-green self-protective coating, which is why the copper skin of the Statue of Liberty remained virtually intact after 100 years of being subjected to high sea winds, driving rains, and beating sun. Close analysis showed that weathering and oxidation of the copper skin had caused only 0.005 inch of wear in a century. For this reason, the copper skin did not need to be significantly rebuilt when the statue was renovated for its centennial. However, high-alloy copper saddles and rivets were used during the restoration to fasten the copper skin to the framing underneath. This was done to ensure the structural integrity of the statue and guard against any galvanic reaction problems. Galvanic action is corrosion caused by an electrical current produced as a result of a chemical reaction between two dissimilar metals.

302SA06.EPS

2.5.6 Nickel and Nickel Alloys

Nickel, a corrosion-resistant metal, is used as an alloying element as well as for plating other metals. Nickel-clad steel and nickel-based alloys containing over 50% nickel are used extensively in industry. They have good corrosion resistance and unique magnetic thermal expansion, thermal conductivity, and modulus of elasticity properties. Many high-strength, high-temperature nickel super alloys have been developed for aerospace, automotive, and hot-die applications. Alloys of nickel and chromium (Nichrome®) are used to make electrical resistance heating elements.

UNS and international standard designations have been assigned to many nickel and nickel-alloys. However, trade names, such as Monel 400®, Inconel 600®, and Hastelloy 22®, are more commonly used to identify them. *Table 15* lists some common nickel alloys and some of their uses.

In general, nickel and nickel-based alloys can be welded by shielded metal arc welding (SMAW), GMAW, and GTAW processes. Welding nickel alloys is very similar to welding austenitic stainless steel, with the following exceptions:

- When welding nickel alloys, the surface oxide melts at a much higher temperature than the base metal. Surfaces should be thoroughly cleaned.

- Embrittlement can be caused by lead, sulfur, phosphorus, and other low-temperature metals. The surface should be free of all contaminants.
- Weld penetration in nickel-based alloys is shallower than with other metals; therefore, it requires the use of very wide bevel and groove angles.

2.5.7 Lead, Tin, and Zinc

Lead, tin, zinc, and their alloys have very low melting points. For this reason, they are usually soldered and not welded. Zinc-plated steels (galvanized) can be welded, but the fumes are toxic and must not be inhaled.

2.5.8 Platinum, Gold, Silver, and Other Precious Metals

The precious metals (osmium, iridium, rhodium, platinum, ruthenium, gold, palladium, silver, and their alloys) are used industrially for electrical contacts, mirror coatings, catalysts, electrodes, high-temperature crucibles, and dental structures and fillings. They are usually brazed, soldered, or welded with special equipment.

Table 15 Common Nickel Alloys and Some of Their Uses

COMMERCIALLY PURE NICKEL												
ALLOY	Ni	C	Mn	Fe	S	Si	Cu	Cr	Al	Ti	Other	TYPICAL PROPERTIES AND USES
A-NICKEL	99.4	0.1	0.2	0.15	0.005	0.05	0.1					CHEMICAL INDUSTRY, ELECTROPLATING
D-NICKEL	95.2	0.1	4.5	0.15	0.005	0.05	0.05					RESISTANCE TO SULFUR ATTACK
E-NICKEL	97.7	0.1	2.0	0.10	—	0.05	0.05					SIMILAR TO D-NICKEL
L-NICKEL	99.4	0.02	0.2	0.15	0.005	0.05	0.1					SEVERE PLASTIC FORMING OPERATIONS
Z-NICKEL	94.0	0.16	0.25	0.25	0.005	0.04	0.05					AGE HARDENABLE, SPRINGS, PUMP RODS, SHAFTS

NICKEL-COPPER ALLOYS												
ALLOY	Ni	C	Mn	Fe	S	Si	Cu	Cr	Al	Ti	Other	TYPICAL PROPERTIES AND USES
MONEL®	66.0	0.12	0.90	1.35	0.005	0.15	31.50					CORROSION RESISTANCE, TOUGHNESS, HIGH STRENGTH
R-MONEL®	66.0	0.18	0.90	1.35	0.005	0.15	31.50					FREE-MACHINING
K-MONEL®	65.3	0.15	0.60	1.00	0.005	0.15	29.50		2.80	0.50		AGE-HARDENABLE, NONMAGNETIC
H-MONEL®	65.0	0.1	0.9	1.5	0.015	3.0	29.50					AGE-HARDENABLE, MACHINABLE
S-MONEL®	63.0	0.1	0.9	2.0	0.015	4.0	30.0					AGE-HARDENABLE, ANTI-GALLING, IMPELLERS, PUMP LINERS

NICKEL-IRON ALLOYS												
ALLOY	Ni	C	Mn	Fe	S	Si	Cu	Cr	Al	Ti	Other	TYPICAL PROPERTIES AND USES
INVAR®	36.0	*	*	63.0	*							VERY LOW EXPANSION COEFFICIENT LENGTH STANDARDS
PLATINITE®	46.0			54.0								GLASS SEAL
PERMALLOY®	78.5			21.5								HIGH-MAGNETIC PERMEABILITY, SUBMARINE TELEGRAPH CABLES

* COMBINED AMOUNTS OF SILICON, MANGANESE, AND CARBON TOTAL LESS THAN 1 PERCENT.

NICKEL-CHROMIUM AND NICKEL-IRON-CHROMIUM ALLOYS												
ALLOY	Ni	C	Mn	Fe	S	Si	Cu	Cr	Al	Ti	Other	TYPICAL PROPERTIES AND USES
INCONEL®	76.4	0.04	0.20	7.20	0.007	0.2	0.10	15.85				HEAT AND CORROSION RESISTANCE, NONMAGNETIC
INCONEL® 702	78.0	0.02	0.05	0.30	0.007	0.15	0.05	15.85	3.0	0.60		AGE-HARDENABLE, OPERATES UP TO 2,400°F
S-INCONEL®	68.0	0.20	1.0	8.5	0.008	5.0	0.5	15.5				AGE-HARDENABLE, ANTI-GALLING
INCONEL®-X-750	72.9	0.04	0.70	6.8	0.007	0.30	0.05	15.0	0.80	2.5	CB-0.9	AGE-HARDENABLE, USED UP TO 1,500°F
NIMONIC® 80A	74.5	0.05	0.55	0.55	0.007	0.20	0.05	20.0	1.30	2.50		AGE-HARDENABLE, SIMILAR TO INCONEL® X
NIMONIC® 90	57.0	0.05	0.50	9.50	0.007	0.20	0.05	20.45	1.65	2.60	CO-17.00	SIMILAR TO NIMONIC® 80A
NICHROME®	80.0	0.05	0.1	0.5	—	0.20	—	20.0				HEATING ELEMENTS
CHROMEL®	90.0							10.0				HEATING ELEMENTS, THERMOCOUPLES
NICOLOY 901	37.35	0.04	0.45	33.75	0.007	0.30	0.10	13.50	0.25	2.50	MO-5.90	AGE-HARDENABLE, GAS TURBINE WHEELS
Ni-O-NEL®	41.35	0.03	0.65	31.65	0.007	0.35	1.80	20.2	0.40	0.50	MO-310	RESISTANT TO CERTAIN HOT ACIDS
60-12 Ni-Cr	62.45	0.55		25.0				12.0				RESISTANCE HEATING ELEMENT
65-15 Ni-Cr	65.0			24.00	0.007			15.00				RESISTANCE HEATING ELEMENT
67-17 Ni-Cr	72.45	0.55		10.0				17.00				HEAT-RESISTING ALLOY

302T15.EPS

3.0.0 ◆ FIELD IDENTIFICATION OF BASE METALS

Many of the common base metals and alloys can be generally identified by simple tests and observation of characteristic metallic features. This is especially true of the ferrous metals, which can be identified by magnetic attraction, surface appearance, structural form or shape, and grinding sparks.

3.1.0 Metal Labeling

Because it is difficult to identify base metal types, most sites have a system to label or color-code the various types of metals being used. Labeling is performed by printing the ASTM number on the material (shape or plate) with an indelible marker or paintstick. Color-coding involves assigning a color to a base metal type and then spraying that color on the material. When color-coding is used, a master chart is often posted near the storage area. The master chart identifies the color assigned to each base metal.

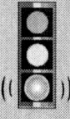

NOTE

Check your site quality standards for the type of material labeling used at your location. Follow these standards to ensure material traceability.

3.2.0 Identification by Magnet

All carbon steels, cast steels, and cast irons are attracted to a magnet. Of the stainless steels, only ferritic and martensitic stainless steels can be magnetized, and these have the bright, corrosion-free surface typical of stainless steel. The largest group of stainless steels—austenitic stainless steels—are nonmagnetic. Nonferrous metals and manganese steels are also nonmagnetic.

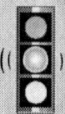

CAUTION

Magnetic testing is sometimes unreliable because of alloys or changes that can occur during welding. Do not rely solely on magnetic testing.

3.3.0 Identification by Appearance

Both hot-rolled and cold-finished carbon steel are milled in standard structural shapes, such as rods, bars, beams, channels, and angles. Unweathered, hot-rolled carbon steel has a blue-gray, smooth-to-scaly surface that oxidizes to a rough, red-brown rust in the weather. The corners of hot-rolled shapes are slightly rounded or chamfered.

Unweathered, cold-rolled carbon steel has a smooth, gray finish that oxidizes to a fine red-brown rust when exposed to the weather. Except for round shapes, the sides and edges are very flat and smooth, and the corners are well formed, not rounded or chamfered.

Cast steel and cast iron look similar on the surface. They both exhibit the shape characteristics typical of castings: curved surfaces and rounded edges, thin and thick regions, reinforcing ribs or raised areas, spokes, and irregular shapes. The cast surfaces are usually slightly textured and can contain mold seams. When broken, cast iron has a grainy appearance, while cast steel has a smooth, gray texture.

There are three types of cast iron: gray, white, and malleable. The exterior surface of all the cast irons has a slightly irregular texture from the sand molds. Malleable iron is sometimes forged. If it has been forged, it will have a smoother appearance and die marks.

Gray cast iron contains a high percentage of carbon and silicon. Gray cast iron is only formed by slow cooling, which forms large crystals and a soft, easily machined metal. During the slow cooling process, the high silicon content forces the carbon to separate out in the form of graphite flakes (free carbon). This gives a fresh fracture the characteristic grainy and gray appearance. If you rub a clean finger or white cloth against a freshly fractured gray cast-iron surface, the graphite leaves a dark smear on the finger or cloth. Gray cast iron can be welded by following special procedures.

White cast iron is chilled rapidly, which forms small crystals and a very hard and brittle metal. The carbon is united with the iron and is not free carbon. Freshly fractured cast iron has an irregular, fine, silvery-white crystalline surface. Because the weldability of white cast iron is poor, it is generally not welded.

Malleable cast iron is white cast iron that has been annealed. The annealing process results in a casting with an outer shell of white cast iron that grades into a core of gray cast iron. The resulting casting is tougher and stronger than gray cast iron and less brittle than white cast iron. Malleable iron can be welded, but special procedures must be used, or it will revert back to white cast iron.

Labeling Material

When cutting a piece of material that is labeled or color-coded, it is very important to make sure that the label or color code is visible on the material being left behind. If possible, make the cut to leave the label or color code. If this is not possible, be sure to mark the material being left to ensure that others will be able to correctly identify materials needed. It is also a good idea to mark the material being removed to ensure that everyone involved will be able to identify the material as that specified for the job.

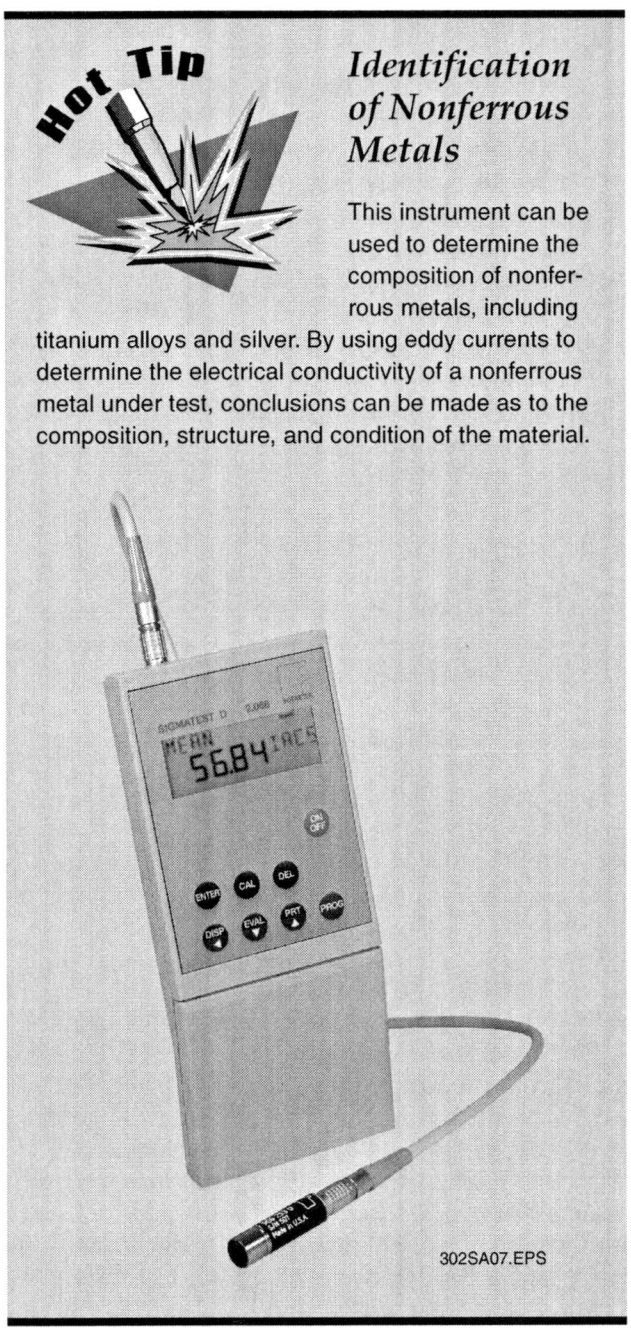

Identification of Nonferrous Metals

This instrument can be used to determine the composition of nonferrous metals, including titanium alloys and silver. By using eddy currents to determine the electrical conductivity of a nonferrous metal under test, conclusions can be made as to the composition, structure, and condition of the material.

302SA07.EPS

3.4.0 Identification by X-Ray Fluorescence Spectrometry

Portable metal analyzers (*Figure 1*) that use X-ray fluorescence (XRF) spectrometry provide a fast and accurate analysis for metal and metal alloy identification. These analyzers work by using ionizing radiation from either a radioisotope or an X-ray tube source in a probe to excite a very small sample area of an unknown metal at an atomic level. In turn, the excitation of the elemental atoms in the sample generates wavelengths of light, called photons, which are referred to as X-ray fluorescence. The wavelengths generated differ for each element and are used to identify the elemental composition of the sample. A crystal detector in the probe determines the energy level (wavelengths) of the X-rays from the depth of their penetration into the crystal. This results in a spectrum of wavelengths with varying levels of intensity. The number of counts or hits for each wavelength determines the intensity level. These intensity counts directly correlate to the percentage of each element in the sample. To identify the sample, the elemental percentages are matched to known percentages for alloys in an alloy library. The unit shown in *Figure 1* has a minimum stored library of 227 alloys and can store an unlimited number of custom-made alloy signatures in a library created by the user.

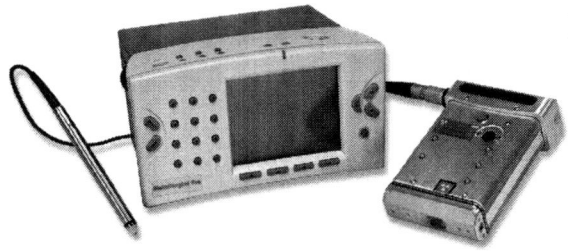

ANALYZER AND PROBE

ANALYZER IN USE   302F01.EPS

Figure 1 ◆ Typical metal analyzer.

Spark Tests

In the past, spark tests were sometimes used by metal workers to identify various types of iron, steel, and steel alloy. The test consisted of observing the characteristics of the spark shower given off when iron or steel was ground by a power grinder. This method required a great deal of skill and training on the part of the metal worker to interpret the spark shower. Because of the highly accurate metal analyzers available today, the spark test is used only on rare occasions to quickly distinguish between low and high carbon (tool) steels.

Determining the Weight and Density of Materials

One cubic inch of pure water at 62°F weighs 0.0361 pound. If you know the density (specific gravity) of a material, you can determine the weight of a cubic inch of the material by multiplying its density by 0.0361. For example, the density of aluminum is 2.71. To calculate the weight for one cubic inch of aluminum, multiply 2.71 × 0.0361 = 0.0978 lb.

To obtain the weight in pounds per cubic foot (lb/ft³) of a material, you multiply its density by 62.355. For example, one cubic foot of aluminum weighs 168.98 lbs (2.71 × 62.355 = 168.98).

If you know the weight per cubic foot of a material, you can calculate the density of the material by multiplying the weight of one cubic foot of the material by 0.01604. For example, one cubic foot of aluminum weighs 168.98 lbs. The density of aluminum is 2.71 (168.98 lb/ft³ × 0.01604 = 2.71).

4.0.0 ◆ PHYSICAL CHARACTERISTICS OF METALS

Physical characteristics are distinctive for every metal. They include the following variable traits:

- Density (specific gravity)
- Electrical conductivity
- Thermal conductivity
- Thermal expansion
- Melting point
- Corrosion resistance

4.1.0 Density

Density is the mass (weight) of a specific volume of metal. Density may also be expressed as specific gravity or specific density. Specific gravity is the weight of a metal compared to the weight of an equal volume of water. The higher the density of a material, the heavier it will be.

Light metals, such as aluminum and magnesium, have low densities (low weights for a given volume) while ferrous metals such as carbon steel and stainless steel have much higher densities. An iron or steel item weighs nearly three times as much as an aluminum item of the same size. *Table 16* lists the densities of some common metals.

4.2.0 Electrical Conductivity

Electrical conductivity is a metal's ability to conduct electricity. All metals can conduct electricity to some degree. Silver is one of the best conductors of electricity, but it is too expensive to use for commercial power transmission. Copper is not quite as good as silver, but for the same diameter conductor or wire, it is a better conductor than aluminum. However, for conductors weighing the same per unit of length, aluminum will carry more current than copper because it has a larger cross section.

The electrical conductivity of a metal is calculated from its measured resistance to current flow (its resistivity). Resistivity is the inverse (opposite) of conductivity. The lower the resistivity value, the higher the conductivity value and the more current a metal can carry for its size. The resistivity of any metal increases and its conductivity decreases as its temperature rises. This means that metals with

Table 16 Densities of Some Common Metals

Metal	Density (lb/in³)
Aluminum	2.71
Copper	8.96
Gold	19.32
Iron or steel	7.87
Magnesium	1.74
Manganese	7.43
Molybdenum	10.20
Nickel	8.90
Tin	7.30
Titanium	4.54
Tungsten	19.30
Zinc	7.13

higher electrical resistances build up heat from welding current faster. They also do not transfer heat away as efficiently, resulting in higher distortion and stresses in the weld zone. *Table 17* lists the electrical resistivity of some common metals.

4.3.0 Thermal Conductivity

Thermal conductivity is a metal's ability to conduct heat. Metals that are good conductors of electricity are also good conductors of heat. Also, as temperature increases, a metal's resistance to the flow of heat increases. Because a metal with poor heat conductivity does not carry the weld heat away from the weld zone efficiently, the temperature rises and distortion is greater. Silver and copper are excellent heat conductors, iron and nickel are fair conductors, and titanium and manganese are poor conductors.

4.4.0 Thermal Expansion

Thermal expansion is the change in size that occurs in a material as its temperature changes. Solids expand in all dimensions when heated and contract from all dimensions when cooled. The increase in unit length when a solid is heated 1° is called its coefficient of linear expansion. The larger the coefficient of expansion, the greater the dimensional increase of the solid for each degree of temperature increase. Large dimensional changes cause distortion and stresses in weldments. *Table 18* lists the coefficients of linear expansion for a variety of materials. *Figure 2* shows the expansion of common metal pipes in inches per 100 feet for a range of temperatures.

4.5.0 Melting Point

The melting point of a solid is the temperature at which the solid converts into a liquid. *Table 19* lists the melting points for some common metals.

4.6.0 Corrosion Resistance

Metals and alloys are corroded by different agents at varying rates. Corrosion is important to consider because it can severely reduce the tensile strength of a metal. When selecting a metal for a particular use, its ability to resist corrosion by the atmosphere and by gaseous or liquid chemicals may be an important factor. Certain metals may not be reactive to specific chemicals. Iron is often used to handle strong caustics. Other metals produce a protective oxide coating that retards further corrosion. Both aluminum and lead form weather-resistant oxide coatings in the atmosphere. The stainless steels and nickel-based

Table 17 Densities of Some Common Metals

Metal	Electrical Resistivitiy (Microhms/cm³)
Aluminum	2.92
Copper	1.67
Gold	2.19
Iron or steel	9.71
Magnesium	4.46
Manganese	185.00
Molybdenum	5.17
Nickel	6.84
Titanium	80.00
Zinc	5.91

Table 18 Coefficients of Linear Expansion for Some Common Metals

Metal	Millionths of an Inch/°F
Aluminum	13.3
Copper	9.2
Iron	6.5
Magnesium	14.0
Manganese	12.0
Molybdenum	2.7
Nickel	7.4
Titanium	4.7
Zinc	22.1

Table 19 Melting Points for Some Common Metals

Metal	Melting Point (°F)
Aluminum	1,215
Copper	1,981
Iron	2,800
Magnesium	1,202
Manganese	2,273
Molybdenum	4,760
Nickel	2,651
Titanium	3,300
Zinc	787

alloys are designed to withstand corrosion from such things as the atmosphere, elevated temperatures, and strong chemicals.

5.0.0 ◆ MECHANICAL PROPERTIES OF METALS

The mechanical properties of metals determine how they will react to or be affected by the external forces of tension, compression, torsion, shear, impacts, and cold shaping (working).

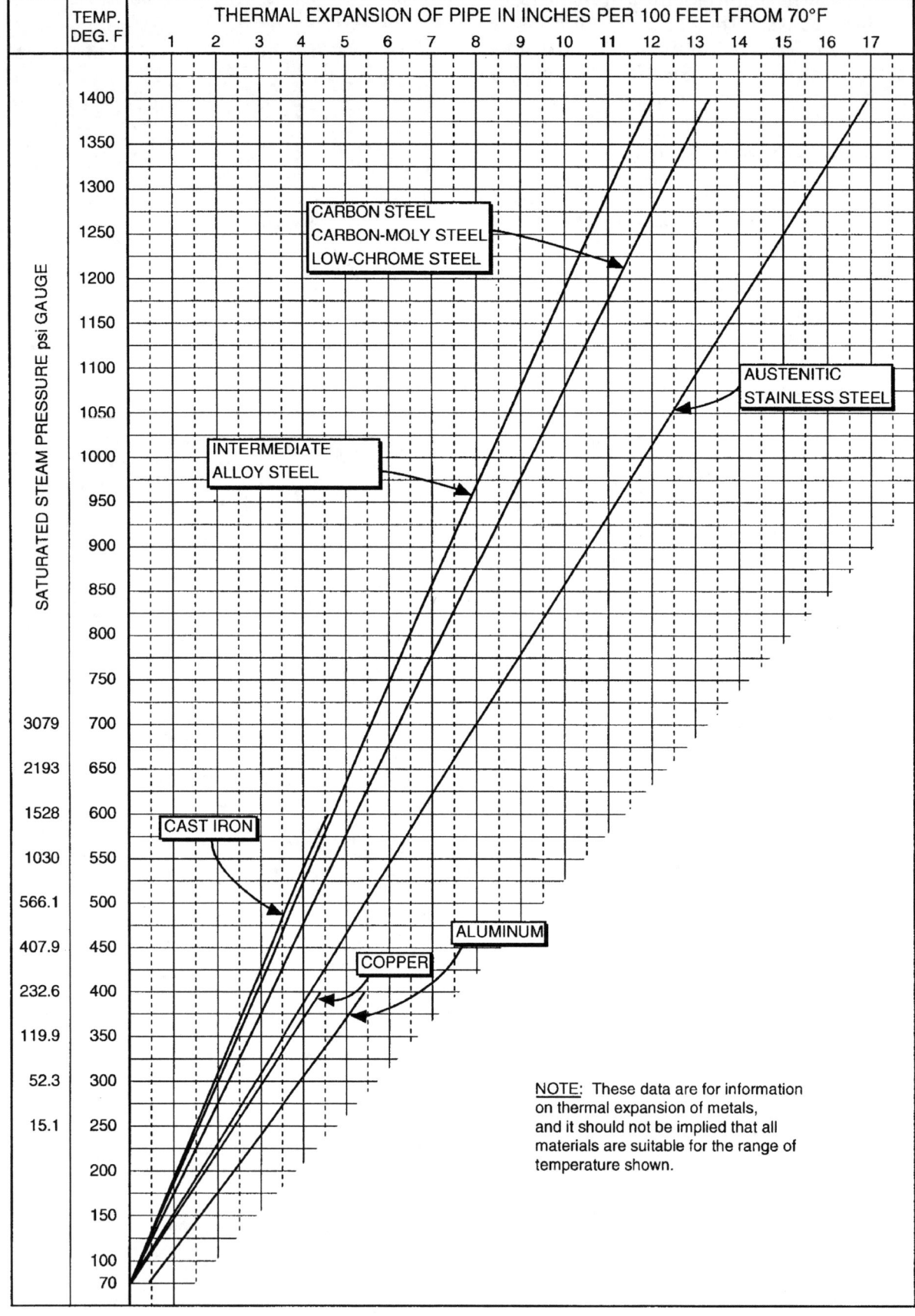

Figure 2 ◆ Expansion of different types of pipe.

5.1.0 Stress-Strain Relationship

Stress is the force or load applied to a specimen that causes it to deform. Strain is the magnitude of deformation caused by stress. Within certain limits, stress increases directly with strain.

The four types of stress are as follows:

- *Tension* – Opposed, in-line forces pulling away from each other at opposite sides of a specimen
- *Compression* – Opposed, in-line forces pushing toward each other against opposite sides of a specimen
- *Torsion* – Opposed, separated, twisting forces acting on the same axis of a specimen
- *Shear* – Opposed, nonintersecting forces acting opposite each other and perpendicular to a specimen

Figure 3 shows the force directions that result in each stress type. Often, several different stress types act on a specimen at the same time.

5.2.0 Elasticity and Elastic Limit

Elasticity is the ability of a material to be strained (deformed) without permanent deformation. As long as a material remains within its elastic range, it will return to its original size and shape when the deforming force (stress) is removed. While within the elastic range, the strain will be directly proportional to the stress that produces it.

Every material has its elastic limit. When the elastic limit is exceeded, the strain is no longer proportional to the stress, and the material will not return to its original size or shape. The material has been distorted. Elastic limits are commonly measured by an extensometer.

5.3.0 Modulus of Elasticity

Every material strains (deforms) in direct proportion to the applied stress while it remains within its elastic range, but the amount each material strains for a given stress is not the same. The ratio of stress to strain is called the modulus of elasticity. It remains the same for any material under the same proportional conditions. For example, a material that will stretch 1 inch under a 10-pound tension will stretch 2 inches under a 20-pound tension as long as the elastic limit is not exceeded. Similarly, a material that will stretch ¼ inch under a 10-pound tension will stretch ½ inch under a 20-pound tension as long as the elastic limit is not exceeded. The modulus of elasticity of a metal can be used to predict how much the metal will deform under a specified load, providing the elastic limit is not exceeded.

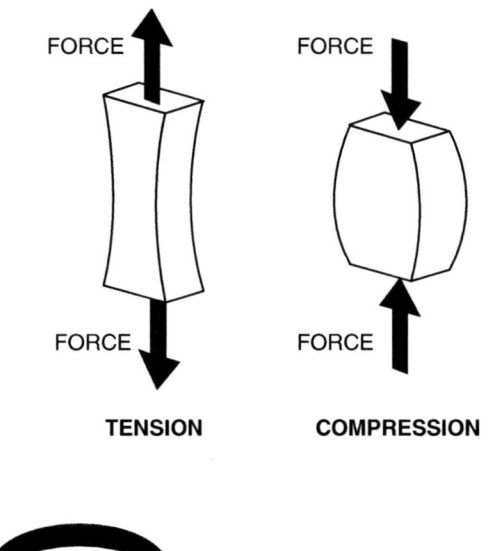

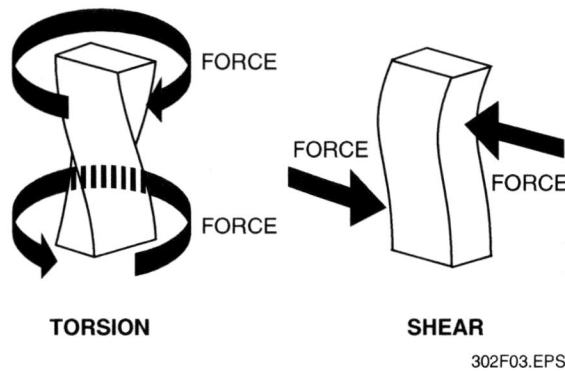

Figure 3 ◆ Force directions that result in each stress type.

5.4.0 Tensile Strength

Tensile strength (yield strength) is the maximum tensile stress a material can withstand without exceeding its elastic limit. It is expressed in force per cross-sectional unit area. The maximum load on a tensile material when it fractures is called the ultimate load. Many engineering designs are based on tensile yield strength.

5.5.0 Ductility

Ductility is the plasticity exhibited by a material under tension loading. Ductility is measured by dividing the change in length of a material by the original length at the point of failure. Ductility is the characteristic that allows metals to be drawn into thinner sections without breaking. Copper and aluminum are examples of common ductile metals.

5.6.0 Hardness

The hardness of a material can be defined as its resistance to indentation. It is related to its elastic and plastic properties. The greater the hardness, the greater the resistance to indentation, penetration,

Machines Used to Test the Properties of Metals

A wide variety of specialized machines are used to test mechanical properties of metals.

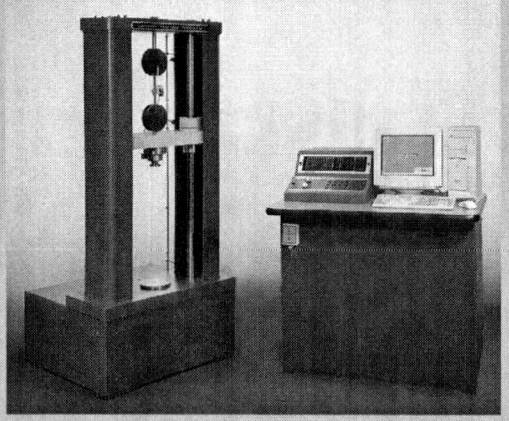

TENSILE TEST MACHINE

LABORATORY MODEL COMPRESSION-STRENGTH TESTING MACHNE

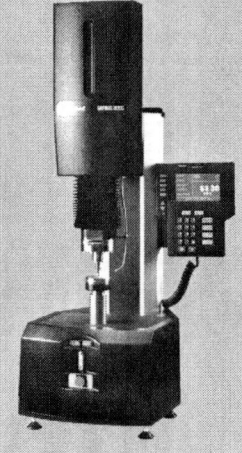

ROCKWELL HARDNESS TESTER

302SA08.EPS

and wear. Hardness is measured by pressing a steel ball or diamond point against a surface with a calibrated force. The diameter or depth of the impression is then measured to determine the hardness. Standard hardness testers include the Brinell, Rockwell, and Vickers instruments. *Figure 4* shows a portable hardness tester suitable for testing on the job site.

The Brinell hardness test is performed by pressing a hardened 10-millimeter steel ball into the metal being tested. Force is applied by standard weights: 500 kilograms for aluminum, copper, and other soft materials; and 3,000 kilograms for

iron, steel, and other hard metals. The diameter of the indentation is measured under a microscope and converted to the Brinell hardness number.

The Rockwell hardness tester uses a ¹⁄₁₆-inch hardened steel ball with a 100-kilogram weight on soft metals and uses a diamond cone and 150-kilogram weight on hard metals. Depth of penetration is read directly from hardness scales on the tester.

The Vickers hardness tester uses a pyramid-shaped diamond penetrator with a variable (1- to 120-kilogram) load. The hardness is read from standard tables by comparing the width of the dent with the pressing load.

302F04.EPS

Figure 4 ◆ Portable hardness tester.

6.0.0 ◆ STRUCTURAL STEEL AND COMMON MILLED SHAPES

Metals are commercially available in many shapes and sizes. They can be shaped by casting, rolling, drawing, forging, cutting, machining, welding, and spinning. Structural steel and other common milled shapes are formed by rolling hot or cold metal between a succession of rollers. The rollers are specially configured for the shapes they are to produce. Structural steel includes a wide range of steel types, classes, and shapes. To ensure uniform standards, the American Society for Testing and Materials (ASTM) specifies the properties of strength, weight, corrosion resistance, and weldability for various steel classifications. *Table 20* lists some common types and classifications of structural steel. Note that in the table the minimum

Table 20 Common Types and Classifications of Structural Steel

STEEL TYPE	ASTM CLASS	MINIMUM YIELD STRESS (ksi)	FORM	REMARKS
Carbon	A 36	36	Plate Shapes Bars Sheets, strips Rivets Bolts, nuts	For buildings and general structures; available in high-toughness grades
	A 529	42	Plate Shapes Bars	For buildings and similar construction
High-strength	A 440	42 to 50	Plate Shapes Bars	Lightweight and superior corrosion resistance
High-strength, low-alloy	A 441	40 to 50	Plate Shapes Bars	Primarily for lightweight welded buildings and bridges
	A 572	42 to 65	Several types Plate Shapes Bars	Lightweight high-toughness for buildings, bridges, and similar structures
Corrosion-resistant, high-strength, low-alloy	A 242	42 to 50	Plate Shapes Bars	Lightweight and added durability; weathering grades available
	A 588	42 to 50	Plate Shapes Bars	Lightweight, durable in high thickness; weathering grades available
Quenched and tempered alloy	A 514	90 to 100	Several types Plate Shapes Bars	Strength varies with thickness and type

302T20.EPS

yield stress is expressed as ksi. This is one of the standard notations used for structural steel. The term ksi is defined as kips per square inch where kips means kilo-pounds (1,000 pounds) of dead load. Hence, a yield stress of 36 ksi means 36,000 pounds per square inch of dead load.

The shape or type of structural steel is identified on drawings with a symbol or abbreviation. Size and dimensions are always given in a specified order. The specification format for designating shape and size is shown with the relevant shape in the following sections.

6.1.0 Plate

Plate is rolled metal with a uniform thickness equal to or greater than ³⁄₁₆ inch. Its identifying symbol is PL and its thickness is given in inches. Its length and width are given in feet.

Example specification: PL ¼" × 4' × 10'

6.1.1 Sheet Metal

Sheet metal is rolled metal with a uniform thickness less than ³⁄₁₆ inch. Thickness is given in inches or indicated by an American Wire Gauge (AWG) number, and length and width are given in feet. Sheet is often rolled into coils.

Example specification: Sheet No. 12 AWG × 4' × 8'

6.1.2 Bars

Bars are rolled to a variety of cross-section shapes and sizes (*Figure 5*). Standard bar shapes include the following:

- *Round* – Round bar is specified by BAR, followed by the bar diameter, a circle with a slash through it, and then the length.
- *Square* – Square bar is specified by BAR, followed by the face width, a square with a slash through it, and then the length.
- *Bar (flat)* – Flat bar is specified by BAR, followed by the wide dimension, the narrow dimension, and then the length.
- *Z-bar* – Z-bar is specified by Z, followed by the **flange** width, the **web** depth, the flange and web thickness, and then the length. The flange and web are always the same thickness.
- *Hexagonal* – Hexagonal bar is specified by HEX, followed by the bar thickness (measured across the flats), and then the length.
- *Octagonal* – Octagonal bar is specified by OCT, followed by the bar thickness (measured across the flats), and then the length.

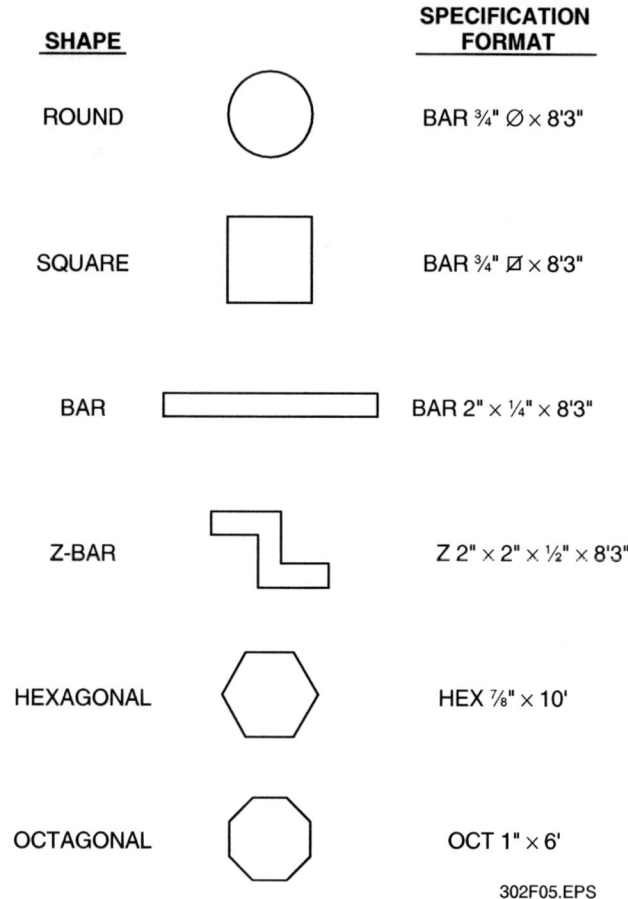

SHAPE	SPECIFICATION FORMAT
ROUND	BAR ¾" ∅ × 8'3"
SQUARE	BAR ¾" ⊡ × 8'3"
BAR	BAR 2" × ¼" × 8'3"
Z-BAR	Z 2" × 2" × ½" × 8'3"
HEXAGONAL	HEX ⅞" × 10'
OCTAGONAL	OCT 1" × 6'

302F05.EPS

Figure 5 ◆ Bar shapes and specification formats.

6.1.3 Angles

Angles are L-shaped bars that have either equal-sized or unequal-sized legs that are always at 90 degrees to each other. *Figure 6* shows both types of angle bar stock and the specification formats for each type. The specification format represents the size and thickness of the angle bar stock legs in inches. Thus, L2 × 2 × ⅜ indicates a 2" × 2" angle of ⅜" thickness. For unequal leg stock, the longest leg is listed first, as shown in *Figure 6*.

6.1.4 Channels

Channels are U-shaped forms made of two flanges connected by a common web. The flanges extend from the same side of the web. The flanges can be of uniform thickness or tapered toward the outer edges. There are several channel variations, each having a unique specification format designation. *Figure 7* illustrates two different channels and their related American Standard Channels specification format designations. The channel designated as C8 × 11.5 represents a channel having a nominal depth of 8" and a weight of 11.5 pounds per foot of length. Similarly, the channel designated C8 × 18.75 has a nominal depth of 8" and a weight of 18.75 pounds

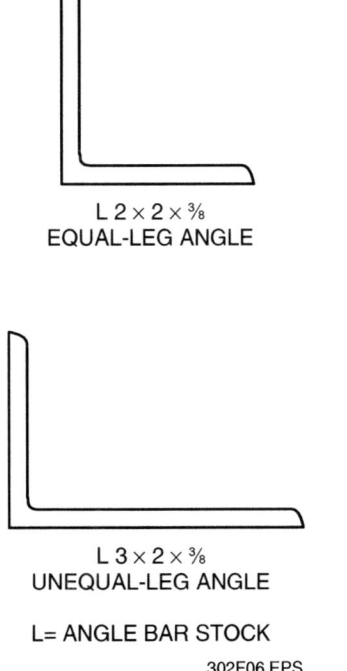

L 2 × 2 × ⅜
EQUAL-LEG ANGLE

L 3 × 2 × ⅜
UNEQUAL-LEG ANGLE

L = ANGLE BAR STOCK

302F06.EPS

Figure 6 ◆ Angle bar stock and specification formats.

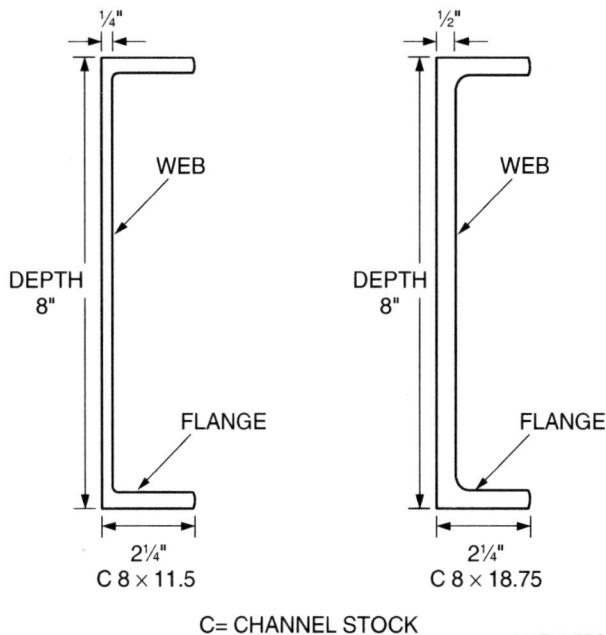

C 8 × 11.5   C 8 × 18.75

C = CHANNEL STOCK

302F07.EPS

Figure 7 ◆ Channels and specification formats.

per foot of length. As shown in the figure, the channel designated as C8 × 11.5 has a flange width of 2¼" and a web thickness of ¼"; while the C8 ×18.75 channel has a flange width of 2¼" and a web thickness of ½". The specific dimensions for the beam depth, flange width, and web thickness for all common channel designations can be found in related tables of dimensions and properties that are provided in most structural engineering and architectural standards reference books.

6.1.5 Beams and Shapes from Beams

Beams (*Figure 8*) are made in I-shaped, H-shaped, and T-shaped cross sections and are made with flat or tapered flanges. In addition to different flange widths and web depths, the thickness of the flanges and webs vary with beam sizes. Also, beam weight (pounds per linear foot) for a beam of a given dimension can be increased by adding thickness to the web and flanges with very little

change in beam width or depth. The specification designator describes the beam type and its nominal depth in inches and the weight of the beam per foot of length. An example specification designator for a beam is S24 × 120, where S24 represents the beam type (S) and nominal depth in inches (24). The 120 is the weight in pounds per foot of length. The specific dimensions for the depth, web thickness, and flange thickness for the various designations of beams are given in related tables of dimensions and properties in most structural engineering and architectural standards books.

Beams are identified with a standard set of symbols. The I-beam and H-beam letter symbols include the following:

- *S (American Standard)* – I-shaped beams with tapered flanges
- *W (Wide flange)* – Wider flanges than S beams with thinner webs and nontapered flanges
- *M* – Similar to W beams but with short flanges
- *HP* – H-shaped beams with nontapered flanges

Miscellaneous Channel Stock

In addition to channels specified in the American Standard Channel format, some channels are specified in the Miscellaneous Channels designation format. Miscellaneous Channels are usually lighter channels than American Standard Channels. The only difference in the specification format is that the format for the Miscellaneous Channels starts with the letters MC instead of C, as used with the American Standard Channels.

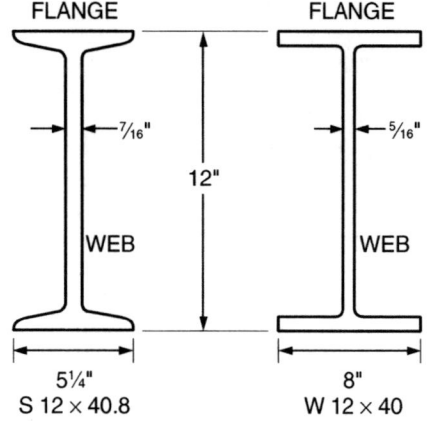

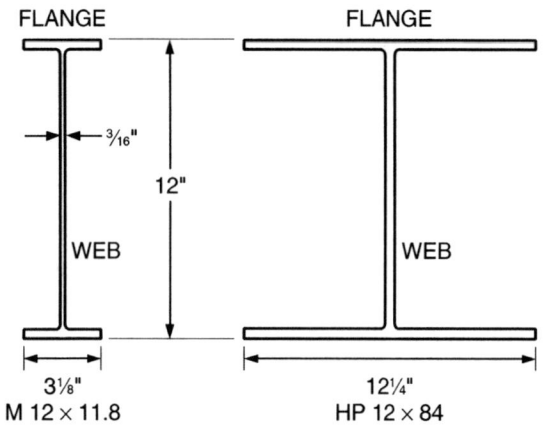

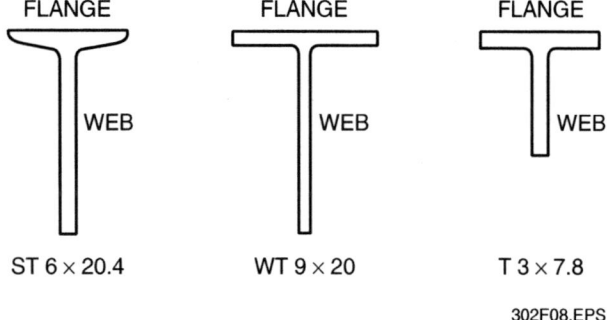

Figure 8 ◆ Structural beam and T-beam shapes.

302F08.EPS

T-beams include structural tees and tee shapes. Structural tees are made by cutting S, W, and M beams down the center, usually by shearing. The conventional symbol for a structural tee is the original beam letter followed by a T. For example, a tee cut from an S8 × 18.4 beam would be identified as ST4 × 9.2.

T-shapes are rolled into their final tee shape. The symbol for tee shapes is a T, without any other letter. The specification gives the nominal depth, flange width, thickness, and length. An example specification is T3 × 7.8. T-shapes are short pieces, commonly used as connections or supports.

6.1.6 Pipe

Pipe size is specified as a nominal ID (approximate inside diameter) up through 12 inches. Pipe 14 inches and larger is specified by the actual OD (outside diameter). Pipe weight (determined by the wall thickness) is specified by a schedule number or strength. Schedule numbers are 5, 10, 20, 40, 80, and 160. Strengths are specified as standard (STD), extra strong (XS), and extra-extra strong (XXS). Standard falls between Schedules 20 and 40. Extra strong falls between Schedules 40 and 80, Extra-extra heavy is Schedule 160 or higher. Because the OD of a given pipe size is always the same, the ID becomes smaller as the pipe wall thickness increases with higher schedule pipe. Table 21 lists standard pipe sizes and schedules with nominal wall thicknesses.

6.1.7 Reinforcing Bars

Reinforcing bars (rebars), sometimes called rerods, are used for concrete reinforcement. They are available in several grades. These grades vary in yield strength, ultimate strength, percentage of elongation, bend-test requirements, and chemical composition. Reinforcing bars can be coated with different compounds, such as epoxy, for use in concrete where corrosion could be a problem. The ASTM has established standard specifications for

I-Beams

The older American Standard I-beams have been replaced by the American Institute of Steel Construction (AISC) wide flange (W) beams. These wide flange shapes are now used for most beams and columns. The wider flange provides a stronger beam than an older I-beam with the same web width.

Table 21 Standard Pipe Sizes and Schedules with Nominal Wall Thicknesses

Nominal Pipe Size	Outside Diam.	Sched. 10	Sched. 20	Sched. 30	STD	Sched. 40	Sched. 60	XS	Sched. 80	Sched. 100	Sched. 120	Sched. 140	Sched. 160	XXS
1/8	0.405	—	—	—	0.068	0.068	—	0.095	0.095	—	—	—	—	—
1/4	0.540	—	—	—	0.088	0.088	—	0.119	0.119	—	—	—	—	—
3/8	0.675	—	—	—	0.091	0.091	—	0.126	0.126	—	—	—	—	—
1/2	0.840	—	—	—	0.109	0.109	—	0.147	0.147	—	—	—	0.188	0.294
3/4	1.050	—	—	—	0.113	0.113	—	0.154	0.154	—	—	—	0.219	0.308
1	1.315	—	—	—	0.133	0.133	—	0.179	0.179	—	—	—	0.250	0.358
1-1/4	1.660	—	—	—	0.140	0.140	—	0.191	0.191	—	—	—	0.250	0.382
1-1/2	1.900	—	—	—	0.145	0.145	—	0.200	0.200	—	—	—	0.281	0.400
2	2.375	—	—	—	0.154	0.154	—	0.218	0.218	—	—	—	0.344	0.436
2-1/2	2.875	—	—	—	0.203	0.203	—	0.276	0.276	—	—	—	0.375	0.552
3	3.5	—	—	—	0.216	0.216	—	0.300	0.300	—	—	—	0.438	0.600
3-1/2	4.0	—	—	—	0.226	0.226	—	0.318	0.318	—	—	—	—	—
4	4.5	—	—	—	0.237	0.237	—	0.337	0.337	—	0.438	—	0.531	0.674
5	5.563	—	—	—	0.258	0.258	—	0.375	0.375	—	0.500	—	0.625	0.750
6	6.625	—	0.250	—	0.280	0.280	—	0.432	0.432	—	0.562	—	0.719	0.864
8	8.625	—	0.250	0.277	0.322	0.322	0.406	0.500	0.500	0.594	0.719	0.812	0.906	0.875
10	10.75	—	0.250	0.307	0.365	0.365	0.500	0.500	0.594	0.719	0.844	1.000	1.125	1.000
12	12.75	—	0.250	0.330	0.375	0.406	0.562	0.500	0.688	0.844	1.000	1.125	1.312	1.000
14 OD	14.0	0.250	0.312	0.375	0.375	0.438	0.594	0.500	0.750	0.938	1.094	1.250	1.406	—
16 OD	16.0	0.250	0.312	0.375	0.375	0.500	0.656	0.500	0.844	1.031	1.219	1.438	1.594	—
18 OD	18.0	0.250	0.312	0.438	0.375	0.562	0.750	0.500	0.938	1.156	1.375	1.562	1.781	—
20 OD	20.0	0.250	0.375	0.500	0.375	0.594	0.812	0.500	1.031	1.281	1.500	1.750	1.969	—
22 OD	22.0	0.250	0.375	0.500	0.375	—	0.875	0.500	1.125	1.375	1.625	1.875	2.125	—
24 OD	24.0	0.250	0.375	0.562	0.375	0.688	0.969	0.500	1.218	1.531	1.812	2.062	2.344	—
26 OD	26.0	0.312	0.500	0.625	0.375	—	—	0.500	—	—	—	—	—	—
28 OD	28.0	0.312	0.500	0.625	0.375	—	—	0.500	—	—	—	—	—	—
30 OD	30.0	0.312	0.500	0.625	0.375	—	—	0.500	—	—	—	—	—	—
32 OD	32.0	0.312	0.500	0.625	0.375	0.688	—	0.500	—	—	—	—	—	—
34 OD	34.0	0.312	0.500	0.625	0.375	0.688	—	0.500	—	—	—	—	—	—
36 OD	36.0	0.312	0.500	0.625	0.375	0.750	—	0.500	—	—	—	—	—	—
42 OD	42.0	—	—	—	0.375	—	—	0.500	—	—	—	—	—	—

Nominal Wall Thickness (in.)

302T21.TIF

reinforcing bars. These grades appear on bar-bundle tags, in color coding, in rolled-on markings on the bars, and/or on bills of materials. The specifications are as follows:

- *A615, Standard Specification for Deformed and Plain Billet-Steel Bars for Concrete Reinforcement*
- *A616, Standard Specification for Rail-Steel Deformed Bars for Concrete Reinforcement*
- *A617, Standard Specification for Axle-Steel Deformed Bars for Concrete Reinforcement*
- *A706, Standard Specification for Low-Alloy Steel Deformed Bars for Concrete Reinforcement*

Why Doesn't the Pipe Size Match Pipe ID or OD?

When pipe was first made, it was from lead, which required a thick wall to hold the specific pressure for a given ID. When the switch was made from lead pipe to steel pipe, a decision had to be made whether to maintain OD so the steel pipe will fit existing fittings or to change OD and keep the ID size.

The decision was made to maintain OD to provide direct replacement with the existing fittings. At that point, neither nominal ID nor OD corresponded to size.

Steel Pipe

The only difference between black iron pipe and galvanized pipe is the addition of a protective zinc coating on the galvanized pipe.

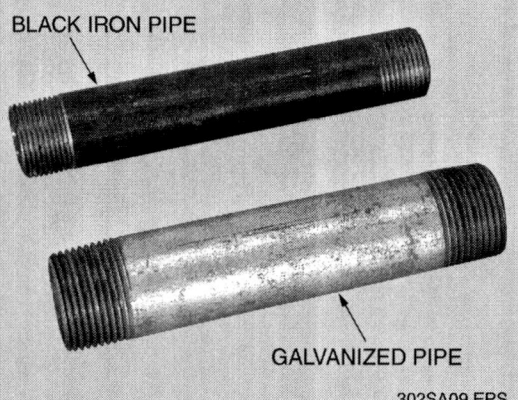

BLACK IRON PIPE

GALVANIZED PIPE

302SA09.EPS

The standard configuration for reinforcing bars is the deformed bar. Different patterns may be impressed upon the bars depending on which mill manufactured them, but all are rolled to conform to ASTM specifications. The purpose of the deformation is to improve the bond between the concrete and the bar and to prevent the bar from moving within the concrete. Plain bars are smooth and round without deformations on them and are used for special purposes, such as for dowels at expansion joints where the bars must slide in a sleeve, for expansion and contraction joints in highway pavement, and for column spirals. Deformed bars are designated by a number in eleven standard sizes (metric or inch-pound) as shown in *Table 22*. The number denotes the approximate diameter of the

Table 22 ASTM Standard Metric and Inch-Pound Reinforcing Bars

Bar Size		Nominal Characteristics					
		Diameter		Cross-Sectional Area		Weight	
Metric	[in.-lb]	mm	[in.]	mm	[in.]	kg/m	[lbs/ft.]
#10	[#3]	9.5	[0.375]	71	[0.11]	0.560	[0.376]
#13	[#4]	12.7	[0.500]	129	[0.20]	0.944	[0.668]
#16	[#5]	15.9	[0.625]	199	[0.31]	1.552	[1.043]
#19	[#6]	19.1	[0.750]	284	[0.44]	2.235	[1.502]
#22	[#7]	22.2	[0.875]	387	[0.60]	3.042	[2.044]
#25	[#8]	25.4	[1.000]	510	[0.79]	3.973	[2.670]
#29	[#9]	28.7	[1.128]	645	[1.00]	5.060	[3.400]
#32	[#10]	32.3	[1.270]	819	[1.27]	6.404	[4.303]
#36	[#11]	35.8	[1.410]	1006	[1.56]	7.907	[5.313]
#43	[#14]	43.0	[1.693]	1452	[2.25]	11.38	[7.65]
#57	[#18]	57.3	[2.257]	2581	[4.00]	20.24	[13.60]

302T22.EPS

WELDING LEVEL THREE — TRAINEE MODULE 29302-03

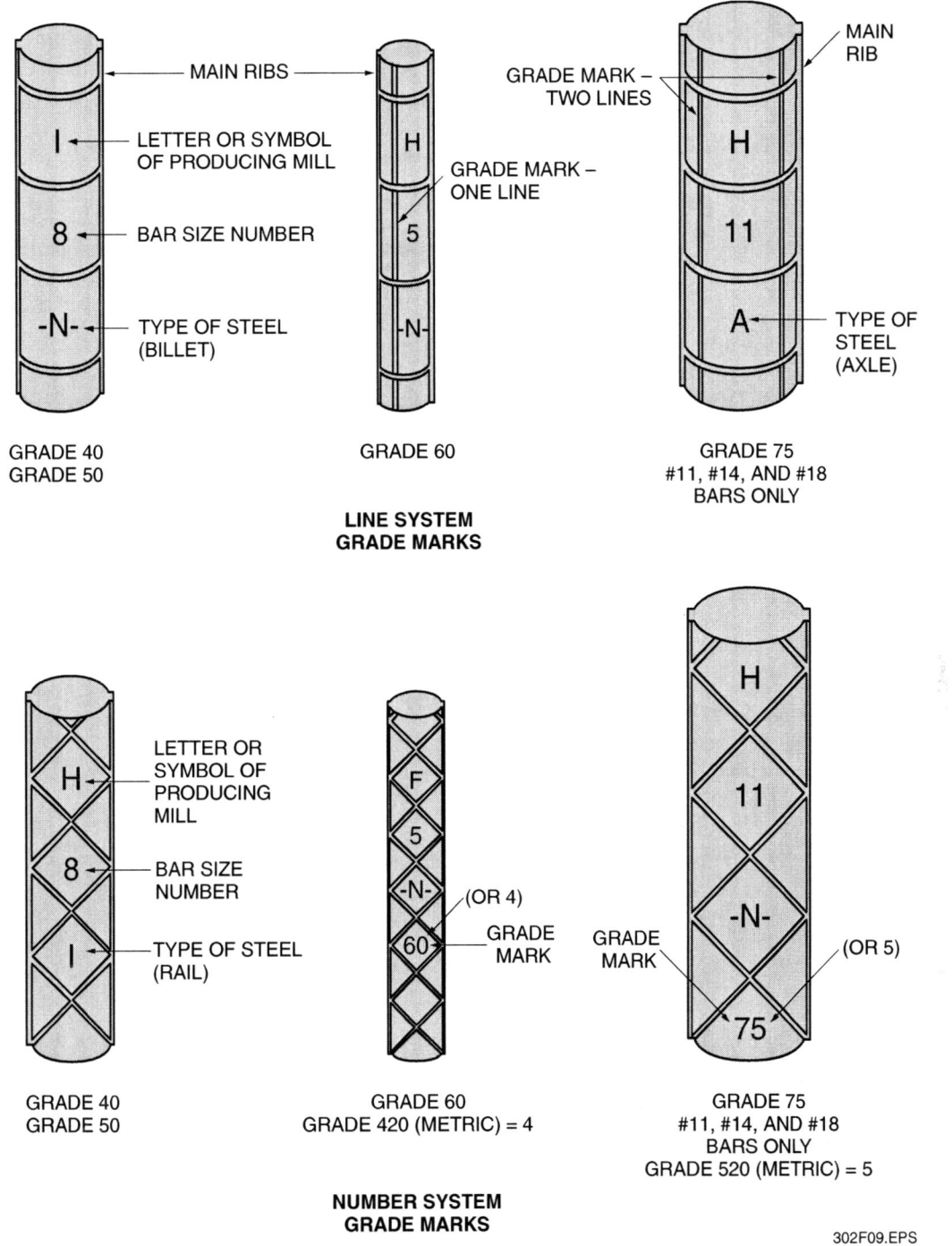

LINE SYSTEM GRADE MARKS

MAIN RIBS

LETTER OR SYMBOL OF PRODUCING MILL

BAR SIZE NUMBER

TYPE OF STEEL (BILLET)

GRADE 40
GRADE 50

GRADE MARK – ONE LINE

GRADE 60

MAIN RIB

GRADE MARK – TWO LINES

TYPE OF STEEL (AXLE)

GRADE 75
#11, #14, AND #18
BARS ONLY

NUMBER SYSTEM GRADE MARKS

LETTER OR SYMBOL OF PRODUCING MILL

BAR SIZE NUMBER

TYPE OF STEEL (RAIL)

GRADE 40
GRADE 50

(OR 4)
GRADE MARK

GRADE 60
GRADE 420 (METRIC) = 4

GRADE MARK

(OR 5)

GRADE 75
#11, #14, AND #18
BARS ONLY
GRADE 520 (METRIC) = 5

302F09.EPS

Figure 9 ◆ Reinforcement bar identification.

bar in eighths of an inch or metric (mm). For example, a #5 bar has an approximate diameter of ⅝ inch (0.625). The nominal dimension of a deformed bar (nominal does not include the deformation) is equivalent to that of a plain bar having the same weight per foot. As shown in *Figure 9*, bar identification is accomplished by ASTM specifications, which require that each bar manufacturer roll the following information onto the bar:

- A letter or symbol to indicate the manufacturer's mill
- A number corresponding to the size number of the bar (*Table 22*)
- A symbol or marking to indicate the type of steel (*Table 23*)
- A marking to designate the grade (*Table 24*)

The grade represents the minimum yield (tension strength), measured in kips per square inch (ksi) or megapascals (MPa), that the type of steel used will withstand before it permanently stretches (elongates) and will not return to its original length. Today, Grade 420 is the most commonly used rebar. Rebars are normally supplied from the mill bundled in 60-foot lengths (*Figure 10*) that are then cut in the field to the required length.

6.1.8 Tubing

Tubing is manufactured in square, rectangular, and round cross sections. Round tubing can be distinguished from pipe by its dimensions. Standard round tubing is measured by a nominal OD.

The cross sections of square and rectangular tubing have even outside dimensions and slightly rounded corners. *Figure 11* shows standard tubing shapes and specification formats.

6.1.9 Seamed and Seamless Tubing and Pipe

Both standard and nonstandard shapes can also be produced by other processes. For example, tubing can be produced by extrusion or by rolling and welding. Extrusions can be produced in an almost limitless variety of shapes. When tubing is produced by rolling and welding, the most common welding process used is electronic resistance welding (ERW).

6.1.10 Forged Shapes

Forging is used to produce specialized shapes where high strength is required. An example is forged high-pressure pipe fittings such as flanges and elbows.

6.1.11 Cast Shapes

Casting is another method of producing shapes. Cast iron and steel are used for machine parts, automobile components, and large castings.

6.1.12 Powdered Metals

Commercially prepared powdered metals are used for welding fillers, surface coatings, and mold castings. When cast to make parts, the molded powder is usually sintered (partially welded without melting) to fuse it into a solid metal. This is a common technique for manufacturing complex-shaped parts or porous parts. It is also used with high-temperature metals that cannot be liquid cast.

302F10.EPS

Figure 10 ◆ Reinforcing bars.

Table 23 Reinforcement Bar Steel Types

Symbol/Marking	Type of Steel
A	Axle (ASTM A617)
S or N	Billet (ASTM A615)
I or IR	Rail (ASTM A616)
W	Low-alloy (ASTM A706)

Table 24 Reinforcement Bar Grades

Grade	Identification	Minimum Yield Strength
40 and 50	None	40,000 to 50,000 psi (40 to 50 ksi)
60	One line or the number 60	60,000 psi (60 ksi)
75	Two lines or the number 75	75,000 psi (75 ksi)
420	The number 4	420 MPa (60,000 psi or 60 ksi)
520	The number 5	520 MPa (75,000 psi or 75 ksi)

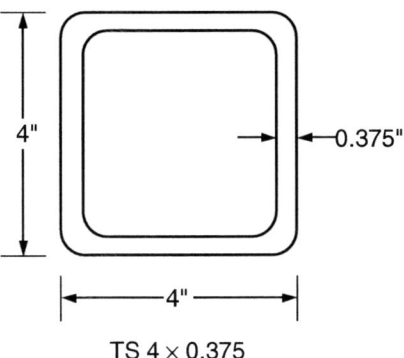

TS 4 × 0.375

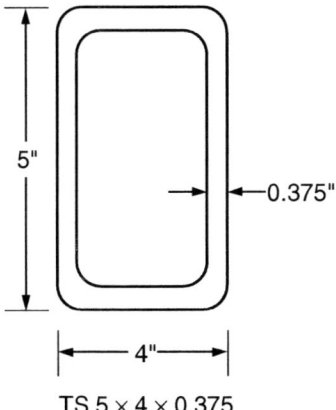

TS 5 × 4 × 0.375

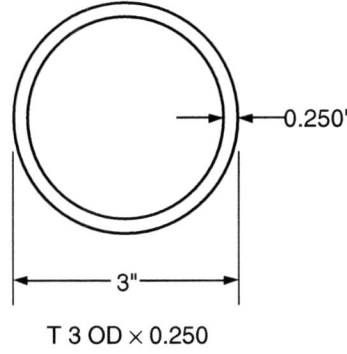

T 3 OD × 0.250

302F11.EPS

Figure 11 ◆ Standard tubing shapes and specification formats.

7.0.0 ◆ METALLURGICAL CONSIDERATIONS FOR WELDING

You must consider many metallurgical variables when determining the effects of welding on both the base metal and the weld. Base metal composition and thickness, filler metal, and joint configuration all affect the weld quality. There are also several processes and design considerations used to change the metal characteristics to produce a more desirable weld. The following sections explain the necessary metallurgical considerations for making acceptable welds:

- Base metal preparations
- Joint design considerations
- Filler metal selection
- Filler metal and electrode selection considerations
- Preheating and interpass temperature control
- Postweld heat treatment

7.1.0 Base Metal Preparations

Because of the defects caused by surface contaminants and the possibility of toxic fumes, all welding codes require surface cleaning. In general, the codes state that the surface to be thermally cut or welded must be clean and free from paint, oil, rust, scale, and other material that would be detrimental to either the weld or the base metal when heat is applied. Cleaning is typically performed by mechanical and/or chemical means.

The most common defect caused by surface contamination is porosity, which occurs when gas pockets or voids appear in the weld metal. When the porosity has a length greater than its width and is approximately perpendicular to the weld face, it is called piping porosity. Piping porosity is formed as the gas pocket floats toward the surface of the weld, leaving a void. The gas pocket is trapped as the weld metal solidifies, but as the next layer of weld is deposited, it continues to float up. Porosity and piping porosity may not be visible on the surface of the weld.

ACR Copper Pipe

Copper pipe (tubing) used in air conditioning and refrigeration systems and equipment is different than the copper pipe used for general plumbing applications. Air conditioning and refrigeration (ACR) copper pipe and fittings are manufactured specifically for use in refrigeration systems. ACR pipe is thoroughly cleaned, dried, capped, and sometimes charged with nitrogen to help prevent contamination of air conditioning and refrigeration systems. Another important difference is that the nominal size of ACR pipe is expressed by its OD, whereas copper pipe used in general plumbing applications is expressed in terms of its ID.

7.1.1 Chemical Cleaning

Chemicals are used to clean oil and grease off metal surfaces that require high-quality welds. These chemicals are very strong, which enables them to do their job quickly. Because they are so strong, they can be very hazardous if mishandled.

There are several different types of chemicals to clean metal. The type of chemical used depends on the base metal to be cleaned. Using the wrong chemical can cause a reaction with the base metal that results in burning, pitting, or discoloration of the surface. The cleaning chemical should be selected specifically for the type of base metal and the application.

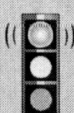 **WARNING!**
Never use a chemical unless you have been specifically and recently trained in its use. Mishandling cleaning chemicals can cause severe burns to the skin, lungs, and eyes. Some chemicals can also cause death or blindness. Always use approved solvents. Procedures and methods for using, storing, and disposing of most chemicals are given in the material safety data sheets (MSDS) for the specific chemical being used. These are available at your shop or job site.

7.1.2 Mechanical Cleaning

Mechanical cleaning is the most common method of removing surface contamination. Tools used for mechanical cleaning include hand tools, power tools, and special sandblasting equipment.

 **WARNING!**
When performing mechanical cleaning, be sure to wear safety glasses and a face shield for protection from the flying particles produced during the cleaning operation. In addition, special clothing is required for sandblasting.

7.2.0 Joint Design Considerations

Welded joints are selected primarily for the safety and strength required for the service conditions. When selecting the joint, groove, and weld to use, there are many considerations to take into account. The following sections explain those considerations.

7.2.1 Codes and Welding Procedure Specifications

A welding code is a detailed listing of the rules and principles that apply to specific welded products. Codes ensure that safe and reliable welded products will be produced and that the weldments will be reasonably safe. Clients specify which codes to follow when placing orders or letting contracts. These orders or contracts impose severe penalties for not conforming to the code(s) specified. In addition, when codes are specified, the use of these codes is mandated with the force of law by one or more government jurisdictions. Always check the contract, order, or project specification for the code(s) specified.

Codes require that WPSs be written for all critical welds. A WPS is a written set of instructions for producing sound welds. It includes the type of joint to use as well as the groove preparation, if required. Each WPS is written and tested in accordance with a particular welding code or specification. All welding requires that acceptable industry standards be followed, but not all welds require a WPS. If a weld does require a WPS, it must be followed. The requirement to use a WPS is often listed on job blueprints as a note or in the tail of the welding symbol. If you are unsure whether the welding being performed requires a WPS, do not proceed until you check with your supervisor.

7.2.2 Joint Preparation

One of the easiest joint preparations is the square groove joint because it only requires butting the edges of the plate together. Butt joints are welded with a square groove weld that can have a partial or complete joint penetration weld. A partial joint penetration weld has much less strength than a complete penetration weld. *AWS D1.1* imposes restrictions for complete penetration welds. With shielded metal arc welding (SMAW), the maximum base metal thickness is ¼ inch and welding from both sides is required. In addition, a root opening of half the thickness of the base metal is required, and you must gouge the root of the first weld before making the second weld. For gas metal arc welding (GMAW) and flux cored arc welding (FCAW), the maximum base metal thickness is ⅜ inch with the same requirements for welding from both sides and for back-gouging.

Fillet welds also require very little joint preparation. When using a fillet weld on outside corner welds, two types of fit-up can be used: corner to corner or half lap. The corner-to-corner joint is difficult to assemble because neither plate can support the other, and you must be careful not to burn through the corner when welding. The half-lap joint is easier to assemble, requires less weld metal, and has less danger of corner burn-through. The half-lap requires a second weld on the inside of the corner. If a half-lap fit-up is used, you must make allowances in the plate dimensions for the lap. Several methods of joint preparation are explained in the following:

- *Single- and double-groove welds* – When possible, you should use a double-groove joint in place of a single groove. The double groove requires half the weld metal that the single groove does. Also, welding from both sides reduces distortion because the forces of distortion work against each other. *Figure 12* shows single- and double-groove welds.

- *U- and J-groove preparation* – On thick joints, U- and J-grooves often require less weld metal than V or bevel joints. The major disadvantage of U- and J-grooves is that they require more preparation time. For this reason, V and bevel joints are used much more frequently than J- or U-grooves.

- *Groove angles and root openings* – The purpose of the groove angle is to allow electrode access to the root of the weld. If the groove angle is larger than necessary, it requires additional weld metal to fill. This increases the time and cost to complete the weld and increases the chance of distortion. If the groove angle is too small, it can result in weld discontinuities. The root preparation is sized to control melt-through. Increasing or decreasing the root preparation (root opening

and root face) results in excess melt-through or insufficient root penetration. As a general rule, as the groove angle decreases, the root opening increases to compensate. Root faces are used with open-root joints but not when metal backing strips are used (except for aluminum). *Figure 13* shows groove angles and root openings.

- *Open-root welds* – For open-root welds on plate, the groove angle is 60 degrees, and the maximum size of the root opening and root face is ⅛ inch. For open root welds on pipe, the groove angle is 60 or 75 degrees, depending on the code or specifications used. The maximum size of the root opening and root faces is referred to in the WPS. *Figure 14* shows open-root joint preparation.

7.3.0 Filler Metal Selection

Because there are so many different electrodes and filler metals available, you must be able to distinguish between them and select the correct

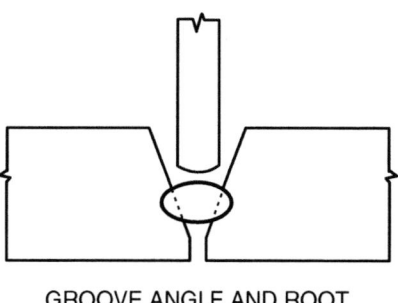

GROOVE ANGLE AND ROOT
OPENING TOO SMALL

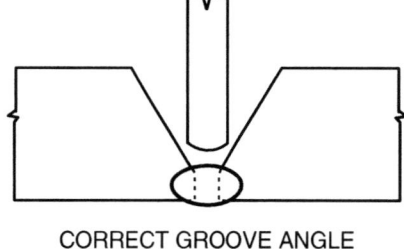

CORRECT GROOVE ANGLE
AND ROOT OPENING

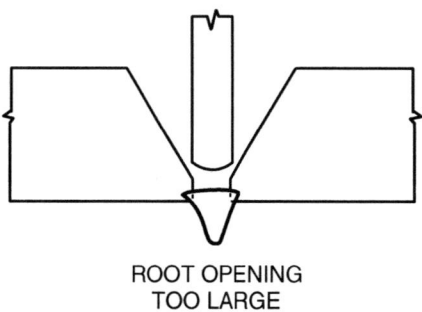

ROOT OPENING
TOO LARGE

302F13.EPS

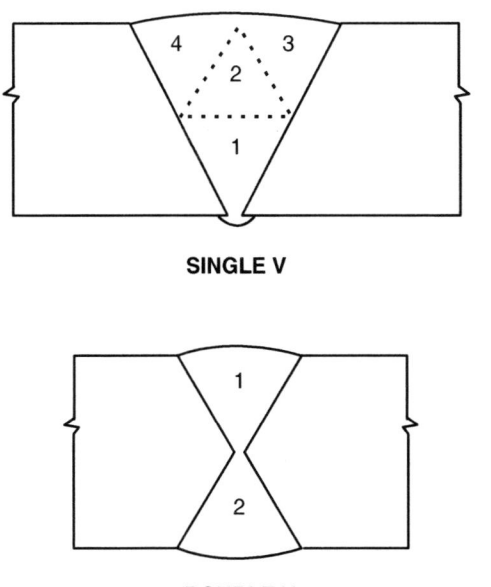

SINGLE V

DOUBLE V

302F12.EPS

Figure 12 ◆ Single- and double-groove welds.

Figure 13 ◆ Groove angles and root openings.

Joint and Surface Preparation

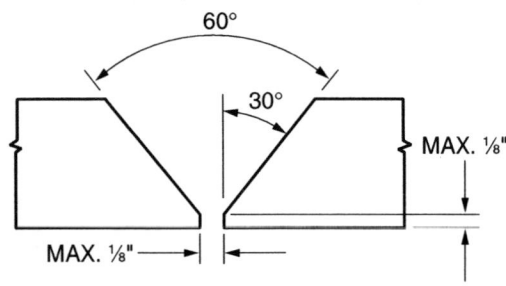

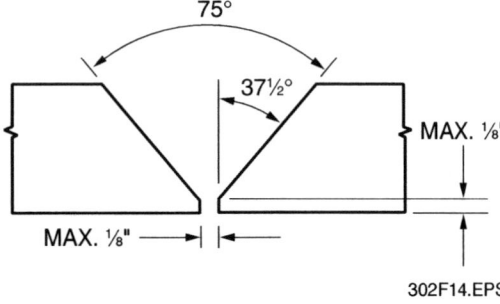

302F14.EPS

Figure 14 ◆ Open-root joint preparation.

electrode and filler metal for the job. Electrodes and filler metals are often classified into one of four groups, based on their general characteristics. The groups are as follows:

- Fast-freeze
- Fast-fill
- Fill-freeze
- Low-hydrogen

7.3.1 Fast-Freeze Group

Fast-freeze electrodes are all-position electrodes that provide deep penetration. However, the arc has a lot of fine weld spatter. The weld bead is flat with distinct ripples and a light slag coating that can be difficult to remove. The arc is easy to control, making fast-freeze electrodes good for vertical and overhead welding.

7.3.2 Fast-Fill Group

Fast-fill electrodes have iron powder in the flux, which gives them high deposition rates but

requires that they be used only for flat welds and horizontal fillet welds. These electrodes have shallow penetration with excellent weld appearance and almost no spatter. The heavy slag covering is easy to remove.

7.3.3 Fill-Freeze Group

Fill-freeze electrodes, also called fast-follow electrodes, are all-position electrodes but are most commonly used for downhill or flat welding. They have medium deposition rates and penetration and are excellent for sheet metal. The weld bead ranges from smooth and ripple-free to even with distinct ripples. The medium slag covering is easy to remove.

7.3.4 Low-Hydrogen Group

Low-hydrogen electrodes are designed for welding high-sulphur, phosphorus, and medium- to high-carbon steels, which have a tendency to develop porosity and underbead cracking. These problems are caused by hydrogen, which is absorbed during welding. Low-hydrogen electrodes are available in fast-fill and fill-freeze characteristics. They produce dense, X-ray-quality welds with excellent notch toughness and ductility.

7.4.0 Filler Metal and Electrode Selection Considerations

If there is a WPS for the weld that is to be made, the filler metal will be specified. It will be given as an AWS classification and/or manufacturer's standard and will sometimes include a manufacturer's name.

When selecting filler metals and electrodes not specified in a WPS, you must consider the several factors explained in the following sections.

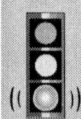

NOTE

The filler metal specified in the WPS must be used.

7.4.1 Base Metal Type

The filler metal should be compatible with the base metal chemical composition and mechanical properties. You can obtain the chemical composition and mechanical properties of the base metal by referring to the base metal mill specifications; then compare the mill specifications to various tables that list base metal and filler metal acceptability.

A weld made on mild steel with a high-tensile electrode, such as E12018, is weaker than one made with E7018. The high-tensile weld is so strong that when stress is placed on the weld, it does not give. It transfers all stress to the heat-affected zone (HAZ) along the weld. This can cause cracking. A weld made with E7018 will give along with the base metal, distributing the stress across the weld, making a stronger joint. The opposite occurs when mild-steel electrodes are used to weld high-strength steel. All the stress occurs in the weld because the surrounding base metal is so much stronger. This causes the weld to fail prematurely.

7.4.2 Base Metal Thickness and Condition

Always use an electrode smaller in diameter than the thickness of the metal to be welded. For sheet metal, 3/32-inch E6013 is recommended because it has shallow penetration. For very thick sections, always use a low-hydrogen electrode for its ductile qualities. To minimize the joint preparation, you can use a deep penetration E6010 or E7010 electrode for the root pass. Before welding, you should always clean the base metal surface. Surface corrosion, plating, coatings, and paint should be removed.

7.4.3 Welding Position and Joint Design

The welding position is important because some electrodes, such as E7024 and E7028, can only be used in the flat and horizontal fillet weld position. In addition, you should never attempt to weld in vertical or overhead positions with electrodes larger than 3/16 inch except for low hydrogen electrodes. The recommended limitation for low hydrogen is 5/32 inch.

The type of joint to be welded is also important to consider when selecting an electrode. When lap and T joints can be positioned flat, they are usually welded with E7024 or E7028. You should weld butt joints that need deep penetration using E6010. If the joint has poor fit-up, use a shallow-penetrating electrode such as E6012 or E6013. If you use low-hydrogen electrodes for open-root joints, back-gouging may be required. To eliminate the need for back-gouging and to make the root pass easier to run, use metal backing, or put the root in with E6010 and then fill out with low-hydrogen.

7.5.0 Preheating and Interpass Temperature Control

Preheating and interpass temperature control procedures are sometimes performed to preserve weldment strength, ductility, and weld quality. Proper preheating and interpass temperature control can reduce shrinkage, prevent excessive hardening, and prevent underbead cracking. Interpass temperature control is also required during certain welding procedures to prevent or reduce localized hardening and shrinkage. If certain metals cool too quickly between weld passes, undesirable changes occur in the base and weld metals, and the heat from successive passes will not reverse these changes. Preheating and interpass temperature heat treatment of metals is described in detail in the Welding module 29301-03, *Preheating and Postweld Heat Treatment of Metals*.

7.6.0 Postweld Heat Treatment

Postweld heat treatment (PWHT) is the heating of a weldment or assembly after welding is completed. It can be used on most types of base metals. Postweld heat treatment may be necessary to reduce residual stresses and to improve corrosion resistance, dimensional stability, fatigue resistance, and impact and low-temperature strength. Postweld heat treatment equipment and techniques include furnaces, banding, and localized heat treatment. Postweld heat treatments include stress relieving, annealing, normalizing, and tempering. Postweld heat treatment of metals is described in detail in the Welding module 29301-03, *Preheating and Postweld Heat Treatment of Metals*.

Summary

Selecting the correct cleaning method and proper filler material and making proper welds depend on your ability to identify base metals and understand the classification systems used for them. You must also understand the characteristics and the physical and mechanical properties of the metal.

On occasion, you will have to deal with unidentified metals: knowing how to identify metals in the field is an important part of making sound decisions and good welds. Structural metals come in different forms and shapes, and you must be familiar with the requirements of each.

Metallurgical considerations for welding include cleaning, joint design, codes and specifications, filler metals, pre- and postweld heat treatments, and the tempering and finishing processes to which metals are subjected, and you must be familiar with all these considerations.

Review Questions

1. The largest group of the ferrous metals includes the _____.
   a. alloy steels
   b. heavy metals
   c. cast irons
   d. carbon steels

2. Quench-and-tempered steels and chromium-molybdenum steels are examples of _____ steels.
   a. common grade stainless
   b. specialty grade stainless
   c. low-alloy
   d. weathering

3. A lower coefficient of thermal conductivity, higher coefficients of thermal expansion, and higher electrical resistance are characteristics of _____ steel.
   a. carbon
   b. stainless
   c. nickel
   d. chromium

4. The principal alloying element in brasses is _____.
   a. zinc
   b. Muntz metal
   c. beryllium
   d. silicon

5. The three types of cast iron are malleable, white, and _____.
   a. green
   b. blue
   c. gray
   d. ductile

6. Malleable cast iron is white cast iron that has been _____.
   a. chilled rapidly
   b. chilled slowly
   c. forged
   d. annealed

7. A metal's ability to conduct heat is known as _____.
   a. density
   b. electrical conductivity
   c. thermal conductivity
   d. melting point

8. Corrosion is an important consideration because it can severely reduce the _____ of a metal.
   a. thermal conductivity
   b. thermal expansion
   c. ductility
   d. tensile strength

9. The ability of a material to be strained (deformed) without permanent deformation is known as _____.
   a. elasticity
   b. modulus
   c. tensile strength
   d. ductility

10. The resistance to indentation of a material is defined as _____.
    a. ductility
    b. hardness
    c. tensile strength
    d. modulus of elasticity

11. A channel has a designation of C12 × 30. This means it is a(n) _____.
    a. American Standard Channel with a depth of 12 inches
    b. Miscellaneous Channel with a weight of 12 pounds per foot
    c. American Standard Channel with a weight of 12 pounds per foot
    d. Miscellaneous Channel with a weight of 30 pounds per foot

12. The most common defect caused by surface contamination is _____.
    a. corrosion
    b. rust
    c. porosity
    d. toxic fumes

13. The type of chemical used to clean metal depends on the _____.
   a. availability of the chemical
   b. base metal
   c. amount of cleaning required by the metal
   d. availability of safety equipment

14. Fast-fill electrodes have shallow penetration with excellent weld appearance and _____.
   a. a light slag coating
   b. a medium slag covering
   c. almost no spatter
   d. excellent notch toughness and ductility

15. When welding, you should always use an electrode that is _____ the thickness of the metal that is to be welded.
   a. smaller in diameter than
   b. larger in diameter than
   c. the same size as
   d. less than $\frac{3}{32}$ inch in diameter regardless of

Trade Terms Introduced in This Module

American Iron and Steel Institute (AISI): An industry organization responsible for preparing standards for steels and steel alloys that are based upon the SAE code system.

American Society for Testing Materials (ASTM): An organization that developed a code system for identifying and labeling steels.

Austenitizing: Heating a steel into or above the transformation temperature range to achieve partial or complete transformation to austenite grain structure.

Casting: The metal object produced by pouring molten metal into a mold.

Coefficient: A numerical measure of a physical or chemical property that is constant for a system under specified conditions, such as the coefficient of friction.

Ductile: Able to be drawn, stretched, or hammered thin without breaking.

Ferrous: Relating to iron or an alloy that is mostly iron.

Flange: The outer members of a structural beam, joist, or rail.

Heat-treatable low-alloy (HTLA) steels: Alloy steels with carbon in the range of about 0.25% to 0.45% and small amounts of chromium, nickel, and/or molybdenum to enhance hardenability.

High-strength low-alloy (HSLA) steels: Steels designed to meet specific mechanical requirements per ASTM specifications with the addition of alloys such as manganese, chromium, nickel, copper, molybdenum, and vanadium.

Malleable: Able to be hammered or pressed into another shape without breaking.

Metallic: Having characteristics of metal, such as ductility, malleability, luster, and heat and electrical conductivity.

Society of Automotive Engineers (SAE): An automotive society responsible for an early system of classifying carbon steels.

Web: The plate joining the flanges of a girder, joist, or rail.

Wrought: Formed or shaped by hammering or rolling.

Additional Resources

This module is intended to present thorough resources for task training. The following reference works are suggested for both instructors and motivated trainees interested in further study. These are optional materials for continued education rather than for task training.

Metals and How to Weld Them. T.B. Jefferson. Cleveland, OH: The Lincoln Welding Foundation.

Machinery's Handbook. Erik Oberg, Franklin D. Jones, Holbrook L. Horton, and Henry H. Ryffel. New York, NY: Industrial Press, Inc.

Figure Credits

Allegheny Ludlum Corporation	302SA01
Cutco Cutlery Corp.	302SA02
Miller Electric Mfg. Co.	302SA03, 302SA04
Dent-X Corporation USA	302SA05
Cliff Oram	302SA06
Foerster Instruments, Inc.	302SA07
ThermoElectron Corp.	302F01 (top)
Fidel Ytturia	302F01 (bottom)
Tinius Olsen Testing Machine Co., Inc.	302SA08 (top)
Instron Corp.	302SA08 (middle)
Portland Cement Association	302SA08 (bottom)
New Age Testing Instruments, Inc.	302F04
Gerald Shannon	302SA09
Erin Kellem	302F10

The NCCER makes every effort to keep these textbooks up-to-date and free of technical errors. We appreciate your help in this process. If you have an idea for improving this textbook, or if you find an error, a typographical mistake, or an inaccuracy in NCCER's Contren® textbooks, please write us, using this form or a photocopy. Be sure to include the exact module number, page number, a detailed description, and the correction, if applicable. Your input will be brought to the attention of the Technical Review Committee. Thank you for your assistance.

Instructors – If you found that additional materials were necessary in order to teach this module effectively, please let us know so that we may include them in the Equipment/Materials list in the Instructor's Guide.

Write: Curriculum Revision and Development Department
National Center for Construction Education and Research
P.O. Box 141104, Gainesville, FL 32614-1104

Fax: 352-334-0932

E-mail: curriculum@nccer.org

Craft _____ Module Name _____

Copyright Date _____ Module Number _____ Page Number(s) _____

Description _____

_____

_____

_____

(Optional) Correction _____

_____

_____

(Optional) Your Name and Address _____

_____

_____

Gas Metal Arc Welding (GMAW) – Pipe

COURSE MAP

This course map shows all of the modules in the third level of the Welding curriculum. The suggested training order begins at the bottom and proceeds up. Skill levels increase as you advance on the course map. The local Training Program Sponsor may adjust the training order.

WELDING LEVEL THREE

```
┌─────────────────────┐         ┌─────────────────────┐
│    29307-03*        │         │     29308-03*       │
│ GTAW – ALUMINUM PIPE│         │  GMAW – ALUMINUM    │
│                     │         │  PLATE AND PIPE     │
└─────────────────────┘         └─────────────────────┘

          ┌──────────────────────────┐
          │        29306-03          │
          │  GTAW – LOW-ALLOY AND     │
          │  STAINLESS STEEL PIPE     │
          └──────────────────────────┘

          ┌──────────────────────────┐
   YOU    │        29305-03          │
   ARE    │  GTAW – CARBON STEEL PIPE │
   HERE   └──────────────────────────┘

┌─────────────────────┐         ┌─────────────────────┐
│     29303-03        │         │      29304-03       │
│    GMAW – PIPE      │         │     FCAW – PIPE     │
└─────────────────────┘         └─────────────────────┘

          ┌──────────────────────────┐
          │        29302-03          │
          │ PHYSICAL CHARACTERISTICS  │
          │ AND MECHANICAL PROPERTIES │
          │      OF METALS           │
          └──────────────────────────┘

          ┌──────────────────────────┐
          │        29301-03          │
          │ PREHEATING AND POSTWELD   │
          │ HEAT TREATMENT OF METALS  │
          └──────────────────────────┘

          ┌──────────────────────────┐
          │    WELDING LEVEL TWO      │
          └──────────────────────────┘

          ┌──────────────────────────┐
          │    WELDING LEVEL ONE      │
          └──────────────────────────┘

          ┌──────────────────────────┐
          │         CORE             │
          │      CURRICULUM          │
          └──────────────────────────┘
```

303CMAP.EPS

*Please note that Modules 29307-03 and 29308-03 are electives
for those progressing through the *Welding Level Three* program.

Figures

Tables

GMAW – Pipe

Objectives

When you have completed this module, you will be able to do the following:

1. Prepare GMAW equipment for open-root V-groove pipe welds.
2. Identify and explain open-root V-groove pipe weld techniques.
3. Perform open-root V-groove pipe welds using GMAW in the following positions:

   - 1G-ROTATED
   - 2G
   - 5G
   - 6G

Prerequisites

Before you begin this module, it is recommended that you successfully complete the following: Core Curriculum; Welding Levels One and Two; Welding Level Three, Modules 29301-03 and 29302-03.

Required Trainee Materials

1. Pencil and paper
2. Appropriate personal protective equipment

1.0.0 ◆ INTRODUCTION

GMAW is an arc welding process that uses a continuous, consumable solid wire or composite cored wire electrode for the filler metal and shielding gas to protect the weld zone. (See *Figures 1* and *2.)* The GMAW process is commonly used to make welds on carbon, low-alloy, and stainless steel, as well as aluminum and other metals.

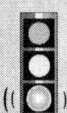

NOTE

Composite cored wire electrodes are a subset of GMAW solid wire carbon steel electrodes as defined in *AWS A5.18.*

Welding pipe with the GMAW process is a fast and effective method of producing high-quality welds. Welding can be continuous, which reduces discontinuities and restarts. With some materials, such as stainless steel, it is common field practice to use GTAW or SMAW for the root pass and GMAW to complete the fill and cover passes.

For the purposes of this training module, the dimensions used are representative of various codes and standards and may not be specific to an exact code or standard. Always refer to the applicable code, standard, or site welding procedure specification (WPS).

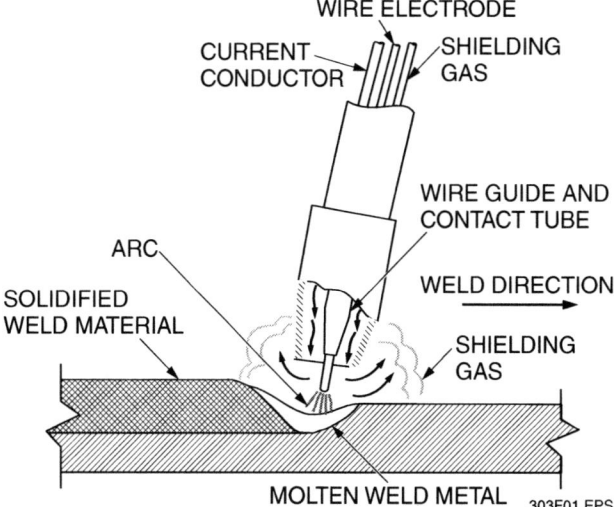

Figure 1 ◆ GMAW process.

WIRE FEEDER

POWER SOURCE

WORKPIECE CABLE

REGULATOR/ FLOWMETER

GAS HOSE

GMAW GUN

303F02.EPS

Figure 2 ◆ Gas shielded GMAW equipment.

2.0.0 ◆ SAFETY SUMMARY

The following is a summary of safety procedures and practices you must observe while cutting or welding. Keep in mind that this is just a summary. Complete safety coverage is provided in the Level One module *Welding Safety*. If you have not completed that module, do so before continuing. Above all, be sure to wear appropriate protective clothing and equipment when welding or cutting.

2.1.0 Protective Clothing and Equipment

- Always use safety goggles with a full face shield or a helmet. The goggles, face shield, or helmet lens must have the proper light-reducing tint for the type of welding or cutting to be performed. Never directly or indirectly view an electric arc without using a properly tinted lens.
- Wear proper protective leather and/or flame retardant clothing along with welding gloves to protect from flying sparks, molten metal, and heat.
- Wear 8" or taller high-top safety shoes or boots. Make sure that the tongue and lace area of the footwear is covered by a pant leg. If the tongue

and lace area is exposed or if the footwear must be protected from burn marks, wear leather spats under the pants or chaps and over the front top of the footwear.

- Wear a solid material (non-mesh) hat with a bill pointing to the rear or, if much overhead cutting or welding is required, a full leather hood with a welding faceplate and the correct tinted lens. If a hard hat is required, use one that allows attachment of both rear deflector material and a face shield.
- If a full leather hood is not worn, wear a face shield and snug-fitting welding goggles over safety glasses for gas welding or cutting. Either the face shield or the lenses of the welding goggles must be an approved shade 5 or 6 filter. For electric arc welding or cutting, wear safety goggles and a welding hood with the correct tinted lens (shade 9 to 14).
- If a full leather hood is not worn, wear earplugs to protect your ear canals from sparks.

2.2.0 Fire/Explosion Prevention

- Never carry matches or gas-filled lighters in your pockets. Sparks can cause the matches to ignite or the lighter to explode, causing serious injury.
- Never perform any type of heating, cutting, or welding until a hot-work permit is obtained and an approved fire watch is established. Most work-site fires caused by these types of operations are started by cutting torches.
- Never use oxygen to blow off clothing. The oxygen can remain trapped in the fabric for a time. If a spark hits the clothing during this period, the clothing can burn rapidly and violently out of control.
- Make sure that any flammable material in the work area is moved or is shielded by a fire-resistant covering. Approved fire extinguishers must be available before attempting any heating, welding, or cutting operations.
- Never release a large amount of oxygen nor use oxygen as compressed air. Its presence around flammable materials or sparks can cause rapid and uncontrolled combustion. Keep oxygen away from oil, grease, and other petroleum products.
- Never release a large amount of fuel gas, especially acetylene. Methane and propane tend to concentrate in and along low areas and can ignite at a considerable distance from the release point. Acetylene is lighter than air but is even more dangerous than methane. When mixed with air or oxygen, it will explode at much lower concentrations than any other fuel.

- To prevent fires, maintain a neat and clean work area, and make sure that any metal scrap or slag is cold before disposal.
- Before cutting or welding containers such as tanks or barrels, check to see if they have contained any explosive, hazardous, or flammable materials, including petroleum products, citrus products, or chemicals that decompose into toxic fumes when heated. As a standard practice, always clean and then fill any tanks or barrels with water, or purge them with a flow of inert gas to displace any oxygen.

2.3.0 Work Area Ventilation

- Make sure to follow confined space procedures before conducting any welding or cutting in the confined space.
- Never use oxygen to ventilate confined spaces.
- Always perform cutting or welding operations in a well-ventilated area. Such operations involving zinc or cadmium materials or coatings result in toxic fumes. For long-term cutting or welding of these materials, always wear an approved, full-face, supplied-air respirator (SAR) that uses breathing air supplied from outside of the work area. For occasional, very short-term exposure, you may use a high-efficiency particulate arresting (HEPA)-rated or metal-fume filter on a standard respirator.
- Make sure confined spaces are ventilated properly for cutting or welding operations.

3.0.0 ◆ WELDING PREPARATION

Before welding can begin, the area has to be readied, the welding equipment must be set up, and the metal to be welded must be prepared. The following sections explain how to set up the equipment for welding.

To practice welding, you need a welding table, bench, or stand. The welding surface must be steel, and provisions must be available for placing weld coupons out of position. *Figure 3* shows a typical welding station.

To set up the area for welding, follow these steps:

Step 1 Check to be sure the area is properly ventilated. Make use of doors, windows, and fans.

Step 2 Check the area for fire hazards. Remove any flammable materials before proceeding.

Step 3 Locate the nearest fire extinguisher. Do not proceed unless the extinguisher is charged and you know how to use it.

303F03.EPS

Figure 3 ◆ Welding station.

Step 4 Position a welding table near the welding machine.

Step 5 Set up flash shields around the welding area.

3.1.0 Practice Pipe Weld Coupons

Pipe weld coupons should be cut from 3" to 12" diameter Schedule 40 or Schedule 80 carbon steel pipe. Each welded joint will require two coupons of the same size and schedule pipe.

NOTE

Most codes allow substitution of carbon steel pipe for stainless steel and low-alloy pipe as a base metal for welder qualification. However, applicable stainless steel or low-alloy filler metals must be used. Always refer to the WPS or site quality standards for the proper qualification information.

Figure 4 shows typical ASME International, American Petroleum Institute (API), and AWS pipe bevel specifications for bevel angles, root faces, and root openings.

To prepare weld coupons for open-root V-groove weld joints, follow these steps:

Step 1 Clean all rust or mill scale from the inside and outside of the carbon steel pipe with a grinder or wire brush. Clean for a distance of ½" from the weld joint.

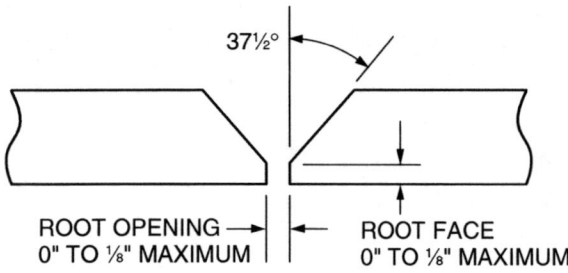

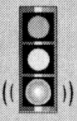

37½°

ROOT OPENING →
0" TO ⅛" MAXIMUM

← ROOT FACE
0" TO ⅛" MAXIMUM

TYPICAL ASME SPECIFICATION

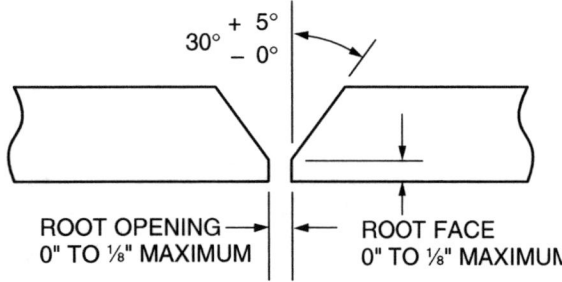

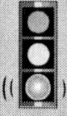

30° $^{+\ 5°}_{-\ 0°}$

ROOT OPENING →
0" TO ⅛" MAXIMUM

← ROOT FACE
0" TO ⅛" MAXIMUM

TYPICAL API AND AWS SPECIFICATION

303F04.EPS

Figure 4 ◆ ASME International, AWS, and API pipe bevel specifications.

Step 2 Bevel the end of the pipe to 30° or 37½° by any acceptable beveling method, such as by using a mechanical cutter or by thermal cutting or grinding.

Step 3 Cut off a section of the beveled pipe end (1½" minimum).

> **NOTE**
>
> For 1G-ROTATED welding, you may have to cut longer coupons to fit the rollers if used.

Follow these steps to prepare the open-root V-groove weld pipe coupon:

Step 1 Check the bevel. There should be no dross, and the bevel angle should be 30° or 37½°.

Step 2 Grind or file a 0" to ⅛" root face on the bevel, as specified by your instructor.

> **NOTE**
>
> Welding codes allow both the root opening and the root face on open-root welds to be from 0" to ⅛" wide. Select and adjust the opening and root face as needed when you start the welding practices.

Step 3 Align the two pipe sections so that the ID (inside diameter) surfaces are even all around. Align small-diameter pipe by clamping both pieces to a piece of angle iron. Align large-diameter pipe with the aid of a pipe alignment jig or by holding a straightedge across the joint, parallel to the pipe axis. The straightedge must be used all around the pipe in case one or both sections are distorted.

Conserve Materials

Pipe for practice welding is expensive and difficult to obtain. Make every effort to conserve and not waste available material. Completely use weld coupons until all surfaces have been welded by cutting the coupon apart and reusing the pieces. Check with the instructor for the appropriate size coupon.

Machined Bevels and Root Faces for GMAW

For V-groove joints, machined bevels and root faces provide the cleanest and most uniform method of preparing materials for satisfactory GMAW welds. This is especially true for stainless steel or aluminum where all contaminants from thermal cutting must be removed. While SMAW, FCAW, or GTAW can tolerate some poor fit-up for root passes, controlling root pass penetration and obtaining good fusion with GMAW is more difficult if the fit-up is poor. In practice, SMAW or GTAW are usually used for the root pass in critical piping applications.

Step 4 Gap the root opening with pieces of filler wire, metal shims, or pieces of bare welding electrode of the correct diameter.

Step 5 When the root opening is correct and the pipe ends are aligned, make the first of four tack welds of no greater than 1".

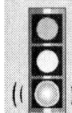

NOTE

Heavy-wall or large-diameter pipe may require longer tacks or more than four tacks.

Hot Tip

Welding Large Diameter Pipe

The practice coupons on large-diameter pipe must be greater than 1½" in length. They need more metal to help absorb the increased heat used to make these welds.

Hi-Lo Gauge

You can use a gauge, such as the one shown here, to check that internal surfaces are even all around.

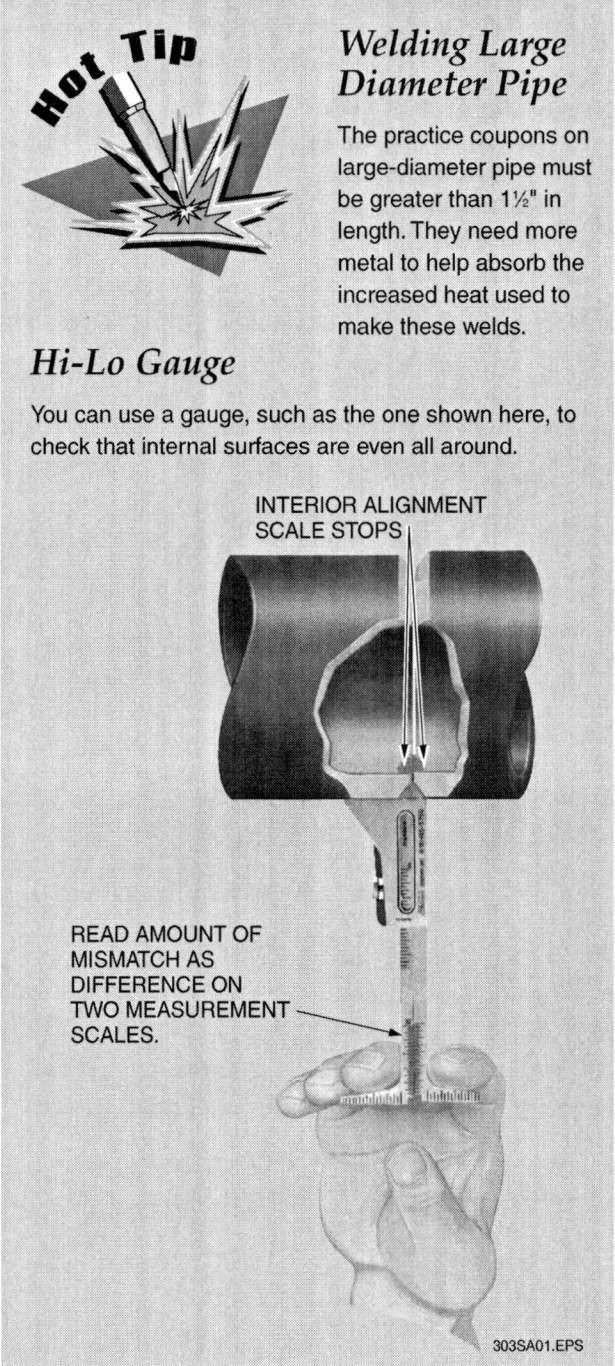

INTERIOR ALIGNMENT SCALE STOPS

READ AMOUNT OF MISMATCH AS DIFFERENCE ON TWO MEASUREMENT SCALES.

303SA01.EPS

Step 6 After the first tack weld, on the opposite side, check the root opening, adjust the gap if necessary, and make the second tack weld.

Step 7 Check the root opening again and weld the third tack midway between the first two tacks.

Step 8 Weld the fourth tack opposite the third tack and midway between the first and second tacks. There should now be four tack welds evenly spaced (90°) around the pipe coupon. This is illustrated by *Figure 5*.

Step 9 Clean and feather the tack welds with the edge of a grinding wheel. Feathering the ends of the tack welds with a grinder helps fuse the tack welds into the root pass.

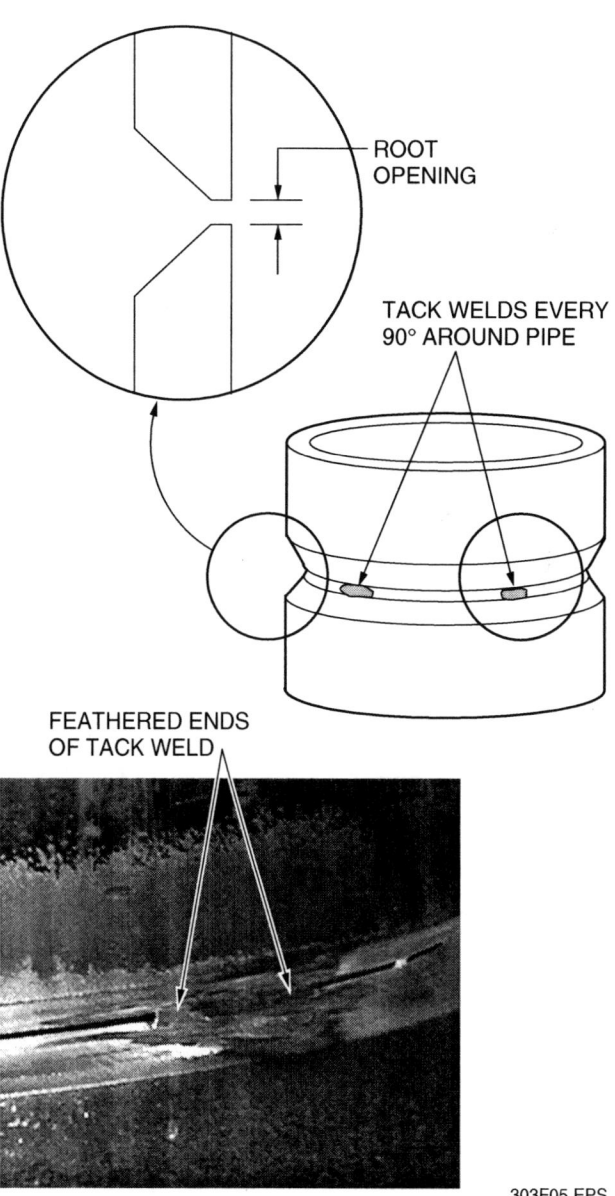

ROOT OPENING

TACK WELDS EVERY 90° AROUND PIPE

FEATHERED ENDS OF TACK WELD

303F05.EPS

Figure 5 ◆ Tacked open-root V-groove weld coupon.

Tack Welds

Upon reaching a tack weld when welding pipe with GMAW, instead of welding over it, you may need to remove the tack weld if it shows evidence of defects caused by improper shielding during the tack weld process. Additionally, welding over a tack weld can cause penetration and fusion problems if you do not pay close attention to the electrode extension.

3.2.0 The Welding Machine

Identify a proper welding machine for GMAW and follow these steps to set it up for use:

Step 1 Verify that the welding machine can be used for GMAW with or without internally controlled gas shielding.

Step 2 Check to be sure that the welding machine is properly grounded through the primary current receptacle.

Step 3 Verify the location of the primary disconnect.

Step 4 Configure the welding machine for gas shielded GMAW parameters as directed by your instructor (*Figure 6*). Configure the gun polarity, and equip the gun with the correct liner material, contact tube for the diameter of the wire being used, and the desired nozzle for the application.

Step 5 In accordance with the manufacturer's instructions, configure and load the wire feeder and gun with the solid or composite electrode wire of the proper diameter as directed by your instructor.

Step 6 Connect the proper shielding gas for the application as described in the previous level and as specified by the wire electrode manufacturer, WPS, site quality standards, or your instructor.

Step 7 Connect the clamp of the workpiece lead to the workpiece.

Step 8 Turn on the welding machine and purge the gun as directed by the gun manufacturer's instructions.

Step 9 Set the initial welding voltage and wire feed speed for the type and size electrode wire as recommended by the manufacturer.

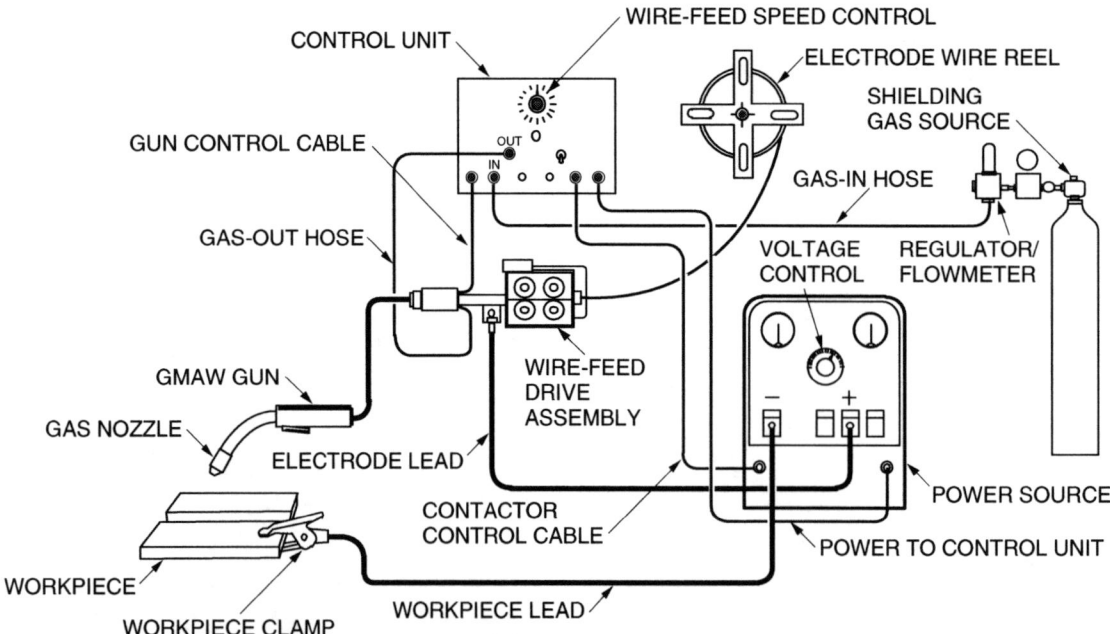

NOTE: THE POLARITY OF THE GUN AND WORKPIECE LEADS IS DETERMINED BY THE TYPE OF FILLER METAL AND APPLICATION.

303F06.EPS

Figure 6 ◆ Configuration diagram of typical GMAW machine.

3.2.1 Welding Voltage

Arc length is determined by voltage, which is set at the power source. Arc length is the distance from the wire electrode tip to the base metal or to the molten pool at the base metal. If voltage is set too high, the arc will be too long and could cause the wire to melt and fuse to the contact tube. A voltage that's too high also causes porosity and excessive spatter. Voltage must be increased or decreased as wire feed speed is increased or decreased. Set the voltage so that the arc length is just above the surface of the base metal. (See *Figure 7*.)

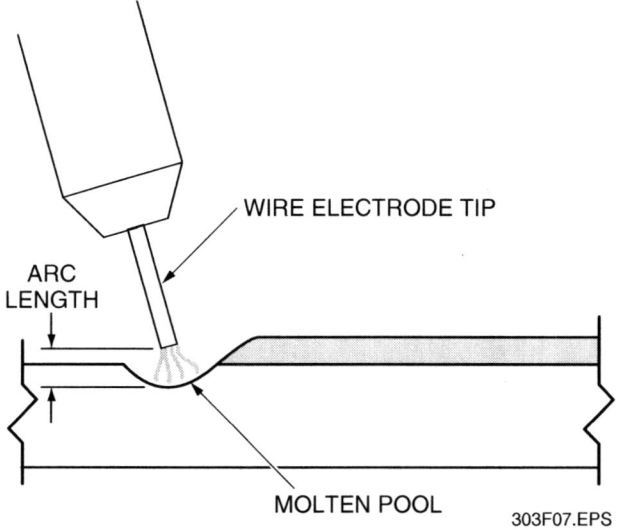

Figure 7 ◆ Arc length.

3.2.2 Welding Amperage

With a standard constant-voltage power source, the electrode feed speed controls the welding amperage after the initial recommended setting. The welding power source provides the amperage to melt the wire electrode while maintaining the selected welding voltage. Within limits, when the wire electrode feed speed is increased, the welding amperage and deposition rate also increase. This results in higher welding heat, deeper penetration, and higher beads. When the wire electrode feed speed is decreased, the welding amperage automatically decreases. With lower welding amperage and less heat, the deposition rate drops, and the weld beads are smaller with less penetration.

Note that some constant-voltage power sources used for GMAW/FCAW provide varying degrees of current modification, such as slope or induction adjustments. Power sources with a slope adjustment allow you to vary the amount of amperage change in relation to the voltage range of the unit. This allows the correct short-circuiting current to be established for certain welding applications and/or wire electrode requirements. Standard constant-voltage GMAW/FCAW units have a current slope that is fixed by the manufacturer for general welding applications and conditions. Other units allow you to adjust the response time for short-circuit current using an induction control. Instead of an instantaneous short-circuit current draw, the current draw can ramp up over a varying amount of time by using adjustable induction in the power source circuits. This smooths the arc characteristics during short-circuiting transfer. Pulse transfer power sources allow you to adjust the peak current for the pulse and the background current between pulses to match specific welding applications and/or wire electrode requirements.

3.2.3 Weld Travel Speed

Weld travel speed is the speed at which the electrode tip passes across the base metal in the direction of the weld. It is measured in inches per minute (ipm). Travel speed has a great effect on penetration and bead size: slower travel speeds build higher beads with deeper penetration while

Wire Electrode Manufacturer's Recommendations

Always obtain and follow the manufacturer's recommendations for shielding gas and for initially setting the welding voltage and wire feed speed parameters. These balanced parameters are critical and are based on the welding position, size, and composition of the solid or composite wire being used.

Arc Voltage

A minimum arc voltage is needed to maintain spray transfer. However, penetration is not directly related to voltage. Penetration will increase with voltage for a time but will actually decrease if the voltage is increased above its optimum.

faster travel speeds build smaller beads with less penetration. Ideally, the welding parameters should be adjusted so that the electrode tip is positioned at the leading edge of the weld puddle during travel.

3.2.4 Gun Position

Gun position for GMAW has greater effect on weld penetration than either voltage or travel speed. The gun position in relation to the direction of the weld is defined the same as for SMAW welding of carbon, stainless, or low-alloy steels.

The gun work angle (*Figure 8*) is defined as a less-than-90° angle between a line perpendicular to the major workpiece surface at the point of electrode contact and a plane determined by the electrode axis and the weld axis. For a T-joint or corner joint, the line is perpendicular to the nonbutting member. For pipe, the plane is determined by the electrode axis and a line tangent to the pipe surface at the same point.

The gun travel angle (see *Figure 8*) is defined as a less-than-90° angle between the electrode axis and a line perpendicular to the weld axis at the point of electrode contact in a plane determined

Travel Speed and Wire Feed Speed

Inexperienced welders are tempted to turn down the wire feed speed if they experience difficulty controlling the weld puddle. Wire electrodes must be run at certain balanced parameters that cannot be changed individually. Voltage, wire feed speed, and travel speed are adjusted and balanced together to control the weld puddle.

by the electrode axis and the weld axis. For pipe, the plane is determined by the electrode axis and a line tangent to the pipe surface at the same point.

- *Neutral angle* – The term *neutral angle* describes the travel angle when the centerline of the gun electrode is exactly perpendicular (90°) to the welding surface. The term is nonstandard for a 0° travel angle. Weld penetration is moderate when the gun electrode is used in this position.

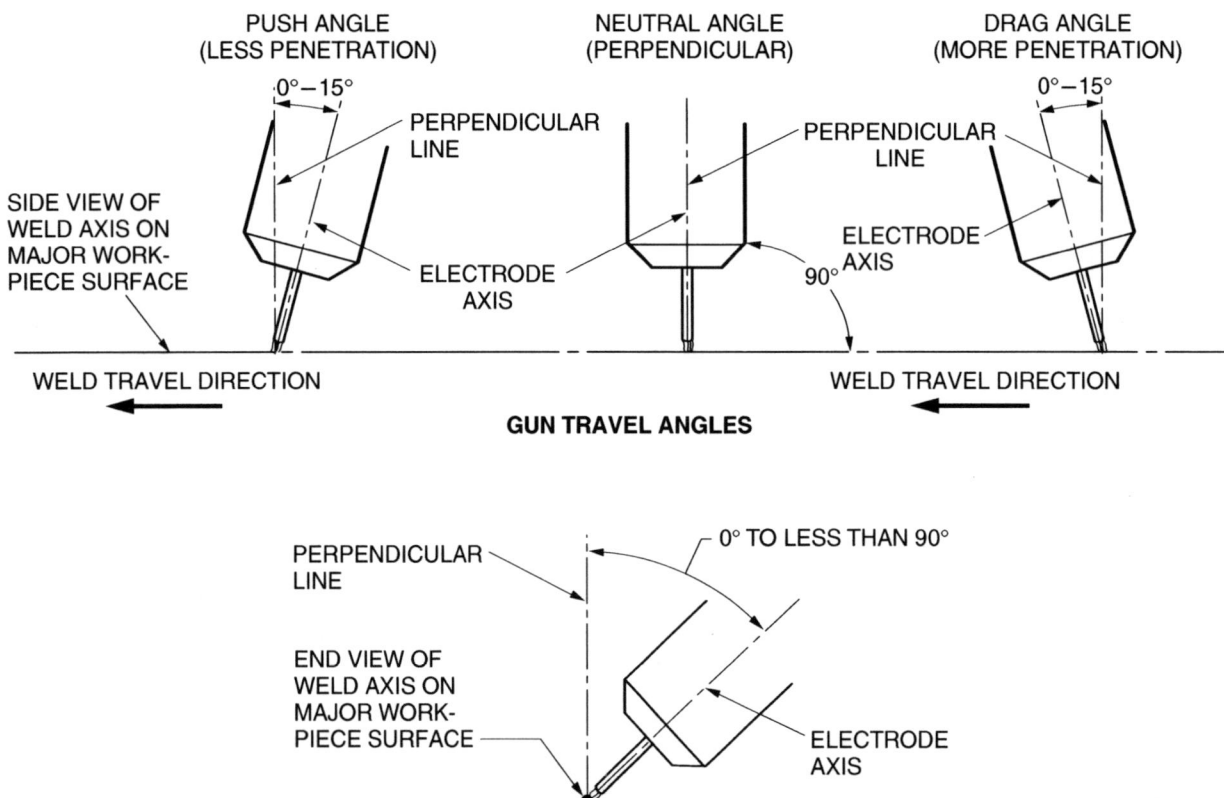

Figure 8 ◆ Gun work and travel angles.

- *Push angle* – A push angle is generally used for welding thin materials or when shallow penetration is required. A push angle is the travel angle created when the welding gun is tilted back so that the electrode tip precedes the gun in the direction of the weld. In this position, the electrode tip and shielding gas are being directed ahead of the weld bead. Welding with this angle produces a smaller weld bead with less penetration. Push angles greater than 15° are not recommended for manual welding.

- *Drag angle* – A drag angle is the travel angle created when the gun is tilted forward so that the electrode tip trails the gun in the direction of the weld. In this position, the electrode tip and shielding gas are being directed back toward the weld bead. Welding with this angle produces a higher and wider weld bead with more penetration. Maximum penetration is achieved with a drag angle of approximately 15° to 25°.

3.2.5 Electrode Extension, Stickout, and Standoff Distance

Electrode extension is the length of the wire that extends beyond the tip of the welding gun's contact tube. Changing the extension influences the welding current for low-conductivity metal wires because it changes the preheating of the wire caused by the resistance of the wire to current flow. As the extension increases, preheating of the wire increases. Because the welding power source is self-regulating, it does not have to furnish as much welding current to melt the wire, so the current output automatically decreases. This results in less penetration and increased deposit rates. Increasing extension is useful for bridging gaps and compensating for mismatch, but it can cause overlap or lack of fusion and a ropy bead appearance. When the extension is decreased, the power source is forced to increase its current output to burn off the wire. Too little extension can cause the wire to weld to the contact tube and can develop porosity in the weld. For high-conductivity metal wires, the preheating effect of wire resistance is minimal, and wire speed, current (if variable), and voltage settings have a more direct effect.

GMAW extension for micro wire (0.030" to 0.045") varies between ¼" to 1", depending on the transfer mode. *Figure 9* shows typical electrode extensions for various GMAW gun configurations and defines other gun nomenclature. Stickout is the distance from the gas nozzle or insulating nozzle to the end of the electrode. Standoff distance is the distance from the gas nozzle or insulating nozzle to the workpiece. Contact tube extension or setback usually depends on the transfer mode for the GMAW application.

GMAW of Stainless Steel

For stainless steel, a push angle is normally used for square or grooved butt joints. Stainless steel is usually welded in the flat and horizontal positions with spray transfer. Vertical and overhead welds usually employ short-circuit transfer for at least the root passes, if not the fill and cover passes. Pulsed spray transfer can also be used for fill and cover passes. Remember that all open joints in stainless steel materials must be backed with copper, stainless steel, or inert gas to prevent air from reaching the weld puddle until the puddle solidifies.

If practicing welding using stainless steel wire electrodes and carbon steel coupons, use the travel angle and transfer methods just outlined along with carbon steel backing rings, backing-gas pipe plugs, or dams, as required. When welding thick-walled pipe, the gun should be moved back and forth in the travel direction and slightly side to side. On thinner materials, only the back and forth motion is necessary.

Arc Blow

When current flows, magnetic fields are created. The magnetic fields tend to concentrate in corners, in deep grooves, or at the ends of the base metal. When the arc approaches these concentrated magnetic fields, it is deflected from the intended path. Arc blow can cause defects such as excessive weld spatter and porosity. If arc blow occurs, try one or both of the following methods to control it:

- Change the position of the workpiece clamp. This will change the flow of welding current, which will affect the way the magnetic fields are created.
- Change the angle of the gun. Reducing or even reversing the angle of the gun can compensate for the arc blow.

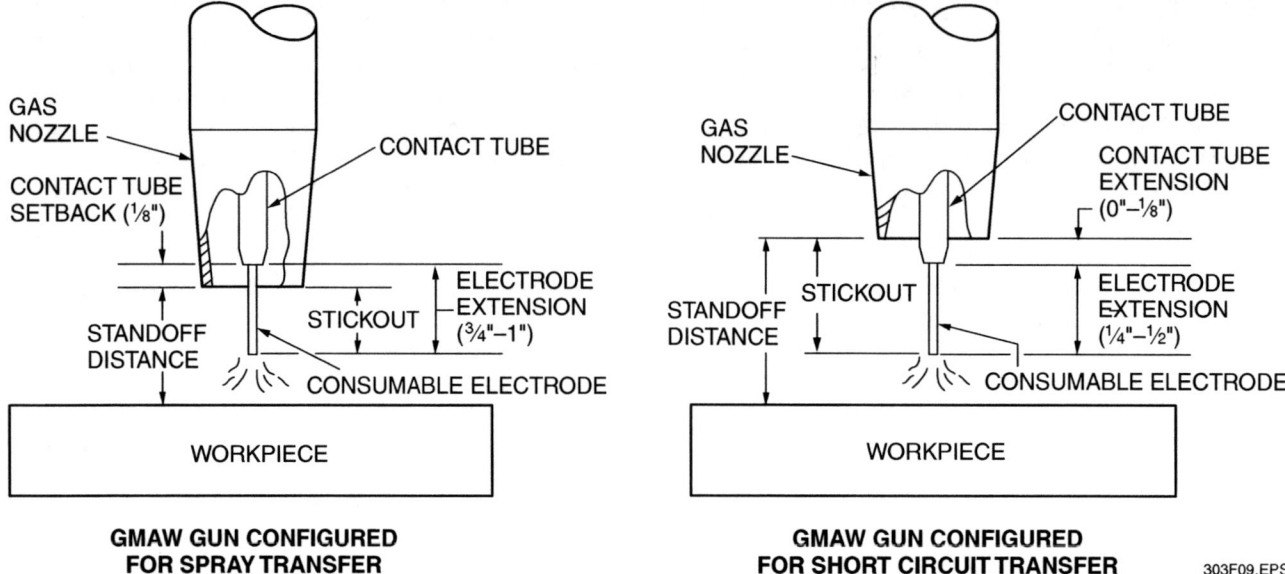

Figure 9 ◆ GMAW gun configurations.

3.2.6 Gas Nozzle Cleaning

As the welding machine is used, weld spatter accumulates on both the gas nozzle and the contact tube. If the gas nozzle is not occasionally cleaned, it will restrict the shielding gas flow, which will cause porosity in the weld.

Clean the gas nozzle with a reamer, round file, or tang of a standard file. After cleaning, the nozzle can be sprayed with or dipped in a special antispatter compound. The antispatter compound helps prevent the spatter from sticking to the nozzle.

4.0.0 ◆ OPEN-ROOT V-GROOVE PIPE WELDS

The open-root V-groove weld is the most common groove weld normally made on carbon, low-alloy, and stainless steel pipe. These welds are typically used for joining medium- and thick-walled pipe used in critical piping systems. Critical piping systems (including pipe, fittings, and welded joints) contain or carry material with the potential to cause long- or short-term catastrophic danger or damage to personnel and/or the environment, should the components fail to contain or carry the material as designed. Welds in critical piping must meet the most stringent code requirements. Non-critical piping is low-pressure piping used for heating and air conditioning, simple water systems, and other service installations. Less-stringent code requirements are used to evaluate non-critical piping welds.

4.1.0 GMAW Welding Techniques

The most difficult part of making an open-root V-groove weld is the root pass. The root pass must have complete penetration and fusion but not an excessive amount of melt-through root reinforcement. The extent of penetration and fusion is significantly important. If penetration is too little, it will create a weak joint. However, if there is too much penetration and excessive root reinforcement is created, the inside diameter of the pipe will be reduced and the flow will be restricted. Incomplete fusion will result in a defective weld. When

Spatter Control

With excessive stickout, more voltage is lost in the electrode length, the arc voltage is lower, and irregular arc action and spatter may occur. Penetration is decreased as stickout is increased.

With shorter stickout and less voltage drop in the wire, higher arc voltage is available. However, with spray transfer, if the stickout is too short, spatter will increase and rapidly build up in the nozzle and on the contact tube. When using short-circuit transfer with contact tube extension, spatter buildup in the nozzle and on the contact tube is reduced.

GMAW is used for the root pass, short-circuiting arc transfer is usually used. Progression of multiple (5G) position welds for root pass welding is usually downhill. The fill and cover passes are usually run uphill. Backing gas must be used for stainless, low-alloy, and nonferrous alloys. *Figure 10* shows approximate modifications to the travel position that may be necessary for multiple (5G) position pipe welds when welding uphill or downhill with GMAW.

After the root pass is run, it should be cleaned and inspected. Check the weld for incomplete fusion and penetration. The root reinforcement should be ⅛" or less.

> **NOTE**
> Some WPSs do not allow grinding of stainless steel welds.

Remove any excess buildup or undercut with a hand grinder by grinding the face of the root pass with the edge of a grinding disk. Use care not to grind through the root pass nor widen the groove. A narrow grinding wheel (not a cutoff wheel) is effective for this purpose.

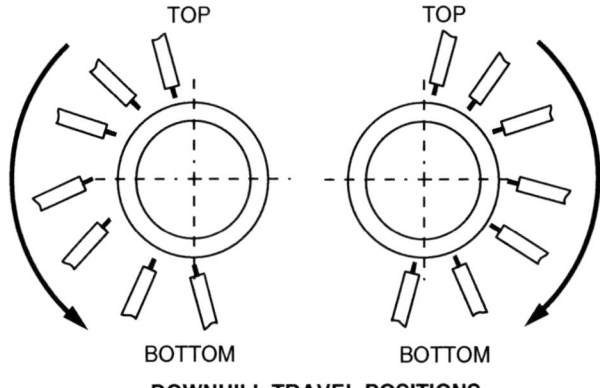

DOWNHILL TRAVEL POSITIONS

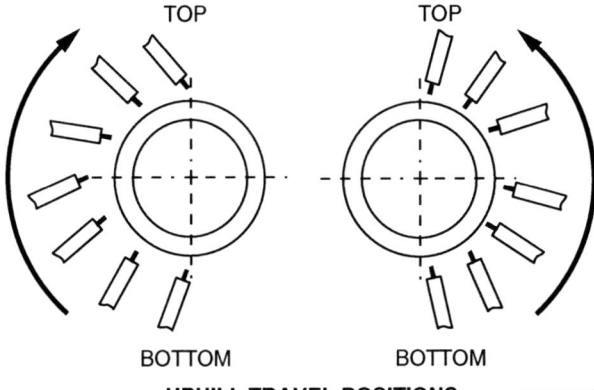

UPHILL TRAVEL POSITIONS   303F10.EPS

Figure 10 ◆ Approximate travel positions.

4.2.0 Pipe Groove Weld Test Positions

Groove welds may be made in all positions on pipe. The weld position is determined by the axis of the pipe. Four standard weld test positions, shown in *Figure 11* and described as follows, are used with pipe:

- *Flat (1G-ROTATED) welding position* – The pipe axis is horizontal, and the pipe is slowly rotated while being welded on the top. The weld beads are flat.

ROTATE PIPE AND DEPOSIT
WELD AT OR NEAR THE TOP

15°
15°

PIPE HORIZONTAL (±15°) AND ROLLED
TO KEEP WELD FLAT

1G-ROTATED POSITION

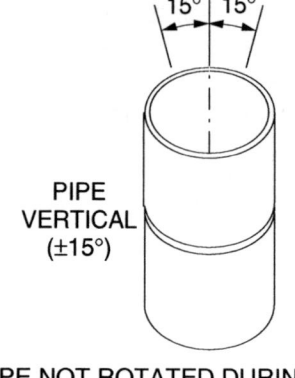

15° | 15°

PIPE
VERTICAL
(±15°)

PIPE NOT ROTATED DURING WELDING

2G POSITION

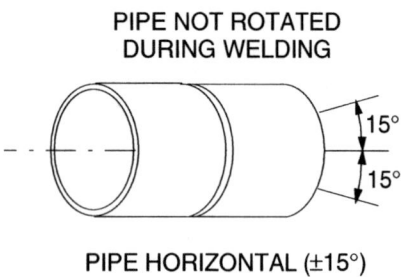

PIPE NOT ROTATED
DURING WELDING

15°
15°

PIPE HORIZONTAL (±15°)

5G POSITION

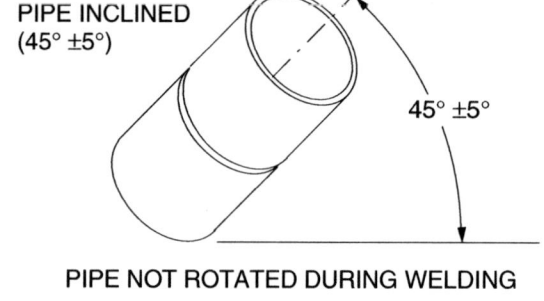

PIPE INCLINED
(45° ±5°)

45° ±5°

PIPE NOT ROTATED DURING WELDING

6G POSITION

303F11.EPS

Figure 11 ◆ Four basic pipe groove weld test positions.

- *Horizontal (2G) welding position* – The pipe axis is vertical, and the pipe rotation is fixed. The weld beads are horizontal.
- *Multiple (5G) welding position* – The pipe axis is horizontal, and the pipe rotation is fixed. The weld beads are flat, vertical, and overhead.
- *Inclined multiple (6G) welding position* – The pipe axis is inclined nominally at 45° ±5°, and pipe rotation is fixed. The weld beads are horizontal, vertical, and overhead.

The pipe axis positions for 1G, 2G, and 5G can vary ±15° from the basic position. The pipe axis position for 6G can vary ±5° from the nominal 45° angle. See *Table 1*.

To destructively test 5G or 6G welds, the test specimens are cut from the four regions illustrated in *Figure 12*: midway between the 12-o'clock and 3-o'clock, 3-o'clock and 6-o'clock, 6-o'clock and 9-o'clock, and 9-o'clock and 12-o'clock positions. (The 12-o'clock position is the top of the coupon groove and 6-o'clock is the bottom.)

4.3.0 Acceptable and Unacceptable Pipe Weld Profiles

Pipe groove welds should be made with slight reinforcement (not exceeding ⅛") and a gradual transition to the base metal at each toe. The root pass should have complete penetration. The root reinforcement on the inside of the pipe ranges from being flush to a maximum of ⅛". Pipe groove welds must not have excessive reinforcement or underfill at the face or root, excessive undercut, or

Table 1 Pipe Groove Weld Test Positions

Weld Test Position	Pipe Axis	Weld Beads	Pipe Position
1G-ROTATED	Horizontal	Flat	Slowly rotated
2G	Vertical	Horizontal	Fixed
5G	Horizontal	Flat, vertical, overhead	Fixed
6G	Inclined 45°	Horizontal, vertical, overhead	Fixed

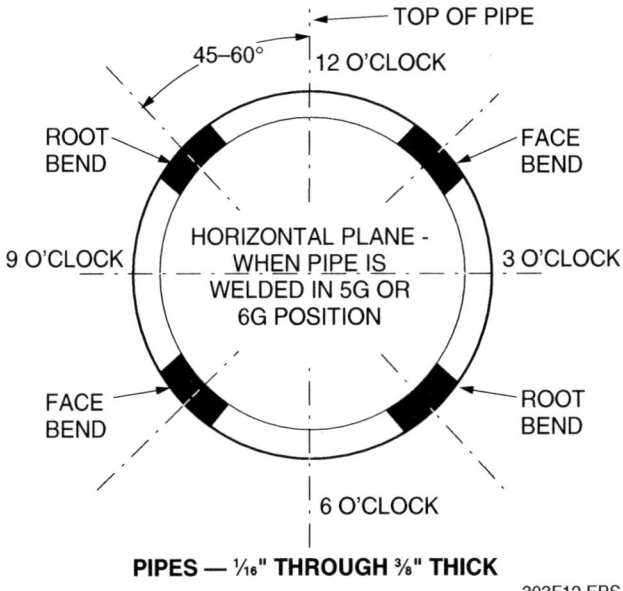

PIPES — 1/16" THROUGH 3/8" THICK

303F12.EPS

Figure 12 ◆ Test specimen regions of 5G or 6G position pipe.

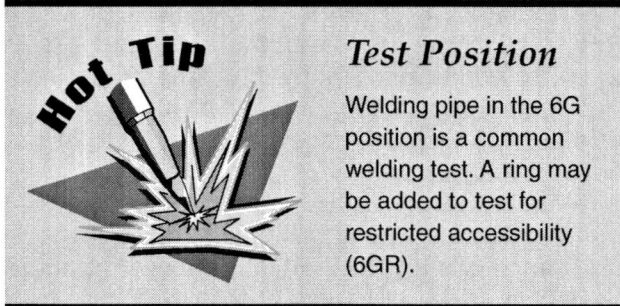

5.0.0 ◆ PRACTICING OPEN-ROOT V-GROOVE WELDS

This section of the module explains how to perform the following open-root V-groove weld positions:

- Flat (1G-ROTATED)
- Horizontal (2G)
- Multiple (5G)
- Inclined multiple (6G)

5.1.0 Flat (1G-ROTATED) Position

Practice the 1G-ROTATED position open-root V-groove pipe welds using a shielding gas and a solid or composite filler wire with a wire diameter as specified by your instructor. For the root pass, keep the gun angle at 90° (0° work angle) to the

overlap. Excessively large cover passes will reduce the pipe's strength. They cause the stresses in the pipe to be concentrated along the sides of the weld and do not permit the pipe to expand and contract in a uniform manner along its length. If a weld has any of these defects, as shown in *Figure 13*, it must be repaired.

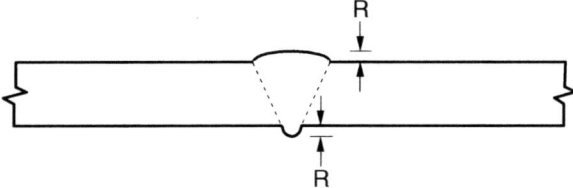

R = FACE AND ROOT REINFORCEMENT PER CODE
NOT TO EXCEED 1/8" MAXIMUM

ACCEPTABLE WELD PROFILE

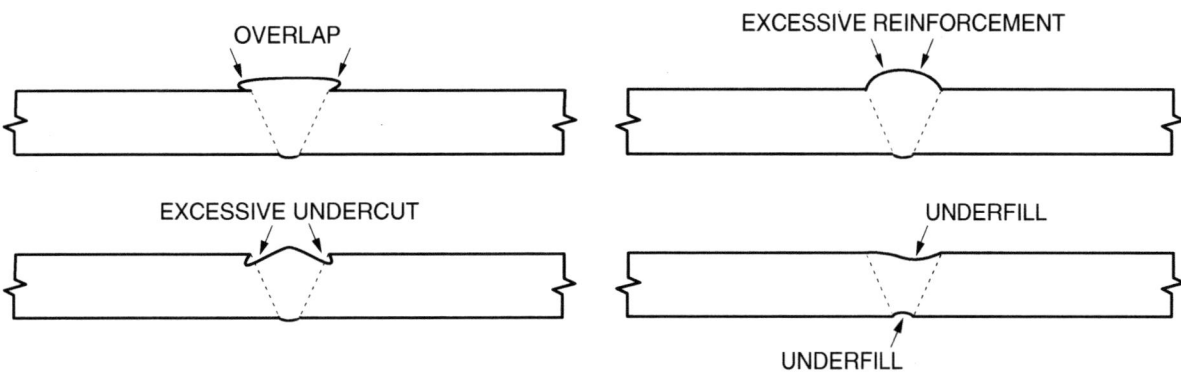

UNACCEPTABLE WELD PROFILES

303F13.EPS

Figure 13 ◆ Acceptable and unacceptable pipe groove weld profiles.

pipe axis with a 10° to 15° drag angle. Use a slight forward and backward motion with a slight side-to-side oscillation to run the root pass and to control penetration, paying particular attention to tie into the tack welds. Clean the face of the root pass with a brush.

When running the fill and cover passes, use stringer or weave beads with a slight side-to-side motion to ensure tie-in at the toes of the weld bead. Try to keep the wire at the leading edge of the weld puddle to ensure proper penetration. Avoid using weave beads for stainless steel pipe. Take particular care at the termination of the weld to fill the crater. Run all passes at or near the top of the pipe as the pipe is rotated.

Follow these steps to practice open-root V-groove pipe welds in the 1G-ROTATED position:

Step 1 Tack weld together the practice pipe weld coupon, as explained earlier.

Step 2 Position the pipe weld coupon horizontally on two sets of rollers at a comfortable welding height. *Figure 14* shows roller supports commonly found in pipe welding shops.

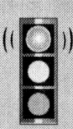

WARNING!

OSHA prohibits the use of homemade jack stands. Only stands manufactured to industry standards are acceptable. Four-legged jack stands are required for pipe over 10" in diameter.

Step 3 Make sure the workpiece clamp is attached directly onto the pipe coupon. This will prevent the welding current from passing through the roller bearings or arcing between the rollers and the pipe coupon.

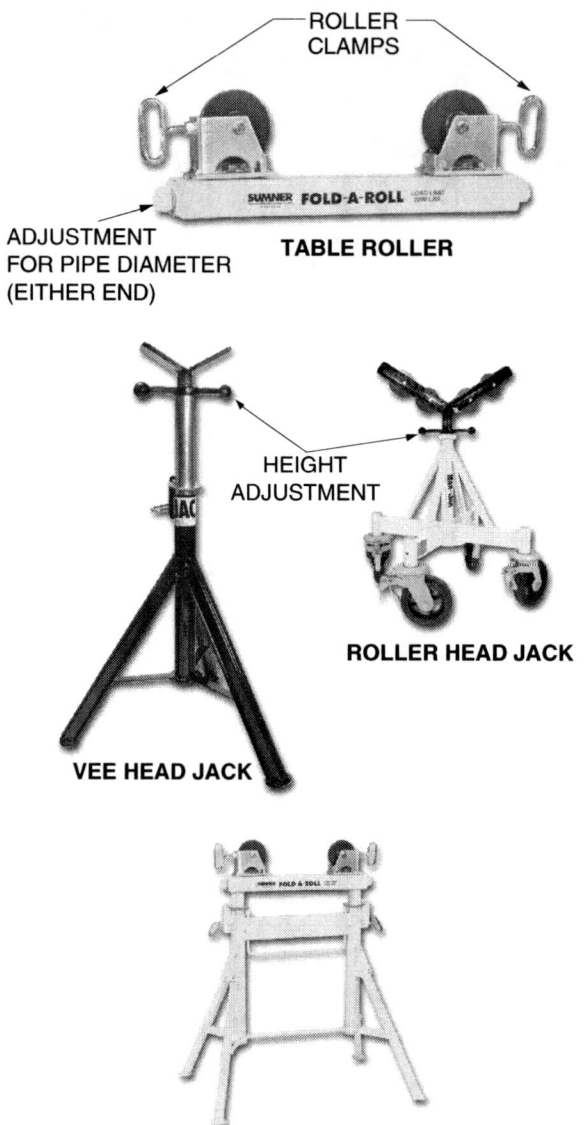

Figure 14 ◆ Pipe roller supports.

Open-Root V-Groove Stainless Steel Pipe Welds

If practicing open-root V-groove stainless steel pipe welds, remember to use push angles in place of drag angles. Also practice with a backing gas and backing-gas pipe plugs or dams.

A BACKING-GAS PIPE PLUG

303SA03.EPS

Fill Pass

Follow these guidelines when running the fill pass:

- Fill craters to preclude cracking.
- Ensure high-strength, high-pressure pipe cleanliness by slightly grinding or filing the crater.
- Stagger the location of the start and stop spots for each weld.

Step 4 Run the root pass using a slight forward and backward motion with a side-to-side oscillation. Position a tack weld at the 11 o'clock position, start the weld bead on the tack weld, and advance toward the 12 o'clock position.

Step 5 Roll the pipe as necessary to keep the weld in the flat position.

Step 6 Brush the root pass to clean the weld.

Step 7 Use the same rolling procedure to make the remaining passes. Use stringer or weave beads as applicable. Pay particular attention at the tie-ins for proper fusion and to prevent excess buildup. Overlap the passes, and clean the weld after each pass. *Figure 15* shows the bead sequences and work angles.

Hot Tip

Powered Pipe Roller

A pipe roller that is operated by an electric motor activated by a foot switch is shown here. These devices are primarily used for large, heavy pipe sections.

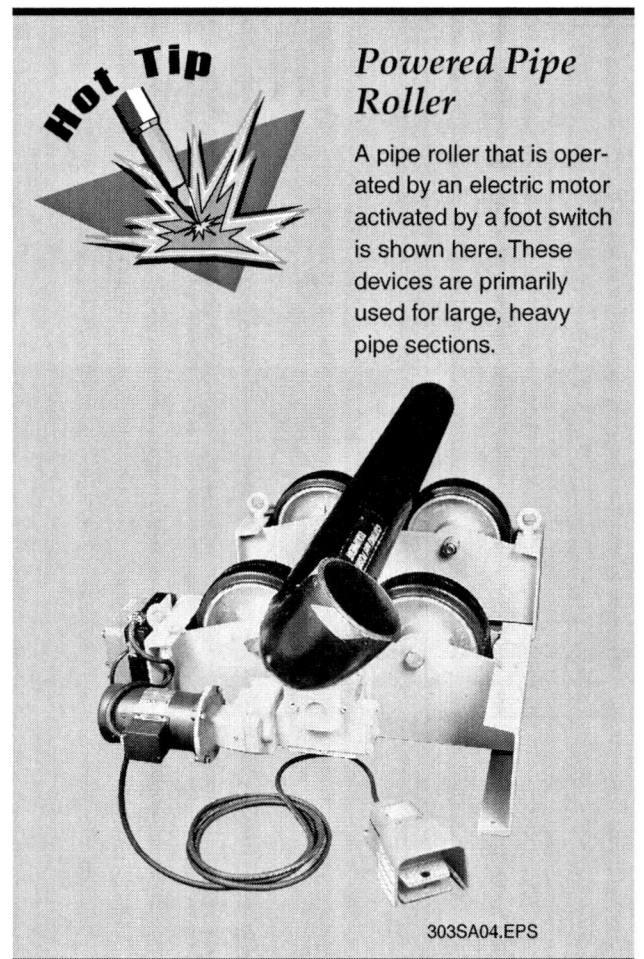

303SA04.EPS

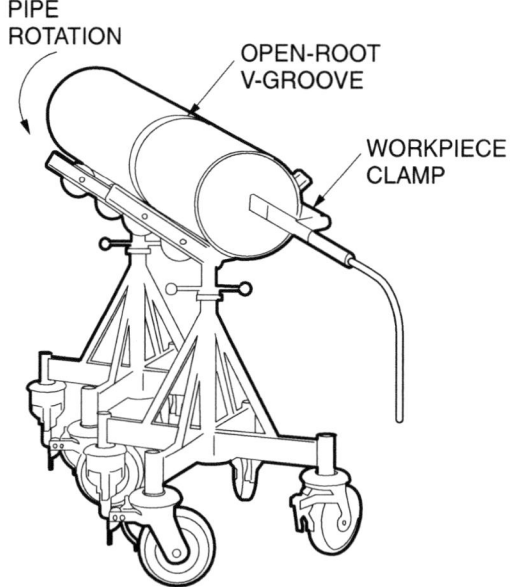

EXAMPLE OF PIPE ROTATION

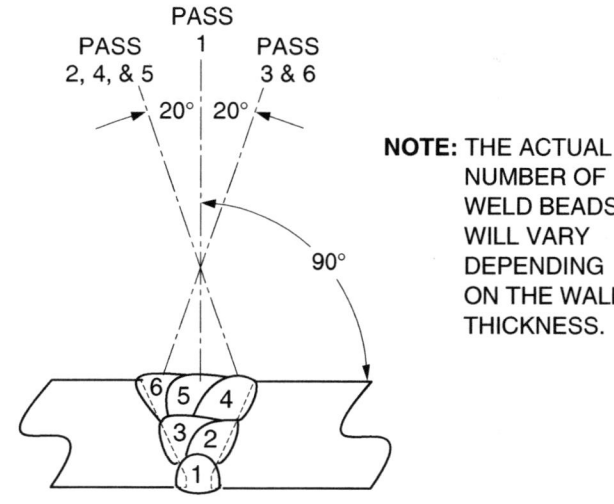

NOTE: THE ACTUAL NUMBER OF WELD BEADS WILL VARY DEPENDING ON THE WALL THICKNESS.

STRINGER BEAD SEQUENCE

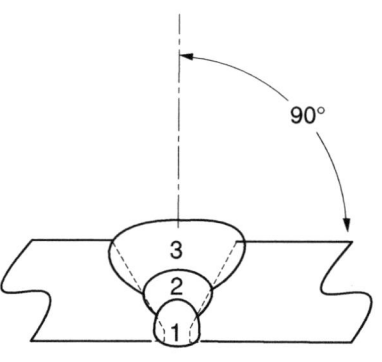

WEAVE BEAD SEQUENCE
(EXCEPT FOR STAINLESS STEEL)

303F15.EPS

Figure 15 ◆ Multiple-pass 1G-ROTATED bead sequences and work angles.

Welding Thick-Walled Pipe

When welding a V-groove joint in thick-walled pipe, it is sometimes difficult to maintain an appropriate electrode extension because of the gun size or the depth/angle of the groove. In these cases, extend the contact tube beyond the nozzle as necessary to maintain a ⅜" electrode extension.

Use of GMAW for a Root Pass

In practice, the WPS or site standards may require that the root pass for critical pipe welds be accomplished using GTAW or SMAW. For non-critical pipe welds, GMAW may be allowed.

Cover Pass

What effect will excessively large cover passes have on the strength of a pipe?

5.2.0 Horizontal (2G) Position

Practice 2G position open-root V-groove pipe welds using a shielding gas and a solid or composite filler wire with a wire diameter as specified by your instructor. The gun should be at a 10° to 15° drag angle and at 90° (0° work angle) to the surface of the pipe for the root pass. To prevent the weld puddle from sagging, the gun work angle can be dropped slightly but not more than 10°. For the root pass, use a slight side-to-side oscillation to control penetration, as shown in *Figure 16*. Pay particular attention to tie into the tack welds.

When running the fill/cover passes, use stringer beads and a slight side-to-side motion to ensure tie-in at the toes of the weld bead. Try to keep the wire electrode at the leading edge of the weld puddle. Take particular care at the termination of the weld to fill the crater.

Figure 17 shows the 2G pipe bead sequences and work angles.

Follow these steps to practice open-root V-groove pipe welds in the 2G position:

Step 1 Tack weld together the practice pipe weld coupon, as explained earlier.

Step 2 Clamp or tack weld the pipe coupon with the axis of the pipe vertical.

Step 3 Run the root pass as shown in *Figure 16*. Start the root pass on a tack weld. Clean the crater between restarts if restarts are necessary. Continue this procedure until the root pass is completed.

Step 4 Brush the root pass to clean the weld.

Step 5 Run the remaining passes at the appropriate work angles. Overlap the passes, and clean the weld after each pass.

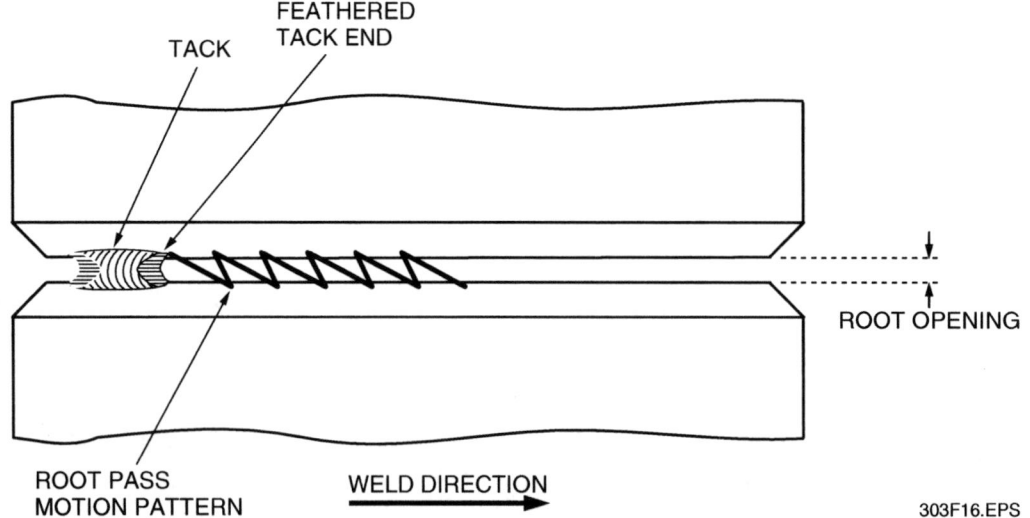

Figure 16 ◆ Root pass motion pattern.

5.3.0 Multiple (5G) Position

Practice the multiple (5G) position open-root V-groove pipe welds using a shielding gas and a solid or composite filler wire with a wire diameter as specified by your instructor. The gun should be at approximately a 10° to 15° drag or push angle for the root, fill, and cover passes, depending on whether your instructor directs you to run downhill or uphill passes. In some cases, the root pass is made downhill and the fill and cover passes are made uphill. Refer to *Figure 10* for the travel position modifications that may be necessary as the passes are made on both sides of the pipe.

When running the fill and cover passes, use stringer or weave beads with a slight side-to-side motion to ensure tie-in at the toes of the weld bead or weave bead. Try to keep the wire electrode at the leading edge of the weld puddle. Pay particular attention at the termination of the weld to fill the crater. *Figure 18* shows the 5G pipe bead sequences and work angles.

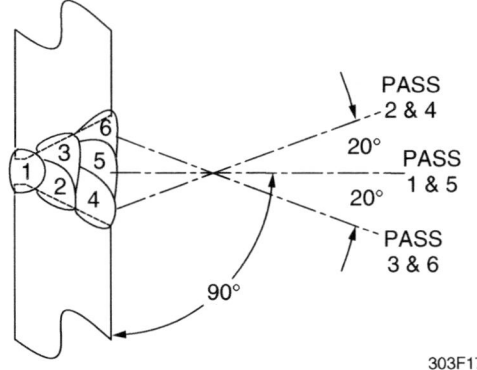

Figure 17 ◆ Multiple-pass 2G bead sequence and work angles.

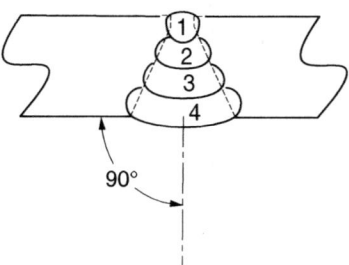

WEAVE BEAD SEQUENCE
(EXCEPT STAINLESS STEEL)

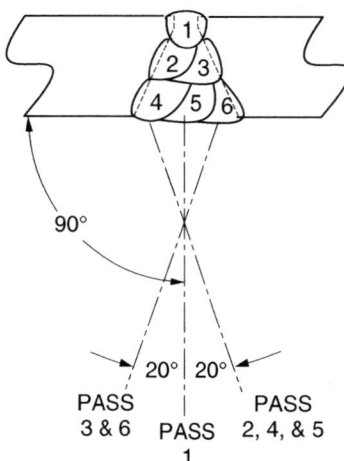

STRINGER BEAD SEQUENCE   303F18.EPS

Figure 18 ◆ Multiple-pass 5G bead sequences and work angles.

> ## Hot Tip
>
> ### Starting Pipe Welds
>
> When starting a pipe weld, briefly use a short electrode extension to create a hot weld that will ensure proper penetration and flatten the weld face. This will make terminating the finishing weld at the starting point easier.

Follow these steps to practice open-root V-groove pipe welds in the 5G position:

Step 1 Tack weld together the practice pipe weld coupon, as explained earlier.

Step 2 Clamp or tack weld the pipe weld coupon into position with the pipe axis horizontal. Be sure to position the tack welds so they are not in the areas from which the test coupons will be cut.

Step 3 Run the root pass with a stringer bead. Modify the travel angles as necessary, as shown in *Figure 10,* for an uphill or downhill pass, as directed by your instructor.

Step 4 Brush the root pass to clean the weld.

Step 5 Run the remaining passes at the appropriate work angles in either an uphill or downhill direction, as directed by your instructor. Overlap the passes, and clean the weld after each pass.

5.4.0 Inclined Multiple (6G) Position

Practice the inclined multiple (6G) (45°) position open-root V-groove pipe welds using a shielding gas and a solid or composite filler wire with a wire diameter as specified by your instructor. The gun should be at approximately a 10° to 15° drag or push angle for the root, fill, and cover passes, depending on whether your instructor directs you to run downhill or uphill passes. In some cases, the root pass is made downhill and the fill and cover passes are made uphill. Refer to *Figure 10* for the travel position modifications that may be necessary as the passes are made on both sides of the pipe.

When running the fill and cover passes, use weave beads or stringer beads and a slight side-to-side motion to ensure tie-in at the toes of the weld bead. Try to keep the wire electrode at the leading edge of the weld puddle. Pay particular attention at the termination of the weld to fill the crater. *Figure 19* shows the 6G pipe bead sequences and work angles.

Follow these steps to practice open-root V-groove pipe welds in the 6G position:

NOTE

If required for training purposes, a restricting ring may be added to the 6G position coupon to form a 6GR position coupon.

Step 1 Tack weld together the practice pipe weld coupon, as explained earlier.

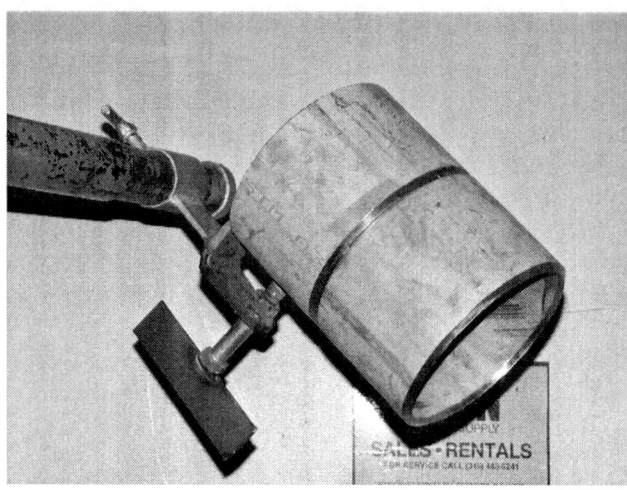

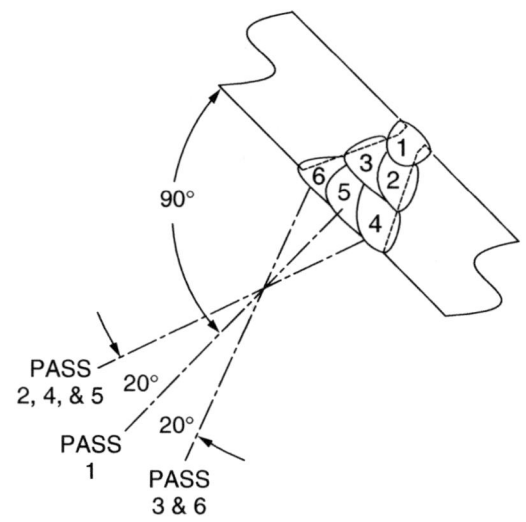

Figure 19 ◆ Multiple-pass 6G bead sequence and work angles.

303F19.EPS

Step 2 Clamp or tack weld the pipe weld coupon into position with the pipe axis inclined 45° to the horizontal plane. Be sure to position the tack welds so that they are not in the areas from which the test specimen will be cut.

Step 3 Run the root pass with a stringer bead. Modify the travel angles as necessary, as shown in *Figure 10,* for an uphill or downhill pass, as directed by your instructor.

Step 4 Brush the root pass to clean the weld.

Step 5 Run the remaining passes at the appropriate work angles in either an uphill or downhill direction as directed by your instructor. Overlap the passes, and clean the weld after each pass.

Summary

The ability to make open-root V-groove welds on pipe in all positions is one of the more difficult skills you must develop as a welder. The open-root V-groove weld is the most common weld joint used for joining medium- and thick-walled pipe. You must be able to set up the equipment, perform the welding, and recognize acceptable welds. Open-root V-groove pipe welds can be made in the 1G-ROTATED, 2G, 5G, and 6G positions. Practice these welds until you can consistently produce acceptable welds.

Review Questions

1. The GMAW process uses _____ wire electrode for filler wire.
   a. stop-and-go
   b. on-demand
   c. manual feed
   d. continuous

2. With stainless steel, it is common to use _____ for the root pass and _____ to complete the fill and cover passes.
   a. GTAW or SMAW; GMAW
   b. GMAW; GTAW or SMAW
   c. GMAW; FCAW
   d. FCAW or SMAW; GMAW

3. When preparing the welding area to practice open-root V-groove pipe welding, the welding table should be positioned near the _____.
   a. door
   b. vent or window
   c. fire extinguisher
   d. welding machine

4. Arc length is determined by _____.
   a. gun position
   b. voltage
   c. amperage
   d. travel speed

5. Within limits, when the wire electrode speed is increased, _____.
   a. the deposition rate drops
   b. the penetration decreases
   c. the welding amperage automatically decreases
   d. the welding amperage and deposition rate increase

6. Weld penetration in GMAW is affected most by _____.
   a. gun position
   b. travel speed
   c. voltage
   d. amperage

7. Wire electrode extension or stickout is usually dependent on the _____ for the GMAW application.
   a. transfer mode
   b. travel angle
   c. voltage
   d. power source

8. When GMAW is used for the root pass in an open-root V-groove weld, the transfer mode selected is usually _____.
   a. spray transfer
   b. short-circuiting transfer
   c. globular transfer
   d. thermal transfer

9. Pipe axis positions for 1G, 2G, and 5G can vary _____ from the basic position for GMAW pipe welds.
   a. ±10°
   b. ±15°
   c. ±20°
   d. ±25°

10. Pipe over 10" in diameter requires the use of _____ jack stands.
    a. two-legged
    b. three-legged
    c. four-legged
    d. five-legged

Performance Accreditation Tasks

The Performance Accreditation Tasks (PATCs) correspond to and support learning objectives in the *AWS Guide for the Training and Qualification of Welding Personnel*.

PATCs provide specific acceptable criteria for performance and help to ensure a true competency-based welding program for students.

The following tasks are designed to evaluate your ability to run open-root V-groove welds with GMAW equipment in the four standard test positions using carbon steel wire of the appropriate diameter and carbon dioxide shielding gas. Perform each task when you are instructed to do so by your instructor. As you complete each task, show it to your instructor for evaluation. Do not proceed to the next task until told to do so by your instructor. For AWS 2G and 5G certifications, refer to *AWS EG3.0-96* for bend test requirements. For AWS 6G certifications, refer to *AWS EG4.0-96* for bend test requirements.

OPEN-ROOT V-GROOVE PIPE WELD IN THE 1G-ROTATED POSITION

Using carbon steel wire of the appropriate diameter, carbon dioxide shielding gas, and stringer beads, make an open-root V-groove weld on carbon steel pipe in the 1G-ROTATED position.

Note: Depending on site procedures or practices, the root pass for the following tasks may be run using another welding process such as GTAW or SMAW. Check with your instructor to determine the welding process to use for the root pass.

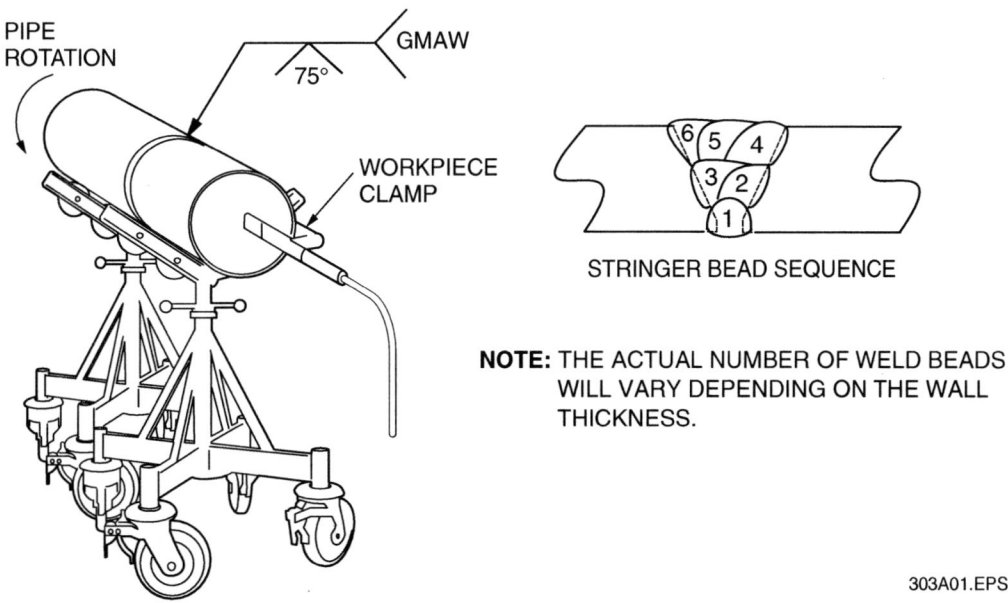

STRINGER BEAD SEQUENCE

NOTE: THE ACTUAL NUMBER OF WELD BEADS WILL VARY DEPENDING ON THE WALL THICKNESS.

303A01.EPS

Criteria for Acceptance

- Uniform appearance on the bead face   _____

- Craters and restarts filled to the full cross section of the weld   _____

- Uniform weld width ±¹⁄₁₆"   _____

- Acceptable weld profile in accordance with the
  ASME Boiler and Pressure Vessel Code, Section IX   _____

- Smooth transition with complete fusion at the toes of the weld   _____

- Complete uniform root reinforcement at least flush with the inside of the
  pipe to a maximum of ⅛"   _____

- No porosity   _____

- No excessive undercut   _____

- No cracks   _____

- No overlap   _____

- No incomplete fusion   _____

OPEN-ROOT V-GROOVE PIPE WELD IN THE 2G POSITION

Using carbon steel wire of the appropriate diameter, carbon dioxide shielding gas, and stringer beads, make an open-root V-groove weld on carbon steel pipe in the 2G position.

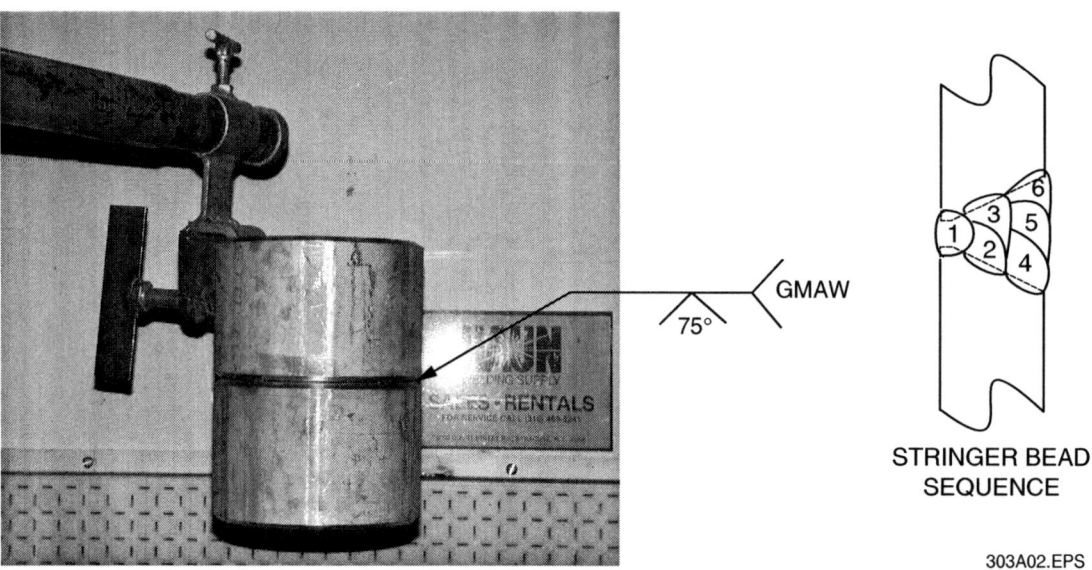

STRINGER BEAD
SEQUENCE

303A02.EPS

Criteria for Acceptance

- Uniform appearance on the bead face
- Craters and restarts filled to the full cross section of the weld
- Uniform weld width ±⅟₁₆"
- Acceptable weld profile in accordance with the *ASME Boiler and Pressure Vessel Code, Section IX*
- Smooth transition with complete fusion at the toes of the weld
- Complete uniform root reinforcement at least flush with the inside of the pipe to a maximum of ⅛"
- No porosity
- No excessive undercut
- No cracks
- No overlap
- No incomplete fusion

OPEN-ROOT V-GROOVE PIPE WELD IN THE 5G POSITION

Using carbon steel wire of the appropriate diameter, carbon dioxide shielding gas, and stringer or weave beads, make an open-root V-groove weld on carbon steel pipe in the 5G position.

GMAW

75°

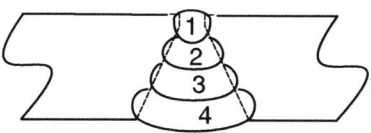

WEAVE BEAD SEQUENCE

STRINGER BEAD SEQUENCE

303A03.EPS

Criteria for Acceptance

- Uniform appearance on the bead face _____
- Craters and restarts filled to the full cross section of the weld _____
- Uniform weld width ±⅟₁₆" _____
- Acceptable weld profile in accordance with the *ASME Boiler and Pressure Vessel Code, Section IX* _____
- Smooth transition with complete fusion at the toes of the weld _____
- Complete uniform root reinforcement at least flush with the inside of the pipe to a maximum of ⅛" _____
- No porosity _____
- No excessive undercut _____
- No cracks _____
- No overlap _____
- No incomplete fusion _____
- No cracks

OPEN-ROOT V-GROOVE PIPE
WELD IN THE 6G (OR 6GR) POSITION

Using carbon steel wire of the appropriate diameter, carbon dioxide shielding gas, and stringer beads, make an open-root V-groove weld on carbon steel pipe in the 6G (or 6GR) position.

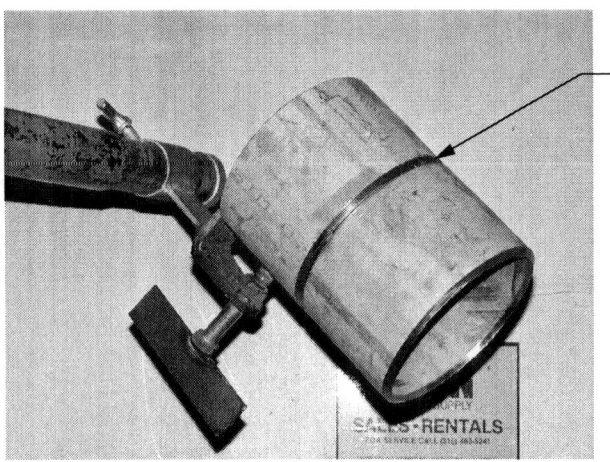

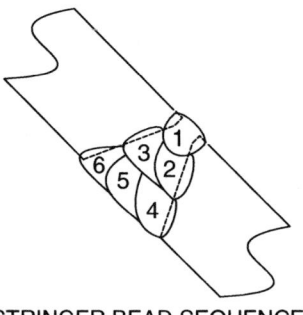

NOTE: IF REQUIRED FOR QUALIFICATION PURPOSES, A RESTRICTING RING MAY BE ADDED TO THE 6G POSITION COUPON TO FORM A 6GR POSITION COUPON.

STRINGER BEAD SEQUENCE　　　　303A04.EPS

Criteria for Acceptance

- Uniform appearance on the bead face _____

- Craters and restarts filled to the full cross section of the weld _____

- Uniform weld width ±⅟₁₆" _____

- Acceptable weld profile and guided bend test in accordance with the *ASME Boiler and Pressure Vessel Code, Section IX* _____

- Smooth transition with complete fusion at the toes of the weld _____

- Complete uniform root reinforcement at least flush with the inside of the pipe to a maximum of ⅛" _____

- No porosity _____

- No excessive undercut _____

- No cracks _____

- No overlap _____

- No incomplete fusion _____

Additional Resources

This module is intended to present thorough resources for task training. The following reference works are suggested for both instructors and motivated trainees interested in further study. These are optional materials for continued education rather than for task training.

ASME – Boiler and Pressure Vessel Code: Section 9, Current Edition. New York, NY: ASME International.

AWS D10.11, Recommended Practices for Root Pass Welding of Pipe Without Backing. Miami, FL: American Welding Society.

MIG Welding Handbook. Florence, SC: L-TEC Welding & Cutting Systems.

Welding of Pipelines and Related Facilities, API Standard 1104, Latest Edition. New York, NY: American Society of Mechanical Engineers International.

Welding Pressure Pipe Lines and Piping Systems. Cleveland, OH: The Lincoln Electric Company.

Welding Skills. R.T. Miller. Homewood, IL: American Technical Publishers.

Figure Credits

Miller Electric Mfg. Co.	303F02
Gerald Shannon	303F03, 303F04, 303F05, 303F17, 303F18, 303F19, 303SA02, 303SA03
Sumner Manufacturing Co., Inc.	303F14
G.A.L. Gage Co.	303SA01
Koike Aronson, Inc.	303SA04

The NCCER makes every effort to keep these textbooks up-to-date and free of technical errors. We appreciate your help in this process. If you have an idea for improving this textbook, or if you find an error, a typographical mistake, or an inaccuracy in NCCER's Contren® textbooks, please write us, using this form or a photocopy. Be sure to include the exact module number, page number, a detailed description, and the correction, if applicable. Your input will be brought to the attention of the Technical Review Committee. Thank you for your assistance.

Instructors – If you found that additional materials were necessary in order to teach this module effectively, please let us know so that we may include them in the Equipment/Materials list in the Instructor's Guide.

Write: Curriculum Revision and Development Department
National Center for Construction Education and Research
P.O. Box 141104, Gainesville, FL 32614-1104

Fax: 352-334-0932

E-mail: curriculum@nccer.org

Craft   Module Name

Copyright Date   Module Number   Page Number(s)

Description

(Optional) Correction

(Optional) Your Name and Address

Flux Cored Arc Welding (FCAW) – Pipe

COURSE MAP

This course map shows all of the modules in the third level of the Welding curriculum. The suggested training order begins at the bottom and proceeds up. Skill levels increase as you advance on the course map. The local Training Program Sponsor may adjust the training order.

WELDING LEVEL THREE

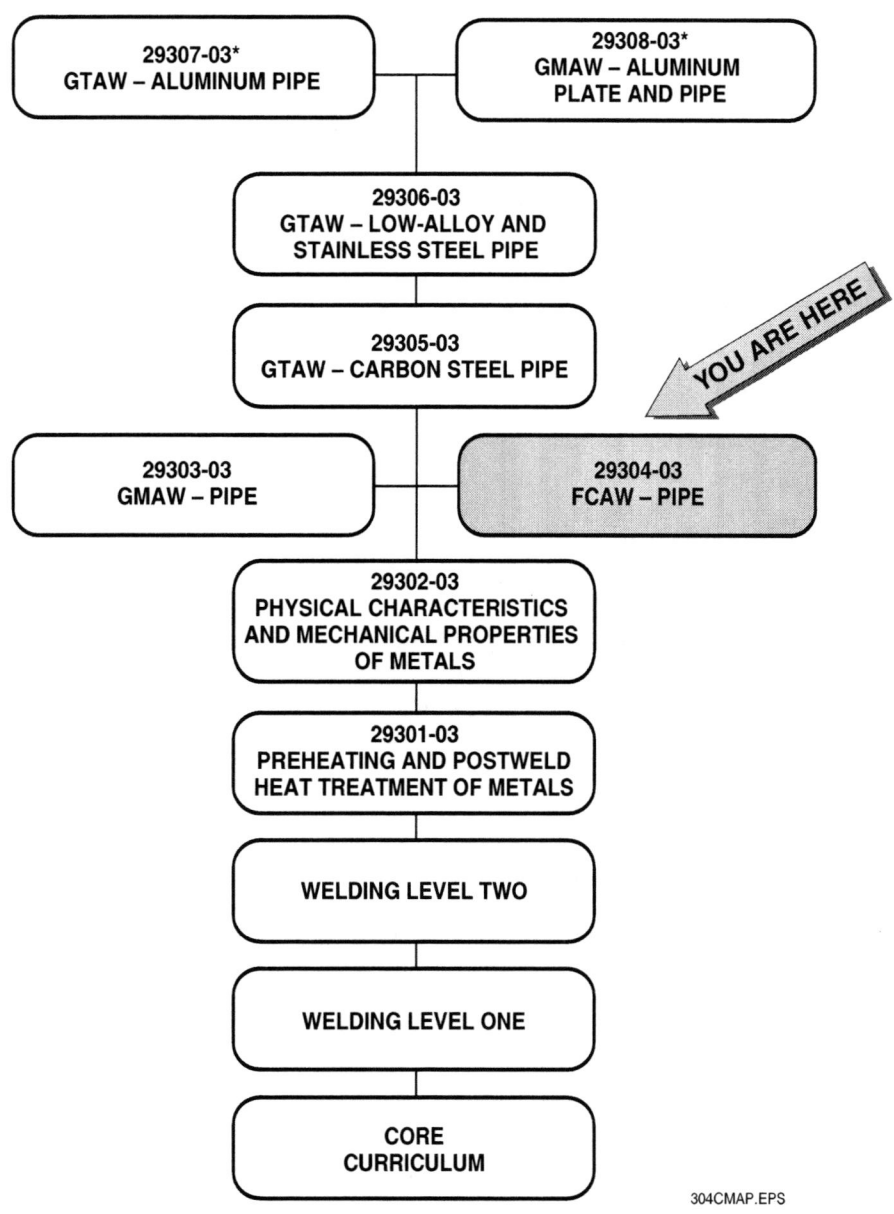

*Please note that Modules 29307-03 and 29308-03 are electives for those progressing through the *Welding Level Three* program.

Figures

Table

Flux Cored Arc Welding (FCAW) – Pipe

Objectives

When you have completed this module, you will be able to do the following:

1. Prepare FCAW equipment for open-root V-groove pipe weld techniques.
2. Identify and explain open-root V-groove pipe welds.
3. Perform open-root V-groove pipe welds using FCAW in the following positions:

   - 1G-ROTATED
   - 2G
   - 5G
   - 6G

Prerequisites

Before you begin this module, it is recommended that you successfully complete the following: Core Curriculum; Welding Levels One and Two; Welding Level Three, Modules 29301-03 and 29302-03.

Required Trainee Materials

1. Pencil and paper
2. Appropriate personal protective equipment

1.0.0 ◆ INTRODUCTION

FCAW is an arc welding process that uses a continuous, consumable flux cored wire as a filler metal to form a weld (see *Figure 1*). Shielding gas protects the molten weld metal from contamination by the atmosphere. The shielding gas can be generated entirely from the flux in the wire core or by both the flux in the wire core and an external shielding gas. In all types of flux cored wires, the flux contains agents for degassing the weld, stabilizing the arc, and forming a protective slag over the weld bead. Other elements are often added to the flux cored wire to alloy with the base metal and improve its chemical and/or mechanical characteristics. The FCAW process is basically limited to welding low-carbon steels, some low-alloy steels, and some stainless steels. FCAW is an excellent process for making groove welds on pipe.

The equipment used for FCAW gas-shielded (FCAW-G) or FCAW self-shielded (FCAW-S) is basically the same except for the components needed for the gas used by the FCAW-G process. Like GMAW, this additional equipment includes guns equipped with a gas nozzle, gas hoses, a gas control solenoid valve, a gas regulator/flowmeter, and a shielding gas cylinder (gas source). *Figures 2* and *3* show typical FCAW equipment.

Wind Resistance

Self-shielded FCAW is particularly useful in situations with high winds because the shielding gas is created inside the welding arc as the flux vaporizes. The molten puddle has a constant supply of shield gas displacing the air from the arc outward and is immediately covered by the residual slag. In contrast, an external shield gas attempts to force gas down from the nozzle to displace the air, but it can be blown away from the weld puddle by strong winds.

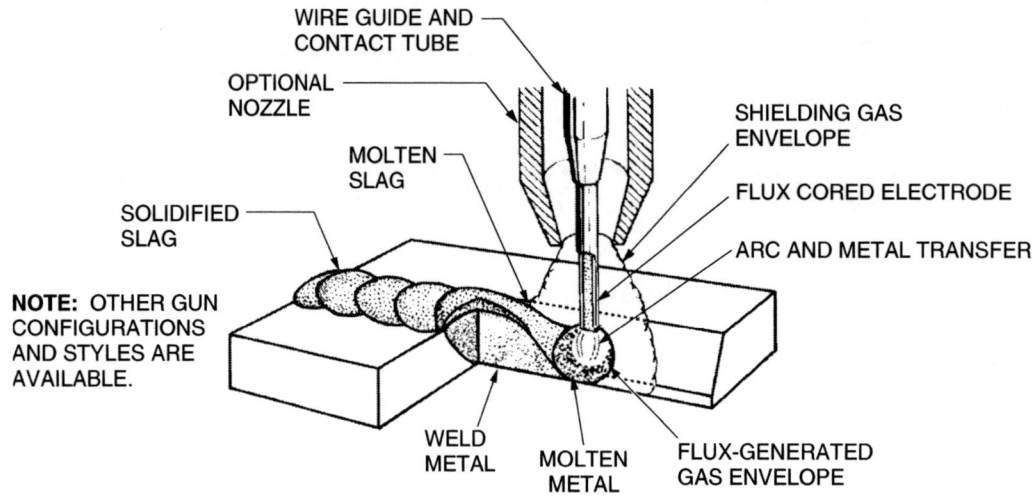

GAS-SHIELDED FCAW

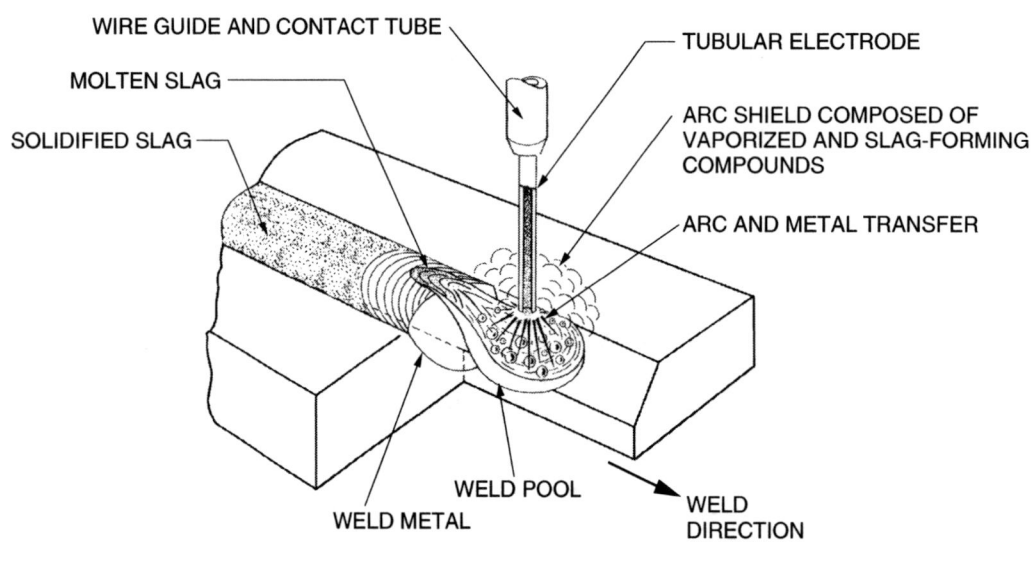

SELF-SHIELDED FCAW

304F01.EPS

Figure 1 ◆ FCAW process.

2.0.0 ◆ SAFETY SUMMARY

The following is a summary of safety procedures and practices you must observe while cutting or welding. Keep in mind that this is just a summary. Complete safety coverage is provided in the Level One module *Welding Safety*. If you have not completed that module, do so before continuing. Above all, be sure to wear appropriate protective clothing and equipment when welding or cutting.

2.1.0 Protective Clothing and Equipment

• Always use safety goggles with a full face shield or a helmet. The goggles, face shield, or helmet lens must have the proper light-reducing tint for the type of welding or cutting to be performed. Never directly or indirectly view an electric arc without using a properly tinted lens.

• Wear proper protective leather and/or flame retardant clothing along with welding gloves to protect from flying sparks, molten metal, and heat.

• Wear 8" or taller high-top safety shoes or boots. Make sure that the tongue and lace area of the footwear is covered by a pant leg. If the tongue and lace area is exposed or the footwear must be protected from burn marks, wear leather spats under the pants or chaps and over the front top of the footwear.

• Wear a solid material (non-mesh) hat with a bill pointing to the rear or, if much overhead cutting

Figure 2 ◆ Gas-shielded FCAW equipment.

Figure 3 ◆ Self-shielded FCAW equipment.

or welding is required, a full leather hood with a welding faceplate and the correct tinted lens. If a hard hat is required, use one that allows attachment of both rear deflector material and a face shield.

- If a full leather hood is not worn, wear a face shield and snug fitting welding goggles over safety glasses for gas welding or cutting. Either the face shield or the lenses of the welding goggles must be an approved shade 5 or 6 filter. For electric arc welding or cutting, wear safety goggles and a welding hood with the correct tinted lens (shade 9 to 14).

- If a full leather hood is not worn, wear earplugs to protect your ear canals from sparks.

2.2.0 Fire/Explosion Prevention

- Never carry matches or gas-filled lighters in your pockets. Sparks can cause the matches to ignite or the lighter to explode, causing serious injury.

- Never perform any type of heating, cutting, or welding until a hot-work permit is obtained and an approved fire watch is established. Most work-site fires caused by these types of operations are started by cutting torches.

- Never use oxygen to blow off clothing. The oxygen can remain trapped in the fabric for a time. If a spark hits the clothing during this period, the clothing can burn rapidly and violently out of control.

- Make sure that any flammable material in the work area is moved or is shielded by a fire-resistant covering. Approved fire extinguishers must be available before any heating, welding, or cutting operations are attempted.

- Never release a large amount of oxygen nor use oxygen as compressed air. Its presence around flammable materials or sparks can cause rapid and uncontrolled combustion. Keep oxygen away from oil, grease, and other petroleum products.

- Never release a large amount of fuel gas, especially acetylene. Methane and propane tend to concentrate in and along low areas and can ignite at a considerable distance from the release point. Acetylene is lighter than air but is even more dangerous than methane. When mixed with air or oxygen, it will explode at much lower concentrations than any other fuel.

- To prevent fires, maintain a neat and clean work area, and make sure that any metal scrap or slag is cold before disposal.

- Before cutting or welding containers such as tanks or barrels, check to see if they have contained any explosive, hazardous, or flammable materials, including petroleum products, citrus products, or chemicals that decompose into toxic fumes when heated. As a standard practice, always clean and then fill any tanks or barrels with water, or purge them with a flow of inert gas to displace any oxygen.

2.3.0 Work Area Ventilation

- Make sure to follow confined space procedures before conducting any welding or cutting in the confined space.
- Never use oxygen to ventilate confined spaces.
- Always perform cutting or welding operations in a well-ventilated area. Such operations involving zinc or cadmium materials or coatings result in toxic fumes. For long-term cutting or welding of these materials, always wear an approved, full-face, supplied-air respirator (SAR) that uses breathing air supplied from outside of the work area. For occasional, very short-term exposure, you may use a high efficiency particulate arresting (HEPA)-rated or metal-fume filter on a standard respirator.
- Make sure confined spaces are ventilated properly for cutting or welding operations.

3.0.0 ◆ WELDING PREPARATION

Before welding can begin, the area has to be readied, the welding equipment must be set up, and the metal to be welded must be prepared. The following sections explain how to set up the equipment for welding.

To practice welding, you need a welding table, bench, or stand. The welding surface must be steel, and provisions must be available for placing weld coupons out of position.

To set up the area for welding, follow these steps:

Step 1 Check to be sure the area is properly ventilated. Make use of doors, windows, and fans.

Step 2 Check the area for fire hazards. Remove any flammable materials before proceeding.

Step 3 Locate the nearest fire extinguisher. Do not proceed unless the extinguisher is charged and you know how to use it.

Step 4 Position a welding table near the welding machine.

Step 5 Set up flash shields around the welding area.

3.1.0 Practice Pipe Weld Coupons

Pipe weld coupons should be cut from 3" to 12" diameter Schedule 40 or Schedule 80 carbon steel pipe. Each welded joint will require two coupons of the same size and schedule pipe.

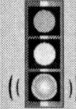

NOTE

Most codes allow substitution of carbon steel pipe for stainless steel and low-alloy pipe as a base metal for welder qualification. However, applicable stainless steel or low-alloy filler metals must be used. Always refer to the welding procedure specifications (WPS) or site quality standards for the proper qualification information.

Figure 4 shows typical ASME International, American Petroleum Institute (API), and American Welding Society (AWS) pipe bevel specifications for bevel angles, root faces, and root openings.

To prepare weld coupons for open-root V-groove weld joints follow these steps:

Step 1 Clean all rust or mill scale from the inside and outside of the carbon steel pipe with a grinder or wire brush. Clean for a distance of ½" from the weld joint.

Step 2 Bevel the end of the pipe to 30° or 37½° by any acceptable beveling method, such as by using a mechanical cutter or by thermal cutting or grinding.

Step 3 Cut off a section of the beveled pipe end (1½" minimum).

NOTE

For 1G-ROTATED welding, you may have to cut longer coupons to fit the rollers, if used.

Follow these steps to prepare the open-root V-groove weld pipe coupon:

Step 1 Check the bevel. There should be no dross, and the bevel angle should be 30° or 37½°.

Conserve Materials

Pipe for practice welding is expensive and difficult to obtain. Make every effort to conserve and not waste available material. Completely use weld coupons until all surfaces have been welded by cutting the coupon apart and reusing the pieces. Check with the instructor for the appropriate size coupon.

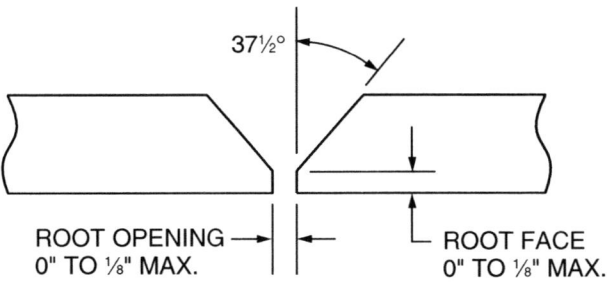

37½°

ROOT OPENING →| |← ROOT FACE
0" TO ⅛" MAX.   0" TO ⅛" MAX.

TYPICAL ASME SPECIFICATION

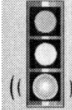
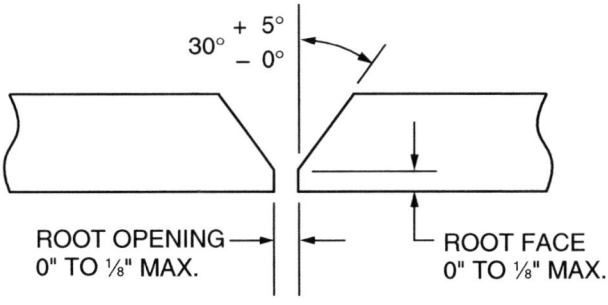

30° +5° / −0°

ROOT OPENING →| |← ROOT FACE
0" TO ⅛" MAX.   0" TO ⅛" MAX.

TYPICAL API AND AWS SPECIFICATION

304F04.EPS

Figure 4 ◆ Pipe bevel specifications.

Step 2 Grind or file a 0" to ⅛" root face on the bevel, as specified by your instructor.

> **NOTE**
> Welding codes allow both the root opening and the root face on open root welds to be from 0" to ⅛" wide. Select and adjust the opening and root face as needed when you start the welding practices.

Step 3 Align the two pipe sections so that the inside diameter (ID) surfaces are even all around. Align small-diameter pipe by clamping both pieces to a piece of angle iron. Align large-diameter pipe with the aid of a pipe alignment jig or by holding a straightedge across the joint, parallel to the pipe axis. The straightedge must be used all around the pipe, in case one or both sections are distorted.

Step 4 Gap the root opening with pieces of filler wire, metal shims, or pieces of bare welding electrode of the correct diameter.

Step 5 When the root opening is correct and the pipe ends are aligned, make the first of four tack welds of no greater than 1".

> **NOTE**
> Heavy-wall or large-diameter pipe may require longer tacks or more than four tacks.

Step 6 After the first tack weld, on the opposite side, check the root opening, adjust the gap if necessary, and make the second tack weld.

Step 7 Check the root opening again and weld the third tack midway between the first two tacks.

Machined Bevels and Root Faces for FCAW

For V-groove joints, machined bevels and root faces provide the cleanest and most uniform method of preparing materials for satisfactory FCAW welds. This is especially true for stainless steel where all contaminants from thermal cutting must be removed. FCAW, SMAW, and GTAW can tolerate some poor fit-up for root passes, unlike GMAW. In practice, GTAW or SMAW are usually used for the root pass in critical piping applications.

Welding Large-Diameter Pipe

The practice coupons on large-diameter pipe must be greater than 1½" in length. They need more metal to help absorb the increased heat used to make these welds.

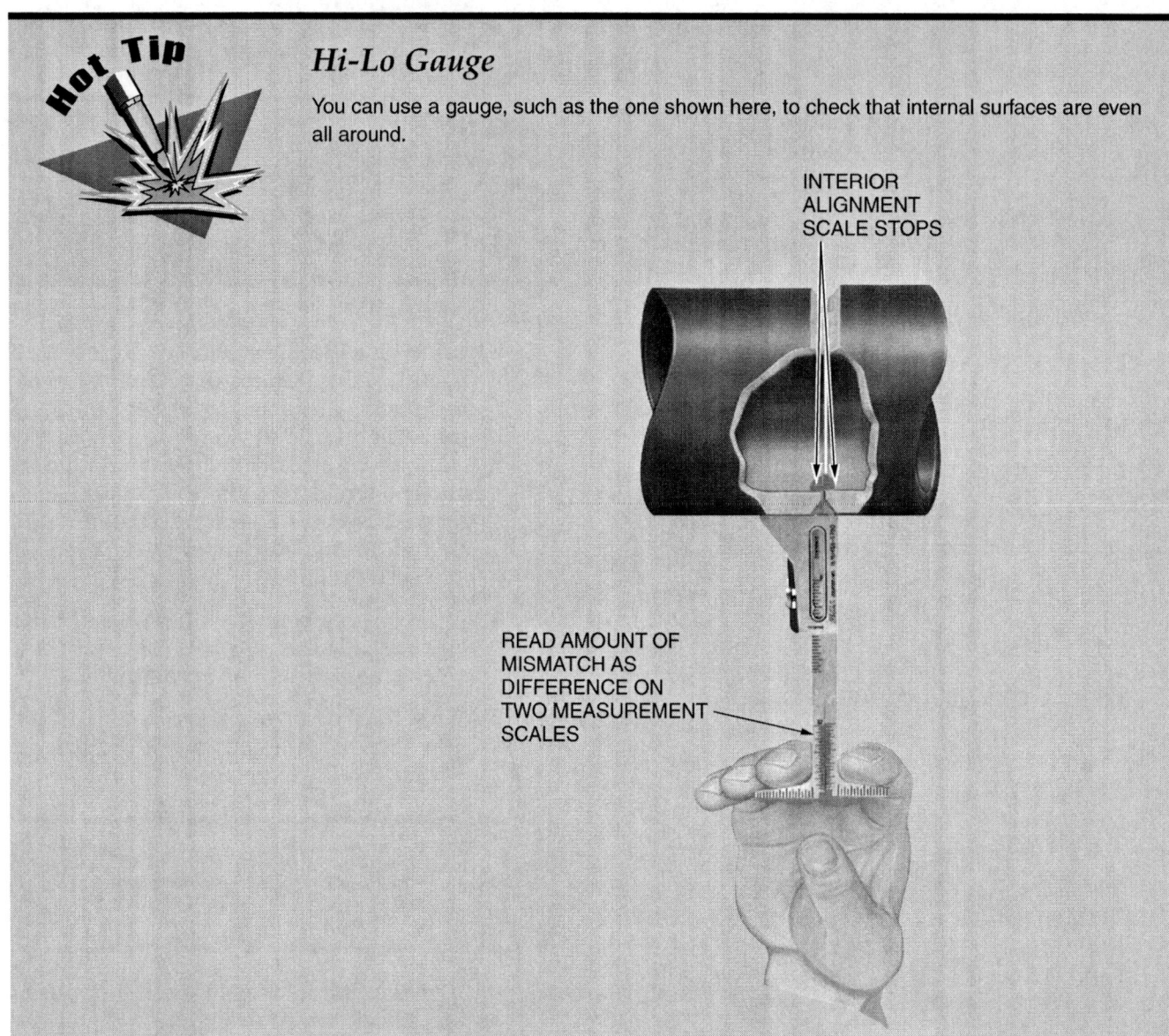

Hot Tip

Hi-Lo Gauge

You can use a gauge, such as the one shown here, to check that internal surfaces are even all around.

INTERIOR
ALIGNMENT
SCALE STOPS

READ AMOUNT OF
MISMATCH AS
DIFFERENCE ON
TWO MEASUREMENT
SCALES

304SA01.EPS

Step 8 Weld the fourth tack opposite the third tack and midway between the first and second tacks. There should now be four tack welds evenly spaced (90°) around the pipe coupon. This is illustrated by *Figure 5.*

Step 9 Clean and feather the tack welds with the edge of a grinding wheel. Feathering the ends of the tack welds with a grinder helps fuse the tack welds into the root pass.

3.2.0 The Welding Machine

Identify a proper welding machine for FCAW and follow these steps to set it up for use.

Step 1 Verify that the welding machine can be used for FCAW with or without internally controlled gas shielding.

Step 2 Check to be sure that the welding machine is properly grounded through the primary current receptacle.

Step 3 Verify the location of the primary disconnect.

Step 4 Configure the welding machine for FCAW-S or FCAW-G parameters as directed by your instructor (*Figure 6*). Configure the gun polarity, and equip the gun with the correct liner material, contact tube for the diameter of the wire being used, and the correct nozzle for the application.

Step 5 In accordance with the manufacturer's instructions, configure and load the wire feeder and gun with the proper diameter flux cored electrode wire as directed by your instructor.

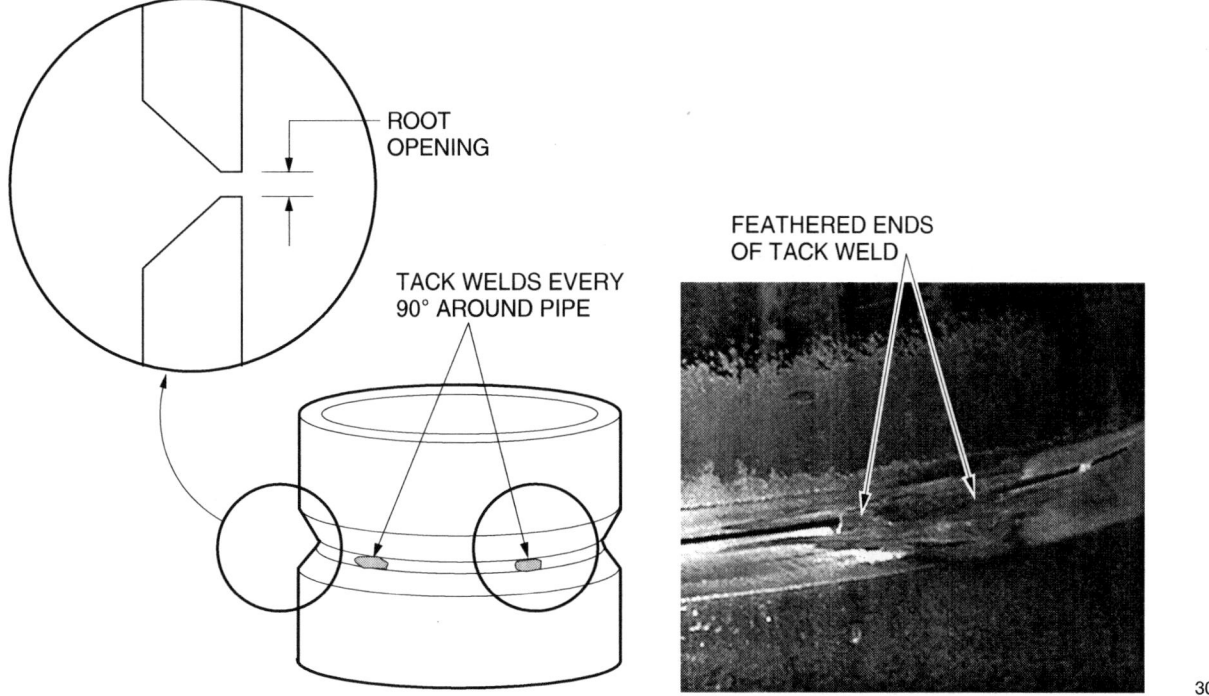

ROOT OPENING

TACK WELDS EVERY 90° AROUND PIPE

FEATHERED ENDS OF TACK WELD

304F05.EPS

Figure 5 ◆ Tacked open-root V-groove weld coupon.

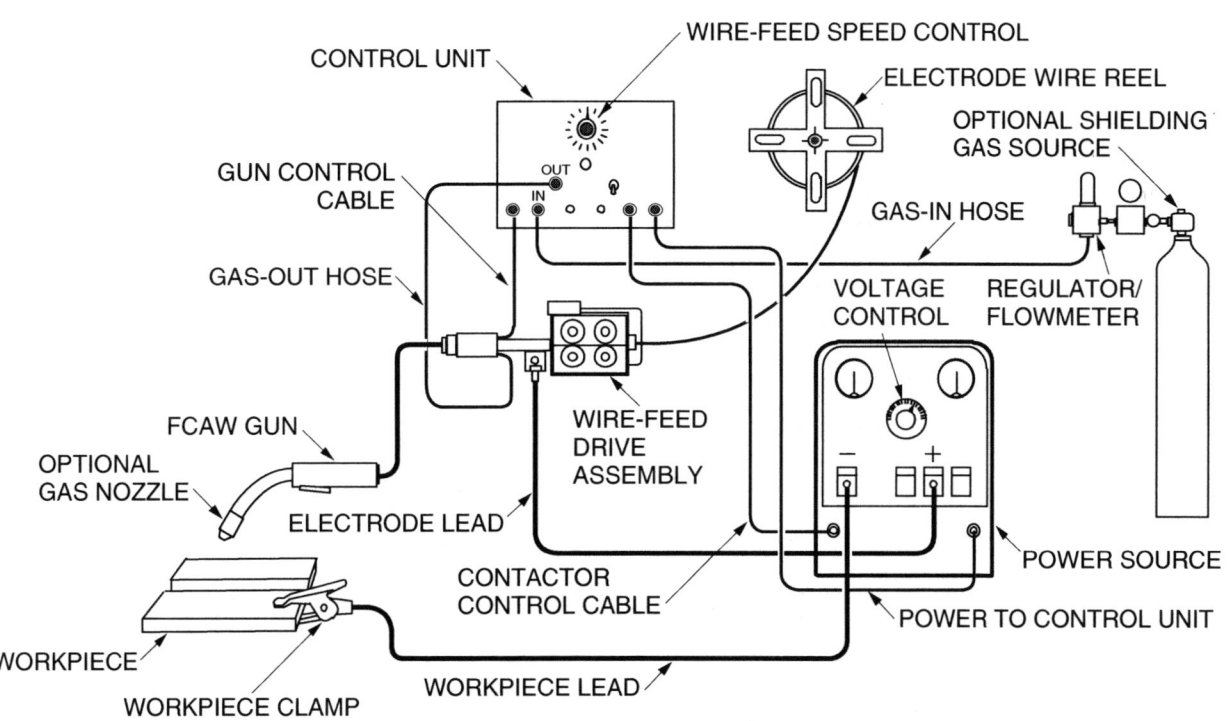

CONTROL UNIT

WIRE-FEED SPEED CONTROL

ELECTRODE WIRE REEL

OPTIONAL SHIELDING GAS SOURCE

GUN CONTROL CABLE

OUT

IN

GAS-IN HOSE

GAS-OUT HOSE

VOLTAGE CONTROL

REGULATOR/ FLOWMETER

FCAW GUN

WIRE-FEED DRIVE ASSEMBLY

OPTIONAL GAS NOZZLE

ELECTRODE LEAD

POWER SOURCE

CONTACTOR CONTROL CABLE

POWER TO CONTROL UNIT

WORKPIECE

WORKPIECE CLAMP

WORKPIECE LEAD

NOTE: THE POLARITY OF THE GUN AND WORKPIECE LEADS IS DETERMINED BY THE TYPE OF FILLER METAL AND APPLICATION.

304F06.EPS

Figure 6 ◆ Configuration diagram of typical FCAW machine.

Step 6 If required, connect the proper shielding gas for the application as described in the previous level and as specified by the wire electrode manufacturer, WPS, site quality standards, or your instructor.

Step 7 Connect the clamp of the workpiece lead to the workpiece.

Step 8 Turn on the welding machine and, if necessary, purge the gun as directed by the gun manufacturer's instructions.

Step 9 Set the initial welding voltage and wire feed speed for the type and size electrode wire as recommended by the manufacturer.

3.2.1 Welding Voltage

Arc length is determined by voltage, which is set at the power source. Arc length is the distance from the wire electrode tip to the base metal or molten pool at the base metal. If voltage is set too high, the arc will be too long and could cause the wire to melt and fuse to the contact tube. A voltage that's too high also causes porosity and excessive spatter. Voltage must be increased or decreased as wire feed speed is increased or decreased. Set the voltage so that the arc length is just above the surface of the base metal (*Figure 7*).

3.2.2 Welding Amperage

With a standard constant-voltage power source, the electrode feed speed controls the welding amperage after the initial recommended setting. The welding power source provides the amperage to melt the wire electrode while maintaining the selected welding voltage. Within limits, when the wire electrode feed speed is increased, the welding amperage and deposition rate also increase. This results in higher welding heat, deeper penetration,

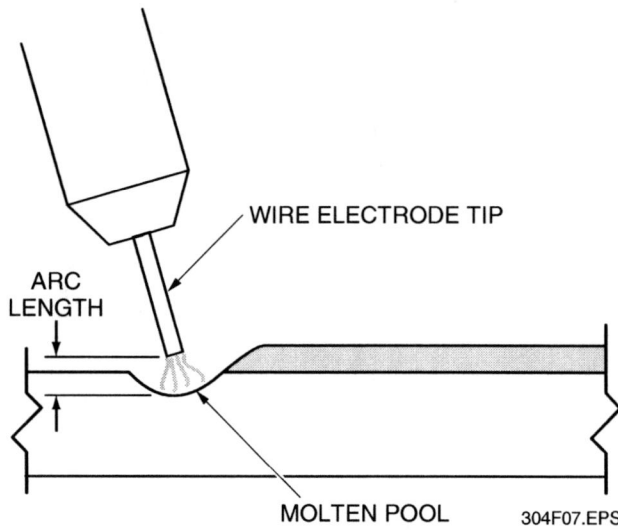

Figure 7 ◆ Arc length.

and higher beads. When the wire electrode feed speed is decreased, the welding amperage automatically decreases. With lower welding amperage and less heat, the deposition rate drops, and the weld beads are smaller with less penetration.

Note that some constant-voltage power sources used for GMAW/FCAW provide some degree of current modification. Standard constant-voltage GMAW/FCAW units have a current slope that is fixed by the manufacturer for general welding applications and conditions. Pulse transfer power sources allow you to adjust the peak current for the pulse and the background current between pulses to match specific welding applications and/or wire electrode requirements.

3.2.3 Weld Travel Speed

Weld travel speed is the speed at which the electrode tip passes across the base metal in the direction of the weld. It is measured in inches per minute

Wire Electrode Manufacturer's Recommendations

Always obtain and follow the manufacturer's recommendations for shielding gas (if required) and for initially setting the welding voltage and wire feed speed parameters. These balanced parameters are critical and are based on the welding position, size, and composition of the flux cored wire being used.

Travel Speed and Wire Feed Speed

Inexperienced welders are tempted to turn down the wire feed speed if they experience difficulty controlling the weld puddle. Wire electrodes must be run at certain balanced parameters that cannot be changed individually. Voltage, wire feed speed, and travel speed are adjusted and balanced together to control the weld puddle.

(ipm). Travel speed has a great effect on penetration and bead size: slower travel speeds build higher beads with deeper penetration while faster travel speeds build smaller beads with less penetration.

3.2.4 Gun Position

Gun position for FCAW has greater effect on weld penetration than either voltage or travel speed. The gun work angle and travel angle in relation to the direction of the weld are defined the same as for SMAW welding. See *Figure 8*.

3.2.5 Electrode Extension, Stickout, and Standoff Distance

Electrode extension is the length of the wire that extends beyond the tip of the welding gun's contact tube. Changing the extension influences the welding current for low-conductivity metal wires because it changes the preheating of the wire caused by the resistance of the wire to current flow. As the extension increases, preheating of the wire increases. Because the welding power source is self-regulating, it does not have to furnish as much welding current to melt the wire, so the current output automatically decreases. This results in less penetration and increased deposit rates. Increasing extension is useful for bridging gaps and compensating for mismatch, but it can cause overlap or lack of fusion and a ropy bead appearance. When the extension is decreased, the power source is forced to increase its current output to burn off the wire. Too little extension can cause the wire to weld to the contact tube and can develop porosity in the weld. For high-conductivity metal wires, the preheating effect of wire resistance is minimal, and wire speed and voltage settings have a more direct effect.

The standard extension for self-shielding FCAW with 0.045" wire and an insulating nozzle is ¾" to 1¼", depending on the welding position. *Figure 9* shows typical electrode extensions for FCAW gun configurations and defines other gun nomenclature. Stickout is the distance from the gas nozzle or insulating nozzle to the end of the electrode. Standoff distance is the distance from the gas nozzle or insulating nozzle to the workpiece.

With shorter stickout and less voltage drop in the wire, higher arc voltage is available. If the stickout is too short, spatter will increase and rapidly build up in the nozzle and on the contact tube.

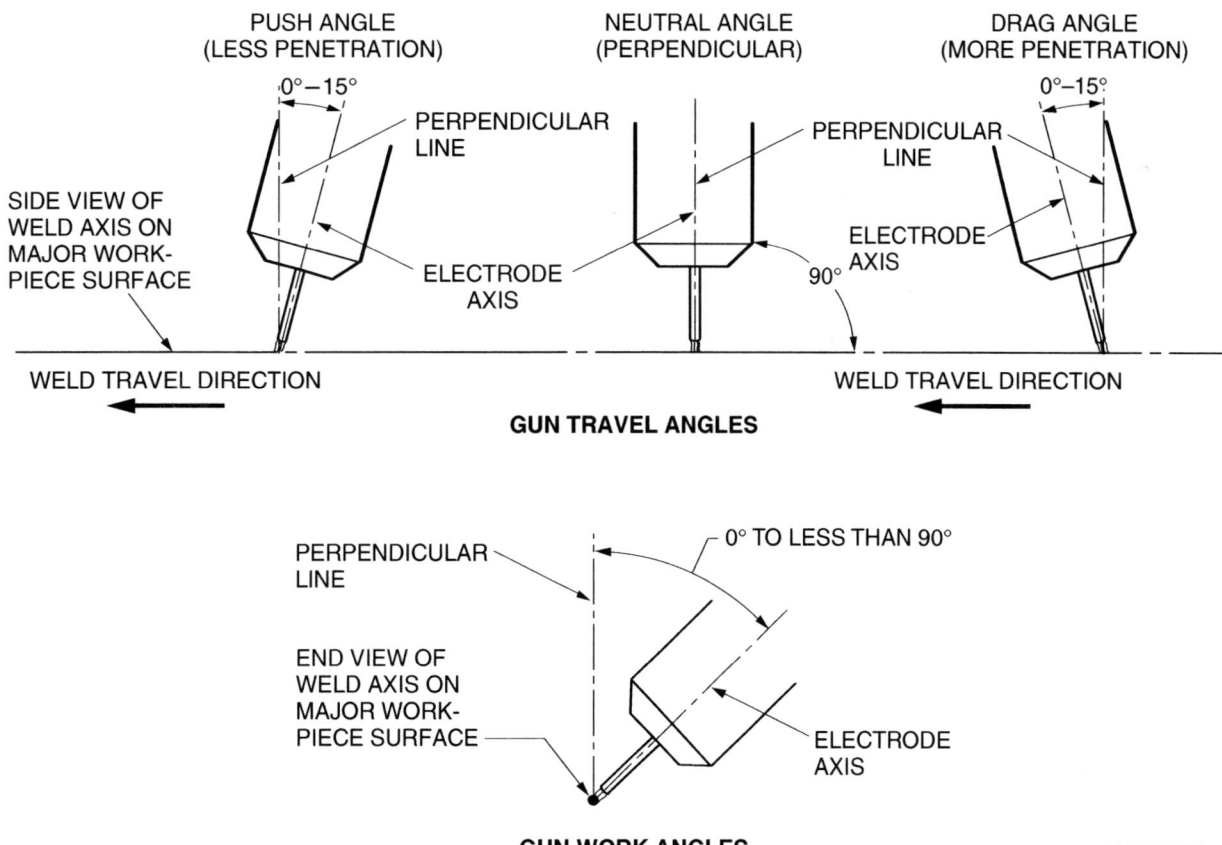

Figure 8 ◆ Gun work and travel angles.

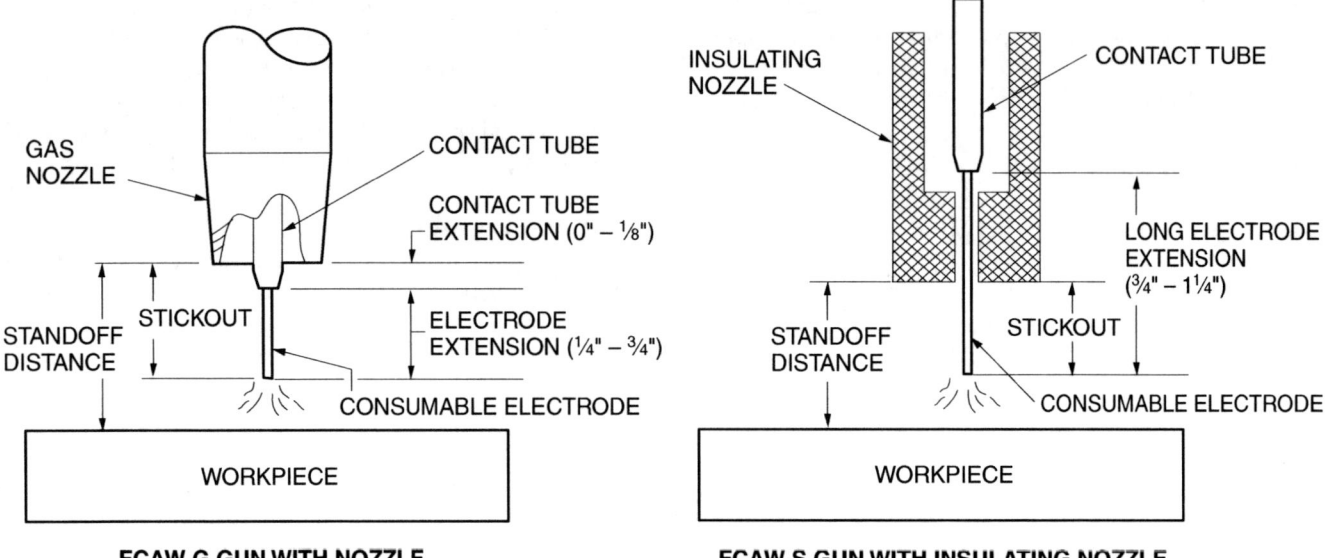

GAS NOZZLE
CONTACT TUBE
CONTACT TUBE EXTENSION (0" – 1/8")
STANDOFF DISTANCE
STICKOUT
ELECTRODE EXTENSION (1/4" – 3/4")
CONSUMABLE ELECTRODE
WORKPIECE
FCAW-G GUN WITH NOZZLE
INSULATING NOZZLE
CONTACT TUBE
LONG ELECTRODE EXTENSION (3/4" – 1 1/4")
STANDOFF DISTANCE
STICKOUT
CONSUMABLE ELECTRODE
WORKPIECE
FCAW-S GUN WITH INSULATING NOZZLE
304F09.EPS

Figure 9 ◆ FCAW gun configurations.

Hot Tip

Spatter Control

With excessive stickout, more voltage is lost in the electrode length, the arc voltage is lower, and irregular arc action and spatter may occur. Penetration is decreased as stickout is increased.

Antispatter Compound

Use only antispatter material specifically designed for GMAW/FCAW-G gas nozzles. Water-based compounds may cause porosity in aluminum and hydrogen embrittlement of hydrogen-sensitive steels. Always refer to MSDS for specific safety considerations.

Miller
ARC GUARD
ANTI-SPATTER GEL
Arcair
PROTEX
EXTRA
ANTI-SPATTER PLUS GUN PROTECTOR

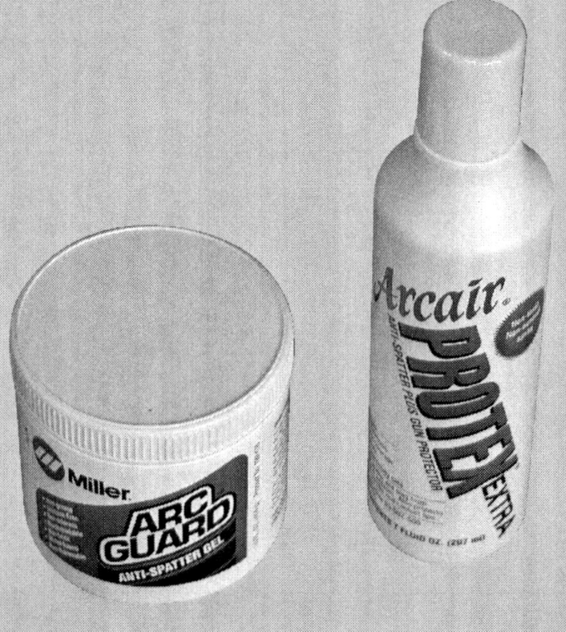

TYPICAL ANTISPATTER GEL AND SPRAY

304SA02.EPS

3.2.6 Gas Nozzle Cleaning

As the welding machine is used, weld spatter accumulates on both the gas nozzle and the contact tube. If the gas nozzle is not occasionally cleaned, it will restrict the shielding gas flow, which will cause porosity in the weld.

Clean the gas nozzle with a reamer, round file, or the tang of a standard file. After cleaning, the nozzle can be sprayed with or dipped in a special antispatter compound. The antispatter compound helps prevent the spatter from sticking to the nozzle.

4.0.0 ◆ OPEN-ROOT V-GROOVE PIPE WELDS

The open-root V-groove weld is the most common groove weld normally made on carbon, low-alloy, and stainless steel pipe. These welds are typically used for joining medium- and thick-walled pipe used in critical piping systems. Critical piping systems (including pipe, fittings, and welded joints) contain or carry material with the potential to cause long- or short-term catastrophic danger or damage to personnel and/or the environment, should the components fail to contain or carry the material as designed. Welds in critical piping must meet the most stringent code requirements. Non-critical piping is low-pressure piping used for heating and air conditioning, simple water systems, and other service installations. Less-stringent code requirements are used to evaluate non-critical piping welds.

4.1.0 FCAW Welding Techniques

The most difficult part of making an open-root V-groove weld is the root pass. The root pass must have complete penetration and fusion but not an excessive amount of melt-through root reinforcement. The extent of penetration and fusion is significantly important. If penetration is too little, it will create a weak joint. However, if there is too much penetration and excessive root reinforcement is created, the inside diameter of the pipe will be reduced and the flow will be restricted. Incomplete fusion will result in a defective weld. Progression of multiple (5G)

position welds for root pass welding is usually downhill. The fill and cover passes are usually run uphill. *Figure 10* shows approximate modifications to the travel position that may be necessary for multiple (5G) position pipe welds when welding uphill or downhill with FCAW.

After the root pass is run, it should be cleaned of slag and inspected. Check the weld for incomplete fusion and penetration. The root reinforcement should be ⅛" or less.

> **NOTE**
> Some WPSs do not allow grinding of stainless steel welds.

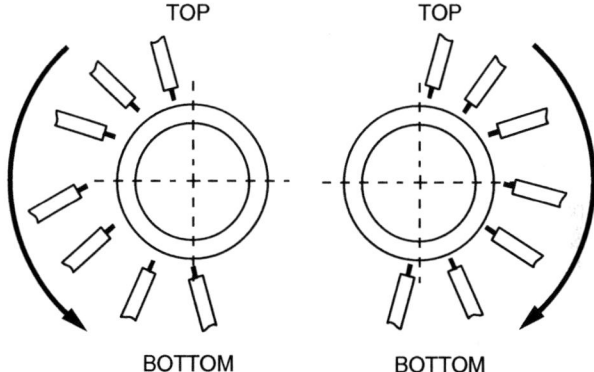

DOWNHILL TRAVEL POSITIONS

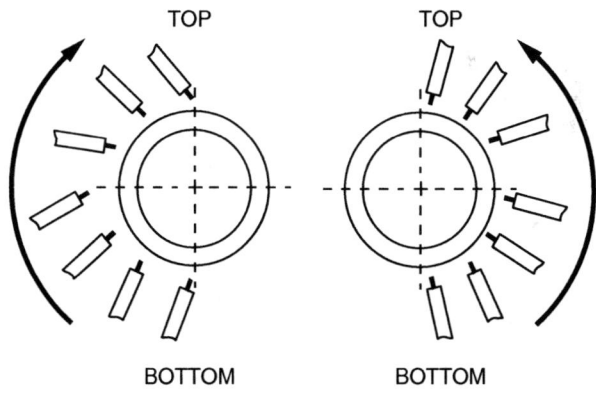

UPHILL TRAVEL POSITIONS

304F10.EPS

Figure 10 ◆ Approximate travel positions.

Arc Strikes

When welding pipe, do not make arc strikes outside the weld groove on the surface of the pipe. An arc strike can cause a hardened spot that can crack as the pipe expands and contracts. An arc strike on the pipe surface is considered a defect, and will require repair or rework.

Remove any excess buildup or undercut with a hand grinder by grinding the face of the root pass with the edge of a grinding disk. Most of the slag along the sides of the weld bead is removed by this grinding action, making it easier to add the remaining passes. Use care not to grind through the root pass or widen the groove. A narrow grinding wheel (not a cutoff wheel) is effective for this purpose.

4.2.0 Pipe Groove Weld Test Positions

Groove welds may be made in all positions on pipe. The weld position is determined by the axis of the pipe. Four standard weld test positions, shown in *Figure 11* and described as follows, are used with pipe:

- *Flat (1G-ROTATED) welding position* – The pipe axis is horizontal and the pipe is slowly rotated while being welded on the top. The weld beads are flat.
- *Horizontal (2G) welding position* – The pipe axis is vertical and the pipe rotation is fixed. The weld beads are horizontal.
- *Multiple (5G) welding position* – The pipe axis is horizontal and the pipe rotation is fixed. The weld beads are flat, vertical, and overhead.
- *Inclined multiple (6G) welding position* – The pipe axis is inclined nominally at 45° ± 5°, and pipe rotation is fixed. The weld beads are horizontal, vertical, and overhead.

The pipe axis positions for 1G, 2G, and 5G can vary ±15° from the basic position. The pipe axis position for 6G can vary ±5° from the nominal 45° angle. See *Table 1*.

To destructively test 5G or 6G welds, the test specimens are cut from the four regions illustrated in *Figure 12*: midway between the 12-o'clock and 3-o'clock, 3-o'clock and 6-o'clock, 6-o'clock and 9-o'clock, and 9-o'clock and 12-o'clock positions. (The 12-o'clock position is the top of the coupon groove and 6-o'clock is the bottom.)

Table 1 Pipe Groove Weld Test Positions

Weld Test Position	Pipe Axis	Weld Beads	Pipe Position
1G-ROTATED	Horizontal	Flat	Slowly rotated
2G	Vertical	Horizontal	Fixed
5G	Horizontal	Flat, vertical overhead	Fixed
6G	Inclined 45°	Horizontal, vertical, overhead	Fixed

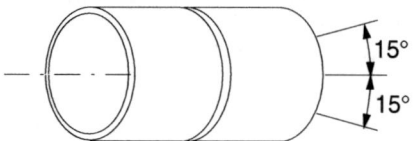

ROTATE PIPE AND DEPOSIT WELD AT OR NEAR THE TOP

PIPE HORIZONTAL (±15°) AND ROLLED TO KEEP WELD FLAT

1G - ROTATED POSITION

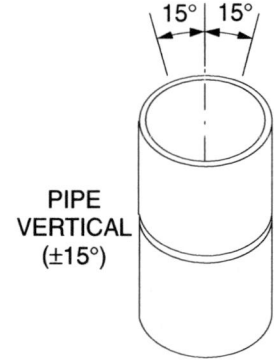

PIPE VERTICAL (±15°)

PIPE NOT ROTATED DURING WELDING

2G POSITION

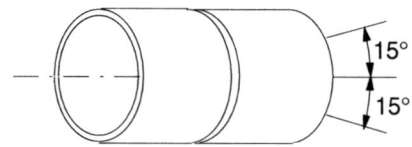

PIPE NOT ROTATED DURING WELDING

PIPE HORIZONTAL (±15°)

5G POSITION

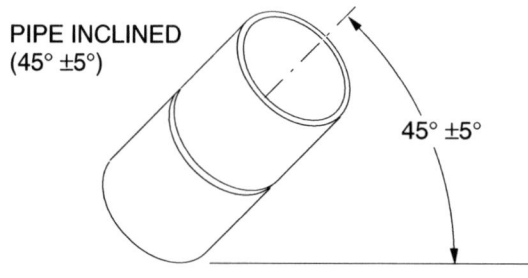

PIPE INCLINED (45° ±5°)

45° ±5°

PIPE NOT ROTATED DURING WELDING

6G POSITION

304F11.EPS

Figure 11 ◆ Four basic pipe groove weld test positions.

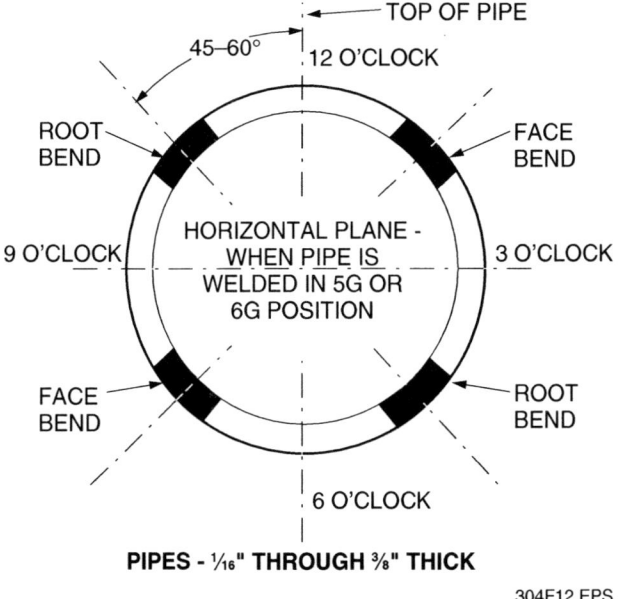

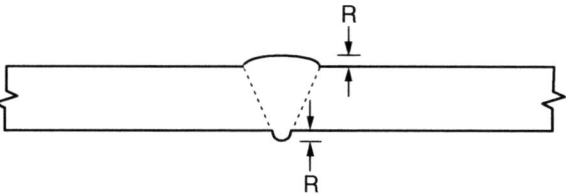

R = FACE AND ROOT REINFORCEMENT PER CODE
NOT TO EXCEED ⅛" MAX.

ACCEPTABLE WELD PROFILE

PIPES - ¹⁄₁₆" THROUGH ⅜" THICK

304F12.EPS

Figure 12 ◆ Test specimen regions of 5G or 6G position pipe.

4.3.0 Acceptable and Unacceptable Pipe Weld Profiles

Pipe groove welds should be made with slight reinforcement (not exceeding ⅛") and a gradual transition to the base metal at each toe. The root pass should have complete penetration. The root reinforcement on the inside of the pipe ranges from being flush to a maximum of ⅛". Pipe groove welds must not have excess reinforcement or underfill at the face or root, excessive undercut, or overlap. Excessively large cover passes will reduce the pipe's strength. They cause the stresses in the pipe to be concentrated along the sides of the weld and do not permit the pipe to expand and contract in a uniform manner along its length. If a weld has any of these defects, as shown in *Figure 13*, it must be repaired.

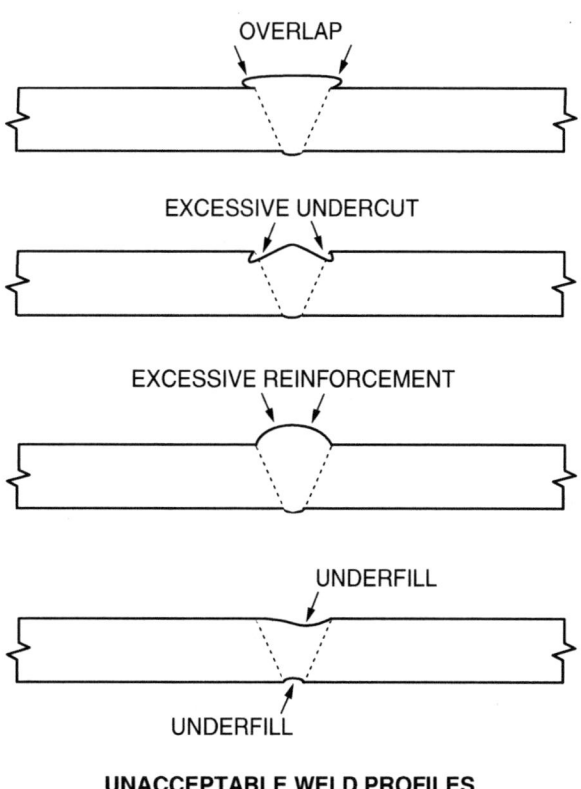

UNACCEPTABLE WELD PROFILES

304F13.EPS

Figure 13 ◆ Acceptable and unacceptable pipe groove weld profiles.

5.0.0 ◆ PRACTICING OPEN-ROOT V-GROOVE WELDS

This section of the module explains how to perform the following open-root V-groove weld positions:

- Flat (1G-ROTATED)
- Horizontal (2G)
- Multiple (5G)
- Inclined multiple (6G)

5.1.0 Flat (1G-ROTATED) Position

Practice the 1G-ROTATED position open-root V-groove pipe welds using a shielding gas (if required) and a flux cored filler wire with a wire diameter as specified by your instructor. Keep the gun angle at 90° (0° work angle) to the pipe axis with a 10° to 20° drag angle. Use a slight side-to-side oscillation to run the root pass and to control penetration, paying particular attention to tie into the tack welds. Clean the face of the root pass with a stainless steel brush.

When running the fill and cover passes, use stringer or weave beads with a slight side-to-side motion to ensure tie-in at the toes of the weld bead. Avoid using weave beads for stainless steel pipe. Take particular care at the termination of the weld to fill the crater. Run all passes at or near the top of the pipe as the pipe is rotated.

Follow these steps to practice open-root V-groove pipe welds in the 1G-ROTATED position:

Step 1 Tack weld together the practice pipe weld coupon, as explained earlier.

Step 2 Position the pipe weld coupon horizontally on two sets of rollers at a comfortable welding height. *Figure 14* shows roller supports commonly found in pipe welding shops.

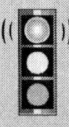

WARNING!
OSHA prohibits the use of homemade jack stands. Only stands manufactured to industry standards are acceptable. Four-legged jack stands are required for pipe over 10" in diameter.

Step 3 Make sure the workpiece clamp is attached directly onto the pipe coupon. This will prevent the welding current from passing through the roller bearings or arcing between the rollers and the pipe coupon.

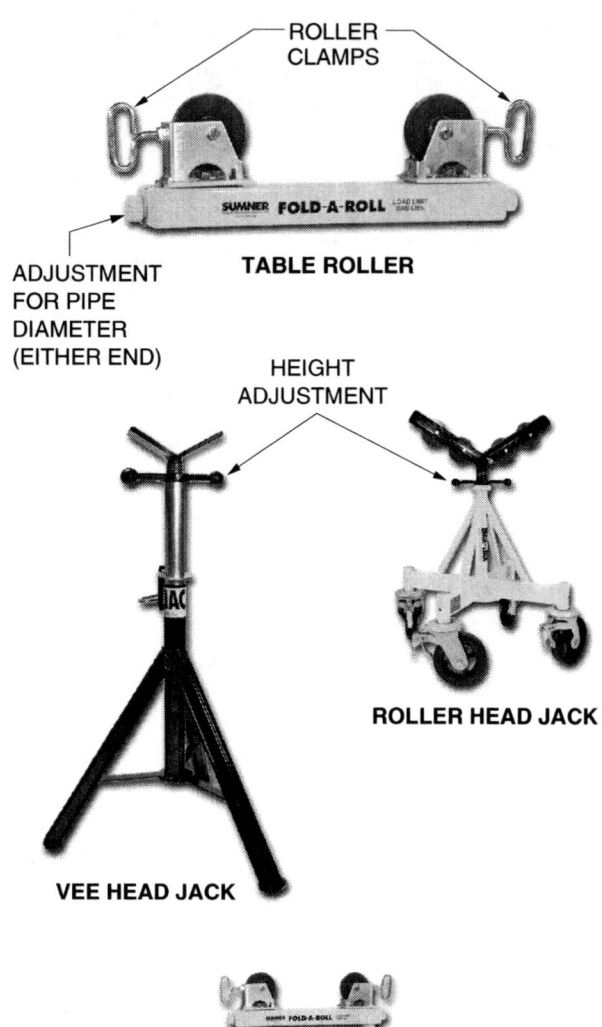

ROLLER CLAMPS

ADJUSTMENT FOR PIPE DIAMETER (EITHER END)

TABLE ROLLER

HEIGHT ADJUSTMENT

ROLLER HEAD JACK

VEE HEAD JACK

FLOOR STAND ROLLER

304F14.EPS

Figure 14 ◆ Pipe roller supports.

CAUTION
Failure to attach the workpiece clamp to the pipe coupon can result in variations in welding current, damage to the roller bearings, and arcing on the rollers and pipe coupon.

Step 4 Run the root pass using a side-to-side oscillation. Position a tack weld at the 11 o'clock position, start the weld bead on the tack weld, and advance toward the 12 o'clock position.

Powered Pipe Roller

A pipe roller that is operated by an electric motor activated by a foot switch is shown here. These devices are primarily used for large, heavy pipe sections.

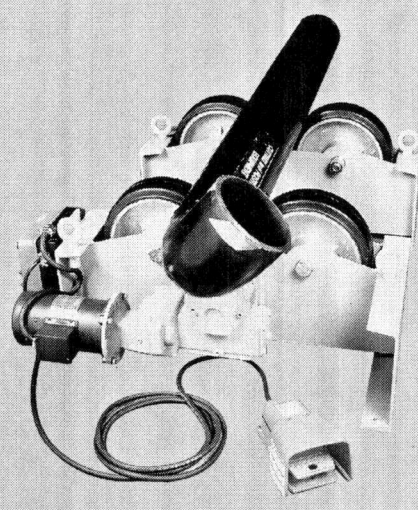

304SA03.EPS

Step 5 Roll the pipe as necessary to keep the weld in the flat position.

Step 6 Chip, grind, file, and/or brush the root pass to remove trapped slag.

Step 7 Use the same rolling procedure to make the remaining passes. Use stringer or weave beads as applicable. Pay particular attention at the tie-ins to prevent slag inclusions or excess buildup. Overlap the passes, and clean the weld after each pass. *Figure 15* shows the bead sequence and work angles.

5.2.0 Horizontal (2G) Position

Practice 2G position open-root V-groove pipe welds using a shielding gas (if required) and a flux cored filler wire with a wire diameter as specified by your instructor. The gun should be

at a 10° to 20° drag angle and at 90° (0° work angle) to the surface of the pipe for the root pass. To prevent the weld puddle from sagging, the gun work angle can be dropped slightly but not more than 10°. For the root pass, use a slight side-to-side oscillation to control penetration, as shown in *Figure 16*. Pay particular attention to tie into the tack welds.

When running the fill/cover passes, use stringer beads and a slight side-to-side motion to ensure tie-in at the toes of the weld bead. (See *Figure 17*.) Take particular care at the termination of the weld to fill the crater.

Follow these steps to practice open-root V-groove pipe welds in the 2G position:

Step 1 Tack weld together the practice pipe weld coupon, as explained earlier.

Step 2 Clamp or tack weld the pipe coupon with the axis of the pipe vertical.

Use of FCAW for a Root Pass

In practice, the WPS or site standards may require that the root pass for critical pipe welds be accomplished using GTAW or SMAW. For non-critical pipe welds, FCAW may be allowed.

Starting Pipe Welds

When starting a pipe weld, briefly use a short electrode extension to create a hot weld that will ensure proper penetration and will flatten the weld face. This will make terminating the finishing weld at the starting point easier.

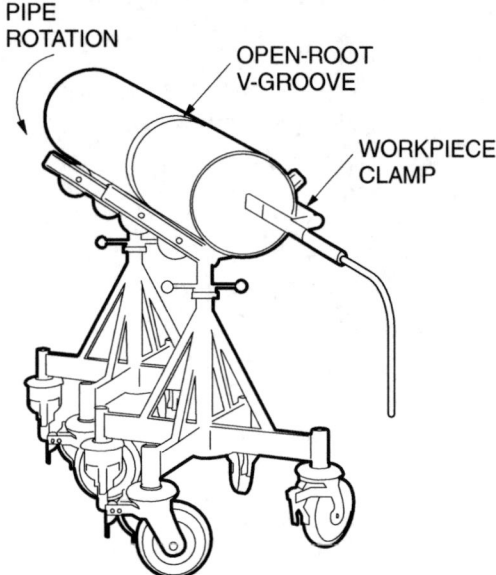

PIPE ROTATION

OPEN-ROOT V-GROOVE

WORKPIECE CLAMP

EXAMPLE OF PIPE ROTATION

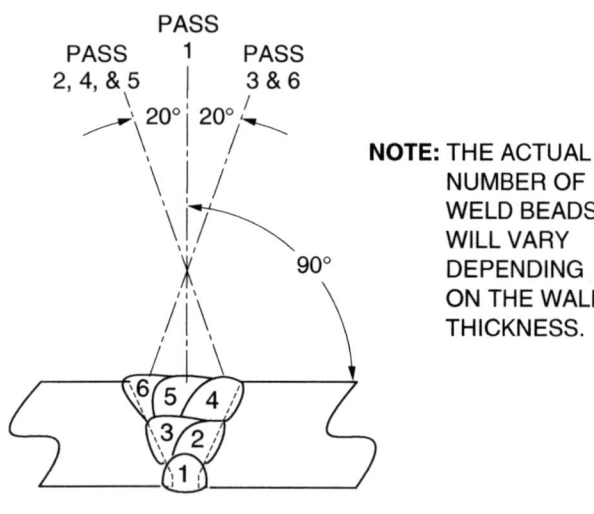

PASS 1

PASS 2, 4, & 5

PASS 3 & 6

20° 20°

90°

NOTE: THE ACTUAL NUMBER OF WELD BEADS WILL VARY DEPENDING ON THE WALL THICKNESS.

STRINGER BEAD SEQUENCE

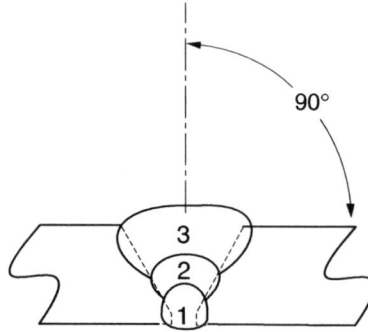

90°

WEAVE BEAD SEQUENCE (EXCEPT FOR STAINLESS STEEL)

304F15.EPS

Figure 15 ◆ Multiple-pass 1G-ROTATED bead sequences and work angles.

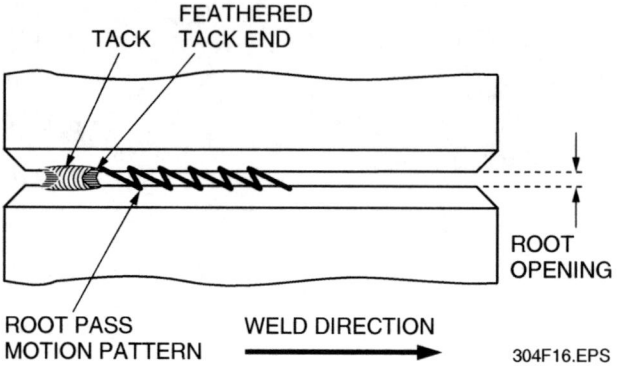

TACK

FEATHERED TACK END

ROOT OPENING

ROOT PASS MOTION PATTERN

WELD DIRECTION

304F16.EPS

Figure 16 ◆ Root pass motion pattern.

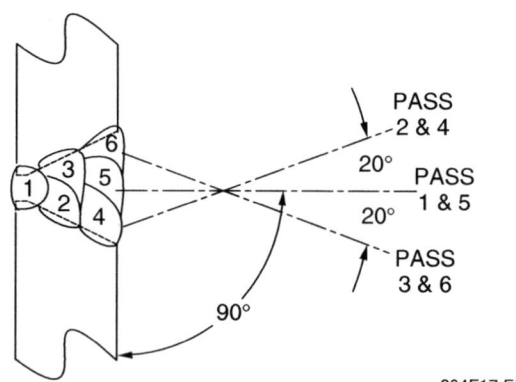

PASS 2 & 4

20°

PASS 1 & 5

20°

PASS 3 & 6

90°

304F17.EPS

Figure 17 ◆ Multiple-pass 2G bead sequence and work angles.

Step 3 Run the root pass as shown in *Figure 16.* Start the root pass on a tack weld. Clean and chip the crater between restarts if restarts are necessary. Continue this procedure until the root pass is completed.

Step 4 Clean, grind, file, and/or brush the root pass to remove trapped slag.

Step 5 Run the remaining passes at the appropriate work angles. Overlap the passes, and clean the weld after each pass.

5.3.0 Multiple (5G) Position

Practice the 5G multiple position open-root V-groove pipe welds using a shielding gas (if required) and a flux cored filler wire with a wire diameter as specified by your instructor. The gun should be at approximately a 10° to 20° drag or push angle for the root, fill, and cover passes, depending on whether your instructor directs you to run downhill or uphill passes. In some cases, the root pass is made downhill and the fill and cover passes are made uphill. Refer to *Figure 10* for the travel position modifications that may be necessary as the passes are made on both sides of the pipe.

When running the fill and cover passes, use stringer or weave beads with a slight side-to-side motion to ensure tie-in at the toes of the weld bead or weave bead. (See *Figure 18*.) Pay particular attention at the termination of the weld to fill the crater.

Follow these steps to practice open-root V-groove pipe welds in the 5G position:

Step 1 Tack weld together the practice pipe weld coupon, as explained earlier.

Step 2 Clamp or tack weld the pipe weld coupon into position with the pipe axis horizontal.

Step 3 Run the root pass with a stringer bead. Modify the travel angles as necessary, as shown in *Figure 10,* for an uphill or downhill pass, as directed by your instructor.

Step 4 Clean, grind, file, and/or brush the root pass to remove trapped slag.

Step 5 Run the remaining passes at the appropriate work angles in either an uphill or downhill direction as directed by your instructor. Overlap the passes, and clean the weld after each pass.

5.4.0 Inclined Multiple (6G) Position

Practice the inclined multiple (6G) (45°) position open-root V-groove pipe welds using a shielding gas (if required) and a flux cored filler wire with a wire diameter as specified by your instructor. The gun should be at approximately a 10° to 20° drag or push angle for the root, fill, and cover passes, depending on whether your instructor directs you to run downhill or uphill passes. In some cases, the root pass is made downhill and the fill and cover passes are made uphill. Refer to *Figure 10* for the travel position modifications that

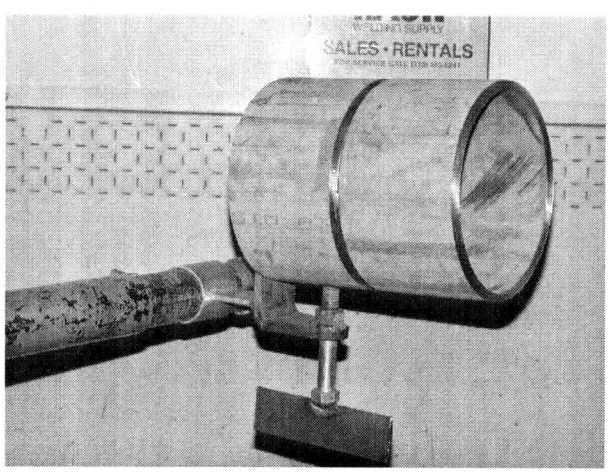

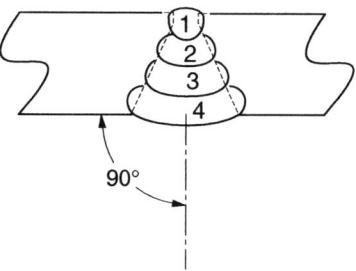

WEAVE BEAD SEQUENCE
(EXCEPT STAINLESS STEEL)

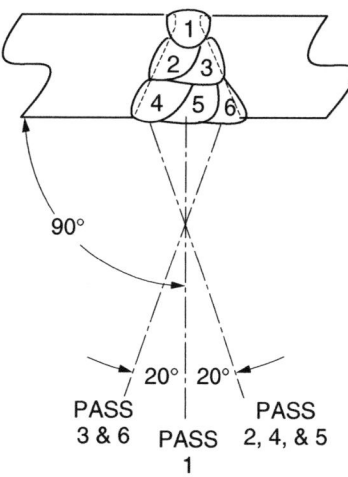

STRINGER BEAD SEQUENCE   304F18.EPS

Figure 18 ◆ Multiple-pass 5G bead sequences and work angles.

may be necessary as the passes are made on both sides of the pipe.

When running the fill and cover passes, use stringer beads and a slight side-to-side motion to ensure tie-in at the toes of the weld bead. (See *Figure 19*.) Pay particular attention at the termination of the weld to fill the crater.

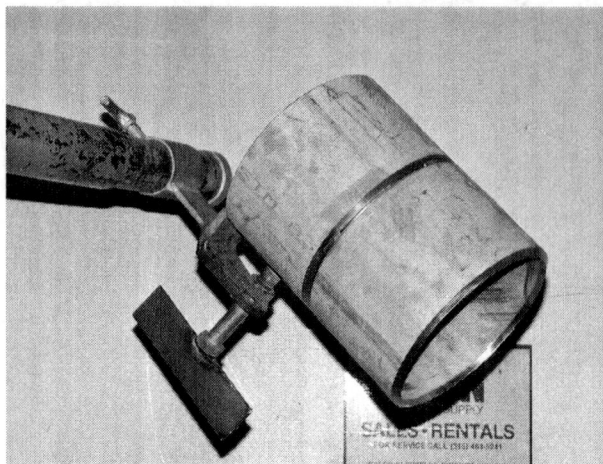

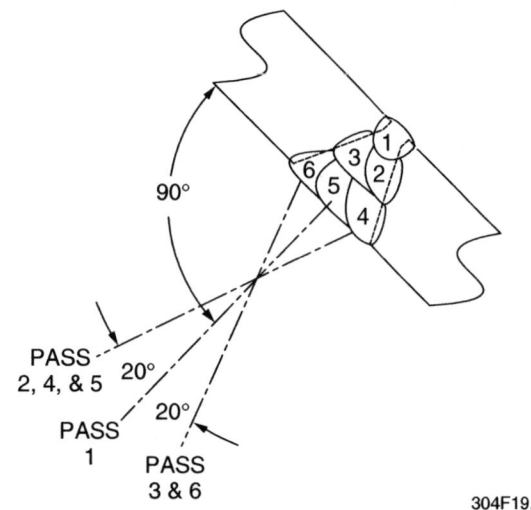

90°

PASS
2, 4, & 5   20°

PASS
1   20°

PASS
3 & 6

304F19.EPS

Figure 19 ◆ Multiple-pass 6G bead sequence and work angles.

Follow these steps to practice open-root V-groove pipe welds in the 6G position:

> **NOTE**
>
> If required for training purposes, a restricting ring may be added to the 6G position coupon to form a 6GR position coupon.

Step 1 Tack weld together the practice pipe weld coupon, as explained earlier.

Step 2 Clamp or tack weld the pipe weld coupon into position with the pipe axis inclined 45° to the horizontal plane.

Step 3 Run the root pass with a stringer bead. Modify the travel angles as necessary, as shown in *Figure 10*, for an uphill or downhill pass, as directed by your instructor.

Step 4 Clean, grind, file, and/or brush the root pass to remove trapped slag.

Step 5 Run the remaining passes at the appropriate work angles in either an uphill or downhill direction as directed by your instructor. Overlap the passes, and clean the weld after each pass.

Summary

The ability to make open-root V-groove welds on pipe in all positions is one of the more difficult skills you must develop as a welder. You must be able to set up the equipment, perform the welding, and recognize acceptable welds. Open-root V-groove pipe welds can be made in the 1G-ROTATED, 2G, 5G, and 6G positions. Practice these welds until you can consistently produce acceptable welds.

Review Questions

1. The FCAW process is basically limited to welding _____.
   a. stainless steels
   b. high-carbon steels and aluminum
   c. stainless steel and aluminum
   d. low-carbon steels, some low-alloy steels, and some stainless steels

2. In FCAW-S, the shielding gas is generated _____.
   a. entirely from the flux in the wire core
   b. entirely from an external shielding gas
   c. from both the flux in the wire core and an external shielding gas
   d. from the molten weld metal

3. The welding codes allow the face on an open-root weld to be from _____ wide.
   a. 0" to ⅛"
   b. 0" to ¼"
   c. ⅛" to ¼"
   d. ⅛" to ½"

4. Feathering the ends of tack welds with a grinder helps _____.
   a. build higher beads with deeper penetration
   b. to increase the deposition rate
   c. fuse the tack welds into the root pass
   d. to adjust the arc length

5. Excessive root reinforcement should be avoided when welding pipe because it _____.
   a. restricts flow within the pipe
   b. decreases penetration
   c. causes porosity in the weld
   d. creates a weak joint

6. Pipe groove welds should be made with slight reinforcement not to exceed _____.
   a. ¹⁄₁₆"
   b. ⅛"
   c. ³⁄₁₆"
   d. ¼"

7. For flux cored arc welding in the 1G-ROTATED position, avoid using _____.
   a. weave beads for aluminum pipe
   b. stringer beads for stainless steel pipe
   c. weave beads for stainless steel pipe
   d. stringer beads for low-carbon steel pipe

8. Pay particular attention to the termination of the weld to _____.
   a. avoid excessive undercut
   b. avoid overlap
   c. blend the weld smoothly over the pipe
   d. fill the crater

9. To prevent the weld puddle from sagging, the gun work angle can be _____.
   a. raised slightly but not more than 10°
   b. dropped slightly but not more than 10°
   c. dropped slightly but not more than 20°
   d. raised slightly but not more than 20°

10. Inclined multiple (6G) position open-root V-groove pipe welds are made with the pipe axis inclined _____ to the horizontal plane.
    a. 10°
    b. 20°
    c. 45°
    d. 75°

Performance Accreditation Tasks

The Performance Accreditation Tasks (PATCs) correspond to and support learning objectives in the *AWS Guide for the Training and Qualification of Welding Personnel.*

PATCs provide specific acceptable criteria for performance and help to ensure a true competency-based welding program for students.

The following tasks are designed to evaluate your ability to run open-root V-groove welds with FCAW-G or FCAW-S equipment in three standard test positions using carbon steel wire of the appropriate diameter. Perform each task when you are instructed to do so by your instructor. As you complete each task, show it to your instructor for evaluation. Do not proceed to the next task until told to do so by your instructor. For AWS 2G and 5G certifications, refer to *AWS EG3.0-96* for bend test requirements. For AWS 6G certifications, refer to *AWS EG4.0-96* for bend test requirements.

OPEN-ROOT V-GROOVE PIPE WELD IN THE 2G POSITION

Using FCAW-G or FCAW-S carbon steel wire of the appropriate diameter and stringer beads, make an open-root V-groove weld on carbon steel pipe in the 2G position. For FCAW-G, use the appropriate shielding gas.

Note: Depending on site procedures or practices, the root pass for the following tasks may be run using another welding process such as GTAW or SMAW. Check with your instructor to determine the welding process to use for the root pass.

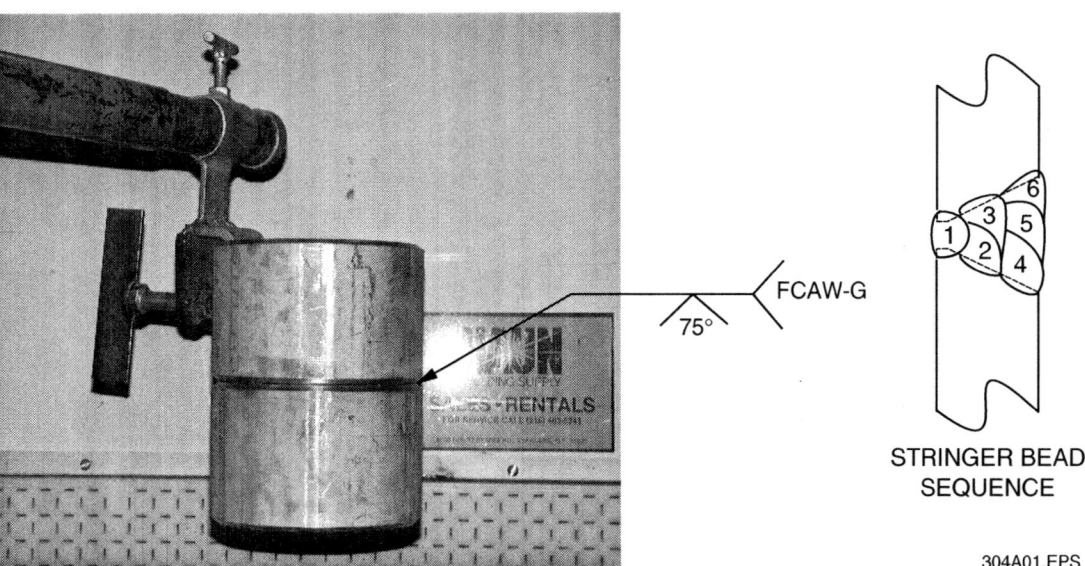

STRINGER BEAD
SEQUENCE

304A01.EPS

Criteria for Acceptance

* Uniform appearance on the bead face   _____

* Craters and restarts filled to the full cross section of the weld   _____

* Uniform weld width ±⅟₁₆"   _____

* Acceptable weld profile in accordance with the
  ASME Boiler and Pressure Vessel Code, Section IX   _____
                                                                          _____

* Smooth transition with complete fusion at the toes of the weld   _____

* Complete uniform root reinforcement at least flush with the inside of the
  pipe to a maximum of ⅛"   _____

* No porosity   _____

* No excessive undercut   _____

* No cracks   _____

* No overlap   _____

* No incomplete fusion   _____

* No pinholes (fisheyes)   _____

* No inclusions   _____

OPEN-ROOT V-GROOVE PIPE WELD IN THE 5G POSITION

Using FCAW-G or FCAW-S carbon steel wire of the appropriate diameter and stringer or weave beads, make an open-root V-groove weld on carbon steel pipe in the 5G position. For FCAW-G, use the appropriate shielding gas.

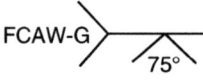

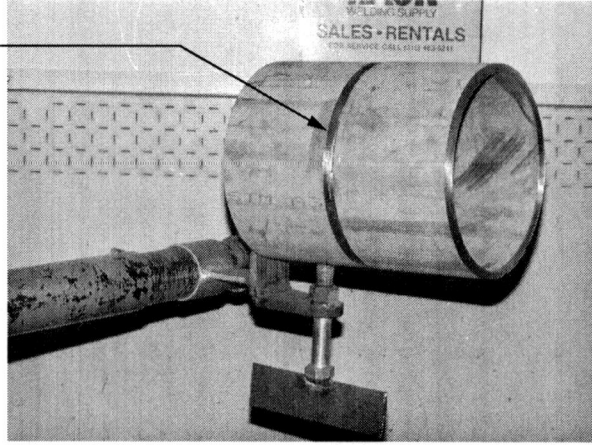

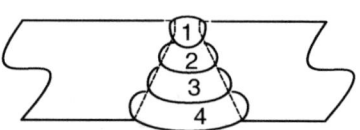

WEAVE BEAD SEQUENCE

STRINGER BEAD SEQUENCE

304A02.EPS

Criteria for Acceptance

- Uniform appearance on the bead face   _____

- Craters and restarts filled to the full cross section of the weld   _____

- Uniform weld width ±¹⁄₁₆"   _____

- Acceptable weld profile in accordance with the
  ASME Boiler and Pressure Vessel Code, Section IX   _____

- Smooth transition with complete fusion at the toes of the weld   _____

- Complete uniform root reinforcement at least flush with the inside of the
  pipe to a maximum of ⅛"   _____

- No porosity   _____

- No excessive undercut   _____

- No cracks   _____

- No overlap   _____

- No incomplete fusion   _____

- No pinholes (fisheyes)   _____

- No inclusions   _____

OPEN-ROOT V-GROOVE PIPE WELD IN THE 6G (OR 6GR) POSITION

Using FCAW-G or FCAW-S carbon steel wire of the appropriate diameter and stringer beads, make an open-root V-groove weld on carbon steel pipe in the 6G (or 6GR) position. For FCAW-G, use the appropriate shielding gas.

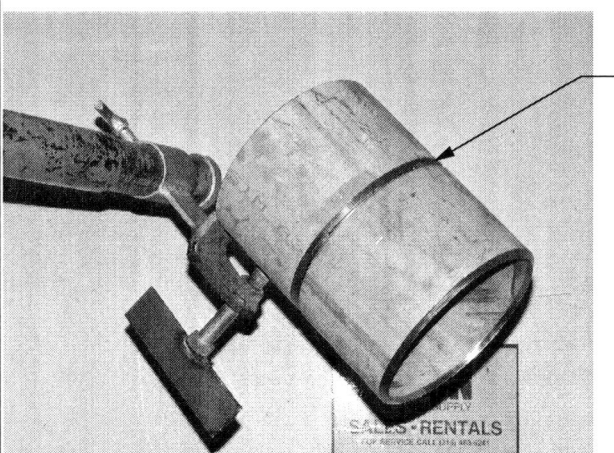

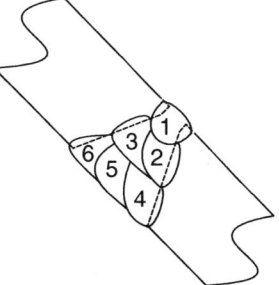

NOTE: IF REQUIRED FOR QUALIFICATION PURPOSES, A RESTRICTING RING MAY BE ADDED TO THE 6G POSITION COUPON TO FORM A 6GR POSITION COUPON.

STRINGER BEAD SEQUENCE

304A03.EPS

Criteria for Acceptance

- Uniform appearance on the bead face　　　　　　＿＿＿＿＿＿

- Craters and restarts filled to the full cross section of the weld　　＿＿＿＿＿＿

- Uniform weld width ±¹⁄₁₆"　　　　　　＿＿＿＿＿＿

- Acceptable weld profile and guided bend test in accordance with the *ASME Boiler and Pressure Vessel Code, Section IX*　　＿＿＿＿＿＿

- Smooth transition with complete fusion at the toes of the weld　　＿＿＿＿＿＿

- Complete uniform root reinforcement at least flush with the inside of the pipe to a maximum of ⅛"　　＿＿＿＿＿＿

- No porosity　　　　　　＿＿＿＿＿＿

- No excessive undercut　　　　　　＿＿＿＿＿＿

- No cracks　　　　　　＿＿＿＿＿＿

- No overlap　　　　　　＿＿＿＿＿＿

- No incomplete fusion　　　　　　＿＿＿＿＿＿

- No pinholes (fisheyes)　　　　　　＿＿＿＿＿＿

- No inclusions　　　　　　＿＿＿＿＿＿

Additional Resources

This module is intended to present thorough resources for task training. The following reference works are suggested for both instructors and motivated trainees interested in further study. These are optional materials for continued education rather than for task training.

ASME – Boiler and Pressure Vessel Code: Section 9, Welding and Brazing Qualifications, Current Edition. New York, NY: ASME International.

AWS D10.11, Recommended Practices for Root Pass Welding of Pipe Without Backing. Miami, FL: American Welding Society.

Dual Shield, Coreshield, Coreweld, Cored Wires, Alloy Rods. Hanover, PA: ESAB Welding and Cutting Products, Filler Metals Division.

Metals Handbook. Metals Park, OH: American Society for Metals.

The Procedure Handbook of Arc Welding. Cleveland, OH: The Lincoln Electric Company.

Welding of Pipeline and Related Facilities, API Standard 1104, Latest Edition. New York, NY: ASME International.

Welding Technology. Gower A. Kennedy. Indianapolis, IN: Macmillian Publishing Company, Inc.

Figure Credits

Miller Electric Mfg. Co.	304F02, 304F03
Gerald Shannon	304F04, 304F05, 304SA02, 304F17, 304F18, 304F19
G.A.L. Gage Co.	304SA01
Sumner Manufacturing Co., Inc.	304F14
Koike Aronson, Inc.	304SA03

The NCCER makes every effort to keep these textbooks up-to-date and free of technical errors. We appreciate your help in this process. If you have an idea for improving this textbook, or if you find an error, a typographical mistake, or an inaccuracy in NCCER's Contren® textbooks, please write us, using this form or a photocopy. Be sure to include the exact module number, page number, a detailed description, and the correction, if applicable. Your input will be brought to the attention of the Technical Review Committee. Thank you for your assistance.

Instructors – If you found that additional materials were necessary in order to teach this module effectively, please let us know so that we may include them in the Equipment/Materials list in the Instructor's Guide.

Write: Curriculum Revision and Development Department
National Center for Construction Education and Research
P.O. Box 141104, Gainesville, FL 32614-1104

Fax: 352-334-0932

E-mail: curriculum@nccer.org

Craft

Module Name

Copyright Date

Module Number

Page Number(s)

Description

(Optional) Correction

(Optional) Your Name and Address

Gas Tungsten Arc Welding (GTAW) – Carbon Steel Pipe

COURSE MAP

This course map shows all of the modules in the third level of the Welding curriculum. The suggested training order begins at the bottom and proceeds up. Skill levels increase as you advance on the course map. The local Training Program Sponsor may adjust the training order.

WELDING LEVEL THREE

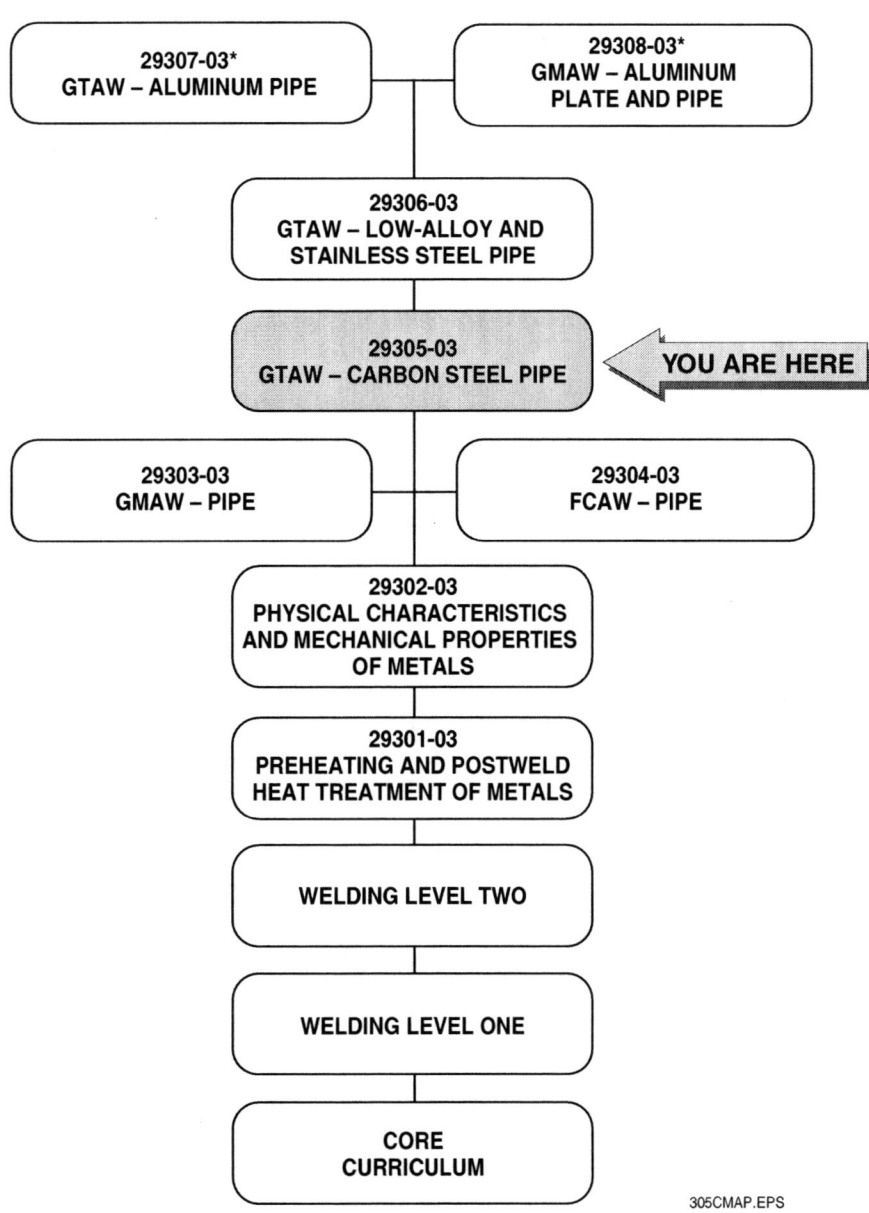

305CMAP.EPS

*Please note that Modules 29307-03 and 29308-03 are electives for those progressing through the *Welding Level Three* program.*

Figures

Gas Tungsten Arc Welding (GTAW) – Carbon Steel Pipe

Objectives

When you have completed this module, you will be able to do the following:

1. Set up GTAW equipment.
2. Identify and explain open-root V-groove pipe weld techniques.
3. Perform open-root V-groove pipe welds using GTAW in the following positions:

   - 1G-ROTATED
   - 2G
   - 5G
   - 6G

Prerequisites

Before you begin this module, it is recommended that you successfully complete the following: Core Curriculum; Welding Levels One and Two; Welding Level Three, Modules 29301-03 through 29304-03 or Modules 29301-03 through 29303-03.

Required Trainee Materials

1. Pencil and paper
2. Appropriate personal protective equipment

1.0.0 ◆ INTRODUCTION

Gas tungsten arc welding (GTAW) is an arc welding process that uses an arc between a tungsten electrode and the base metal to melt the base metal. The electrode, the arc, and the molten base metal are shielded from atmospheric contamination by a flow of inert gas from the torch nozzle. (See *Figure 1.*) The filler metal is a rod of similar composition to the base metal and is usually hand-held and manually fed into the leading edge of the weld puddle. GTAW produces high-quality

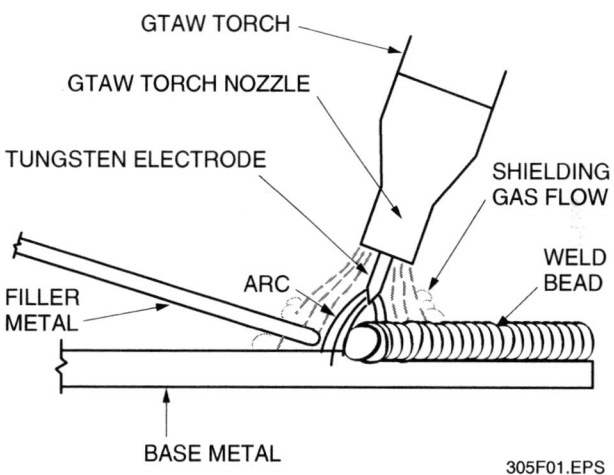

Figure 1 ◆ GTAW process.

welds without slag or oxidation. Because there is no flux, there will be no corrosion due to flux entrapment; therefore, no postweld cleaning is necessary. An exception is the slag left by some flux cored rods, which must be removed.

A major use of GTAW is to make manual high-quality welds in ferrous and nonferrous piping used in both critical and non-critical applications. The GTAW process allows greater control of root penetration and fill than almost any other process. For this reason, GTAW is often used to make the root pass on pipe, even when the fill and cover passes are made with SMAW, GMAW, or FCAW, which all three have higher deposition rates.

This module explains how to set up GTAW equipment and perform open-root V-groove welds on carbon steel pipe with carbon steel filler metal in the 1G-ROTATED, 2G, 5G, and 6G welding positions. The dimensions and specifications in this module are designed to be representative of codes in general and are not specific to any certain code. Always follow the proper codes for your site.

2.0.0 ◆ SAFETY SUMMARY

The following is a summary of safety procedures and practices you must observe while cutting or welding. Keep in mind that this is just a summary. Complete safety coverage is provided in the Level One module *Welding Safety*. If you have not completed that module, do so before continuing. Above all, be sure to wear appropriate protective clothing and equipment when welding or cutting.

2.1.0 Protective Clothing and Equipment

- Always use safety goggles with a full face shield or a helmet. The goggles, face shield, or helmet lens must have the proper light-reducing tint for the type of welding or cutting to be performed. Never directly or indirectly view an electric arc without using a properly tinted lens.
- Wear proper protective leather and/or flame retardant clothing along with welding gloves to protect from flying sparks, molten metal, and heat.
- Wear 8" or taller high-top safety shoes or boots. Make sure that the tongue and lace area of the footwear is covered by a pant leg. If the tongue and lace area is exposed or the footwear must be protected from burn marks, wear leather spats under the pants or chaps and over the front top of the footwear.
- Wear a solid material (non-mesh) hat with a bill pointing to the rear or, if much overhead cutting or welding is required, a full leather hood with a welding faceplate and the correct tinted lens. If a hard hat is required, use one that allows attachment of both rear deflector material and a face shield.
- If a full leather hood is not worn, wear a face shield and snug fitting welding goggles over safety glasses for gas welding or cutting. Either the face shield or the lenses of the welding goggles must be an approved shade 5 or 6 filter. For electric arc welding or cutting, wear safety goggles and a welding hood with the correct tinted lens (shade 9 to 14).
- If a full leather hood is not worn, wear earplugs to protect your ear canals from sparks.

2.2.0 Fire/Explosion Prevention

- Never carry matches or gas-filled lighters in your pockets. Sparks can cause the matches to ignite or the lighter to explode, causing serious injury.

- Never perform any type of heating, cutting, or welding until a hot-work permit is obtained and an approved fire watch is established. Most work-site fires caused by these types of operations are started by cutting torches.
- Never use oxygen to blow off clothing. The oxygen can remain trapped in the fabric for a time. If a spark hits the clothing during this period, the clothing can burn rapidly and violently out of control.
- Make sure that any flammable material in the work area is moved or is shielded by a fire-resistant covering. Approved fire extinguishers must be available before attempting any heating, welding, or cutting operations.
- Always comply with any site requirement for a hot-work permit and/or a fire watch.
- Never release a large amount of oxygen nor use oxygen as compressed air. Its presence around flammable materials or sparks can cause rapid and uncontrolled combustion. Keep oxygen away from oil, grease, and other petroleum products.
- Never release a large amount of fuel gas, especially acetylene. Methane and propane tend to concentrate in and along low areas and can ignite at a considerable distance from the release point. Acetylene is lighter than air but is even more dangerous than methane. When mixed with air or oxygen, it will explode at much lower concentrations than any other fuel.
- To prevent fires, maintain a neat and clean work area, and make sure that any metal scrap or slag is cold before disposal.
- Before cutting or welding containers such as tanks or barrels, check to see if they have contained any explosive, hazardous, or flammable materials, including petroleum products, citrus products, or chemicals that decompose into toxic fumes when heated. As a standard practice, always clean and then fill any tanks or barrels with water, or purge them with a flow of inert gas to displace any oxygen.

2.3.0 Work Area Ventilation

- Make sure to follow confined space procedures before conducting any welding or cutting in the confined space.
- Never use oxygen to ventilate confined spaces.
- Always perform cutting or welding operations in a well-ventilated area. Such operations involving zinc or cadmium materials or coatings result in toxic fumes. For long-term cutting

or welding of these materials, always wear an approved, full-face, supplied-air respirator (SAR) that uses breathing air supplied from outside of the work area. For occasional, very short-term exposure, you may use a high-efficiency particulate arresting (HEPA)-rated or metal-fume filter on a standard respirator.

- Make sure confined spaces are ventilated properly for cutting or welding operations.

3.0.0 ◆ ROOT BACK SIDE PROTECTION

Welding processes that use inert gas to protect the molten puddle on the face of the weld from the atmosphere do not provide protection for the back side of the weld. Although back side protection is usually not required for carbon steel and some low alloys, it becomes essential as the alloy content increases. This is because alloying elements are more susceptible to combining with the oxygen in the atmosphere to form oxides. The oxides are objectionable because they can decrease the corrosion resistance to an unacceptable level and can be pulled into the weld puddle during subsequent welding, causing defects. The most common methods of providing back side protection are the following:

- **Backing gas (purge)**
- Backup flux
- Consumable insert

3.1.0 Backing Gas

Backing gas provides the best and highest-quality protection for welding open-root joints. Argon provides better protection than nitrogen, but nitrogen typically costs one-fourth to one-third less. Because it costs less, nitrogen is used where it will not adversely affect the weld. However, you must use the gas specified by the WPS.

To be effective, the backing gas has to do the following:

- Replace the atmospheric gases to an acceptable level.

- Be contained to provide protection where needed.
- Have a continuous flow to compensate for gas lost during welding.

For short and small-diameter piping, you can plug or cap the ends and any openings to provide containment and to reduce the gas amount and expense. This method of containment is not cost-effective for large piping or long sections. To decrease the volume of piping requiring backing gas replacement, plug the piping near the weld, on both sides, with removable devices called dams. Dams may be inflatable bladders, inflated plastic bags, or foam rubber, or they may be water-soluble dams, which are typically made of water-soluble paper. When you use dams, they must be removed after welding is completed. (See *Figure 2*.) When argon is required as the backing gas, another procedure to reduce the cost is to purge with nitrogen before purging with the argon.

When setting up backing gas, establish the containment, and then seal the weld joint. This is typically done with masking tape. Introduce the backing gas through a tube or hose that passes through or past one of the containment devices or through a nozzle (inlet) that directs the gas through the root opening of the weld. Argon backing gas should always enter at the bottom, and because the gas is typically heavier than the air it is displacing, there should be a vent located near the top as an exit for the atmospheric gases. If the vent hole is in the bottom, the gas will vent out without displacing the atmospheric air in the top section of the pipe. However, if nitrogen is used as the backing gas, the vent should be at the bottom, because nitrogen is lighter than air and will rise to the top.

Check the WPS to determine the backing gas flow rate range. This range is for maintaining the backing gas atmosphere during welding after the pipe has been purged. Higher flow rates can be used for the initial purge to reduce the time it takes to purge the pipe at the onset. An oxygen analyzer is often used to verify that the atmosphere inside the pipe is correct for the welding being performed. When the oxygen analyzer is used, 1% maximum oxygen is the typical criterion for an acceptable welding atmosphere.

Maintaining Proper Backing Gas Flow

As the weld progresses around a pipe, you may have to reduce the backing gas flow to the minimum to prevent blowing the molten weld pool out of the joint and causing holes in the root pass. The backing gas flow should be maintained until at least the hot pass is placed over the root pass and the root pass is completely covered from view.

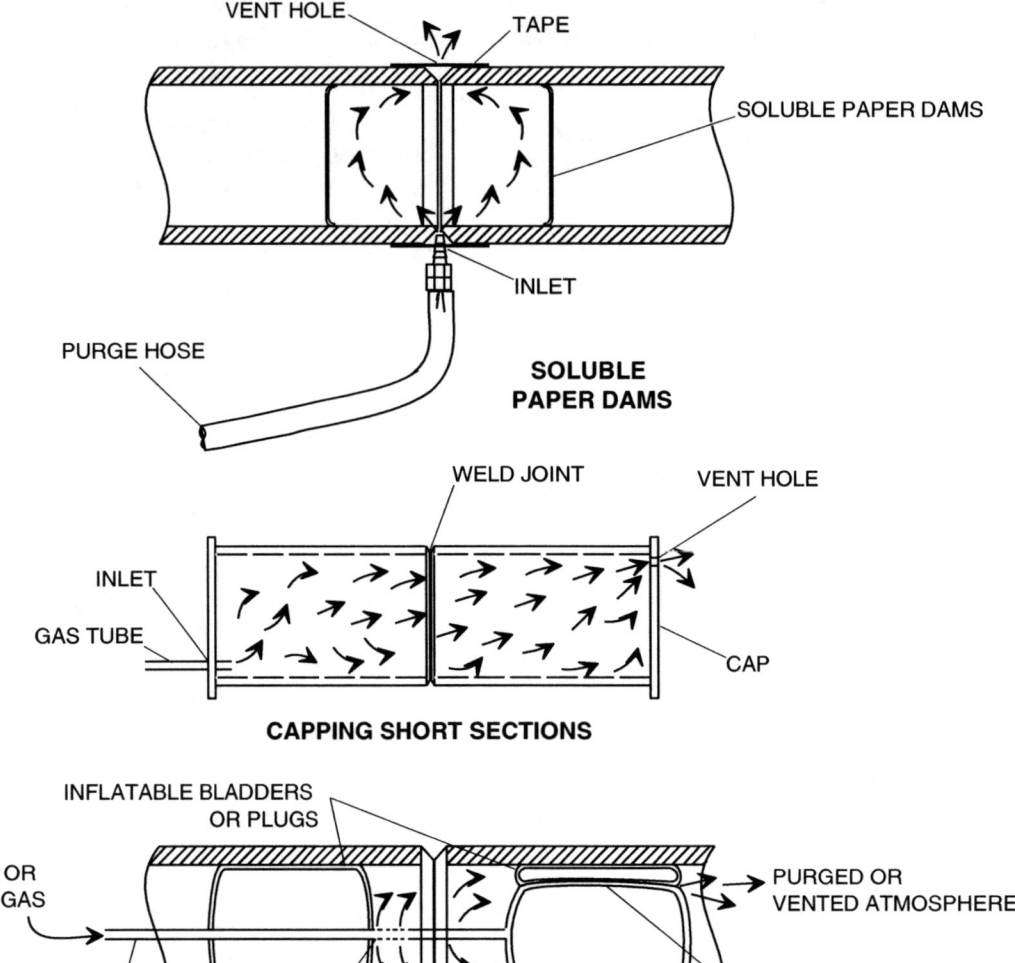

VENT HOLE — TAPE

SOLUBLE PAPER DAMS

INLET

PURGE HOSE

**SOLUBLE
PAPER DAMS**

WELD JOINT   VENT HOLE

INLET

GAS TUBE

CAP

CAPPING SHORT SECTIONS

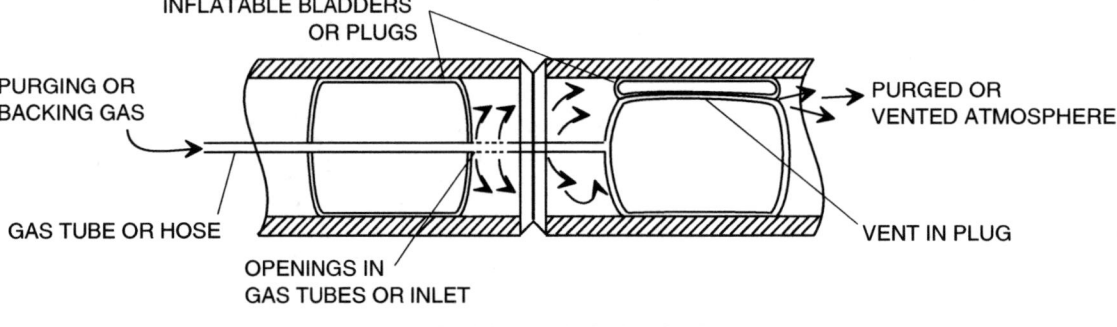

INFLATABLE BLADDERS
OR PLUGS

PURGING OR
BACKING GAS

PURGED OR
VENTED ATMOSPHERE

GAS TUBE OR HOSE

VENT IN PLUG

OPENINGS IN
GAS TUBES OR INLET

PLUGS FOR LONG SECTIONS

305F02.EPS

Figure 2 ◆ Typical pipe-purging methods.

3.2.0 Backup Flux

Using backup flux is another method of providing root protection for open-root welds when weld quality requirements are less critical. Special flux cored rods are available to provide root back side protection during the GTAW process, but typically you will use a dry powder flux that you mix with alcohol (methanol) or acetone to a consistency of thick cream. After the mixture has set for several minutes to allow the alcohol or acetone to react with the flux, you brush an even coat of the mixture onto the back side of the joint. The mixture dries rapidly and leaves a coating of flux. During welding, the flux powder melts, forming a barrier that prevents air from coming in contact with the molten metal. To remove the flux after welding, use a wire brush and hot water.

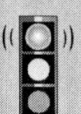

WARNING!
Backup flux may contain silica, fluorides, or other toxic materials. Weld only in well-ventilated spaces. Failure to provide proper ventilation could result in personal injury or death.

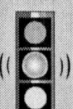

CAUTION
Backup fluxes are formulated for welding specific alloy types. Check the manufacturer's recommendations to be sure you use the correct flux for the alloy being welded. Using the wrong flux could result in weld defects.

3.3.0 Consumable Inserts

Consumable inserts are preplaced filler metal pieces that completely fuse into the root of the joint and become part of the weld. Use consumable inserts when welding critical piping systems that cannot tolerate internal obstructions and also when welding pipe, tube, and other components that require uniformity and a smooth root bead. The insert becomes an integral part of the initial root-weld bead and ensures complete fusion and smooth blending of the insert and adjacent pipe, tube, or fitting ends. Inserts reduce flow restriction within the bore to a minimum and help eliminate root-bead cracking. Except when welding carbon steel pipe, backing gas is usually used with inserts to prevent oxidation of the weld root.

4.0.0 ◆ WELDING PREPARATION

Before welding can begin, the area has to be readied, the welding equipment must be set up, and the metal to be welded must be prepared. The following sections explain how to set up the equipment for welding.

To practice welding, you need a welding table, bench, or stand. The welding surface must be steel, and provisions must be available for placing weld coupons out of position. (See *Figure 3*.)

To set up the area for welding, follow these steps:

Step 1 Check to be sure the area is properly ventilated. Make use of doors, windows, and fans.

Figure 3 ◆ Welding station.

Step 2 Check the area for fire hazards. Remove any flammable materials before proceeding.

Step 3 Locate the nearest fire extinguisher. Do not proceed unless the extinguisher is charged and you know how to use it.

Step 4 Position a welding table near the welding machine.

Step 5 Set up flash shields around the welding area.

4.1.0 Practice Pipe Weld Coupons

Pipe weld coupons should be cut from 3" to 12" diameter Schedule 40 or Schedule 80 carbon steel pipe. Each welded joint will require two coupons of the same size and schedule pipe.

Figure 4 shows typical ASME International, American Petroleum Institute (API), and AWS pipe bevel specifications for bevel angles, root faces, and root openings. To prepare weld coupons for open-root V-groove weld joints, follow these steps:

Step 1 Clean all rust or mill scale from the inside and outside of the carbon steel pipe with a grinder or wire brush. Clean for a distance of ½" from the weld joint.

CAUTION

There is little or no surface cleaning action when using a GTAW torch; therefore, pipe surfaces must be thoroughly cleaned to prevent weld discontinuities.

Step 2 Bevel the end of the pipe to 30° or 37½° by any acceptable beveling method such as by using a mechanical cutter or by thermal cutting or grinding.

Step 3 Cut off a section of the beveled pipe end (1½" minimum).

Conserve Materials

Pipe for practice welding is expensive and difficult to obtain. Make every effort to conserve and not waste available materials. Completely use weld coupons until all surfaces have been welded by cutting the coupon apart and reusing the pieces. Check with the instructor for the appropriate size coupon.

305F03.EPS

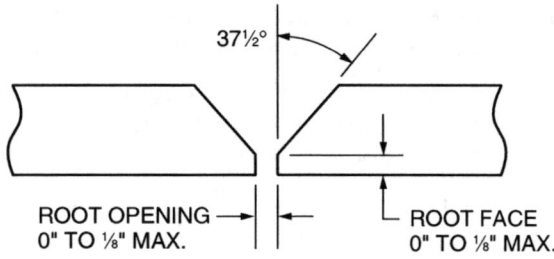

ROOT OPENING →
0" TO ⅛" MAX.

ROOT FACE
0" TO ⅛" MAX.

37½°

TYPICAL ASME AND AWS SPECIFICATION

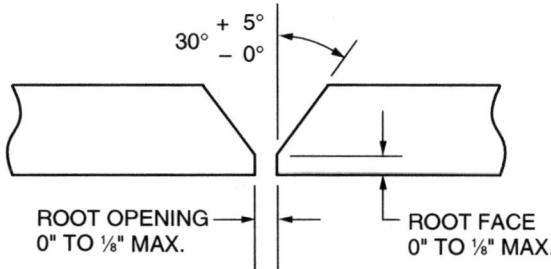

ROOT OPENING →
0" TO ⅛" MAX.

ROOT FACE
0" TO ⅛" MAX.

30° + 5°
− 0°

TYPICAL API SPECIFICATION

305F04.EPS

Figure 4 ◆ Pipe bevel specifications.

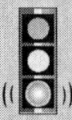
NOTE

For 1G-ROTATED welding, you may have to cut longer coupons to fit the rollers, if used.

Step 4 Check the bevel. There should be no dross, and the bevel angle should be 30° or 37½° with no notches more than ¹⁄₁₆" deep.

Step 5 Grind or file a 0" to ⅛" root face on the bevel as specified by your instructor.

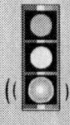
NOTE

Welding codes allow both the root opening and the root face on open root welds to be from 0" to ⅛" wide. Select and adjust the opening and root face as needed when you start the welding practices.

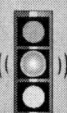
CAUTION

The same root backing required for welding must be used for tack welding coupons. This is because the tack welds become part of the finished root pass. Failure to provide root backing for tack welds when required will result in weld discontinuities.

Step 6 If required, provide for root backing using backup flux, gas backing, or consumable insert.

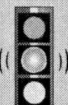
CAUTION

Check your site quality standards and/or WPS to determine the type of backing required. Do not use backup flux unless it is specifically approved at your site. Check with your supervisor if you are unsure of the backing requirements. Failure to follow proper procedures will result in defective welds.

NOTE

If a consumable insert is used, make sure it is secured to one of the pipe sections.

Step 7 Align the two pipe sections so that the inside diameter (ID) surfaces are even all around. Align small-diameter pipe by clamping both pieces to a piece of angle iron. Align large-diameter pipe with the aid of a pipe alignment jig or by holding a straightedge across the joint, parallel to the pipe axis. The straightedge must be used all around the pipe in case one or both sections are distorted.

Step 8 Gap the root opening at 0" to ⅛" with pieces of filler wire, metal shims, or pieces of bare welding electrode of the correct diameter, as directed by your instructor.

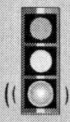
NOTE

Welding codes allow the root opening to be from 0" to ⅛". Adjust the root opening as needed or as directed by your instructor when you start the welding practices.

Step 9 When the root opening is correct and the pipe ends are aligned, make the first of four tack welds of no greater than 1".

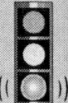
NOTE

Heavy-wall or large-diameter pipe may require longer tacks or more than four tacks.

Welding Large-Diameter Pipe

The practice coupons on large-diameter pipe must be greater than 1½" in length. They need more metal to help absorb the increased heat used to make these welds.

Hi-Lo Gauge

You can use a gauge, such as the one shown here, to check that internal surfaces are even all around when a consumable insert is not used.

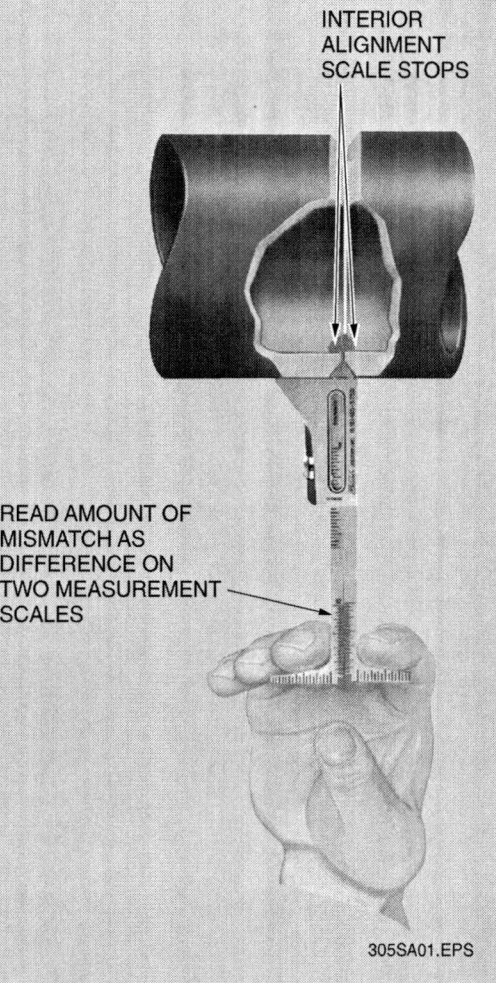

INTERIOR
ALIGNMENT
SCALE STOPS

READ AMOUNT OF
MISMATCH AS
DIFFERENCE ON
TWO MEASUREMENT
SCALES

305SA01.EPS

Step 10 After the first tack weld, on the opposite side, check the root opening, adjust the gap if necessary, and make the second tack weld.

Step 11 Check the root opening again and weld the third tack midway between the first two tacks.

Step 12 Weld the fourth tack opposite the third tack and midway between the first and second tacks. There should now be four tack welds evenly spaced (90°) around the pipe coupon. This is illustrated by *Figure 5*.

Step 13 Feather the tack welds with the edge of a grinding wheel. Feathering the ends of the tack welds with a grinder helps fuse the tack welds into the root pass.

4.2.0 The Welding Machine

Identify a proper constant-current welding machine for GTAW and follow these steps to set it up for use.

Step 1 Verify that the welding machine can be used for GTAW with or without internally controlled gas shielding.

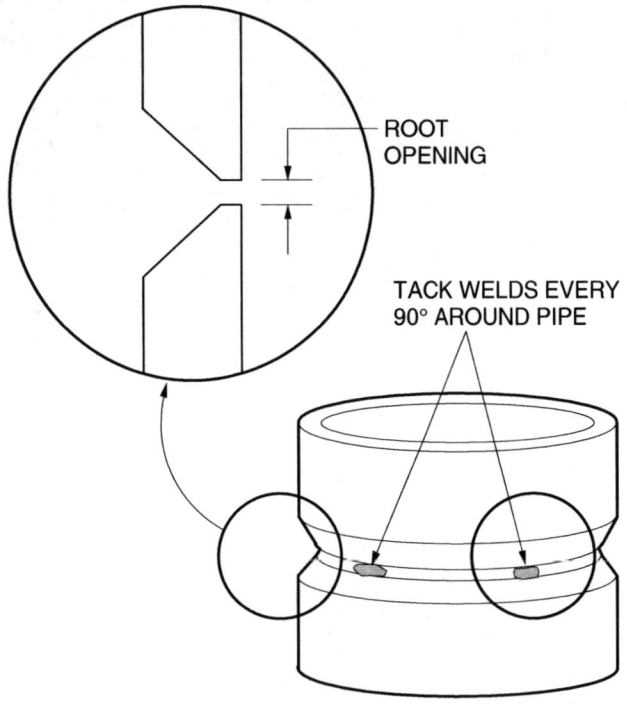

ROOT OPENING

TACK WELDS EVERY 90° AROUND PIPE

FEATHERED ENDS OF TACK WELD

305F05.EPS

Figure 5 ◆ Tacked open-root V-groove weld coupon.

Step 2 Identify an air-cooled or water-cooled GTAW torch. Make sure it is compatible with the welding machine and any cooling unit.

Step 3 Check to be sure that the welding machine is properly grounded through the primary current receptacle.

Step 4 Verify the location of the primary disconnect.

Step 5 Configure the welding machine for GTAW parameters as directed by your instructor (*Figure 6*). Configure the torch polarity and equip the torch with the correct diameter and type of tungsten (³⁄₃₂" or ⅛" EWTh-2) for the filler metal used for the application.

Step 6 Connect the proper shielding gas for the application as described in the previous level and as specified by the filler metal manufacturer, WPS, site quality standards, or your instructor.

Step 7 Connect the clamp of the workpiece lead to the workpiece.

Step 8 Turn on the welding machine and, if necessary, purge the torch as directed by the manufacturer's instructions.

4.3.0 Filler Metals

Filler metals are selected to be compatible with the base metal to be welded. The WPS or site standards specify the filler metal type and size to use.

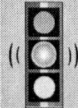

CAUTION

When the WPS or site quality standards specify a filler metal, it must be used to prevent defective welds.

For the welding exercises in this module, you should use ³⁄₃₂" and/or ⅛" carbon steel filler metals of classification ER70S-2 or equivalent to weld the carbon steel pipe coupons. Remove only a small number of filler metals at a time. Keep the remainder in the package to keep them clean. Before using the filler metal, check it for burned ends or contamination such as corrosion, dirt, oil, or grease, which can all cause weld discontinuities. Clean the filler metal with a clean, oil-free rag and chemical cleaner, and snip any burned ends. If the filler metal cannot be cleaned, do not use it.

CAUTION

Do not use RG-45 and RG-60 filler metals. They will cause porosity because they are not deoxidized rods.

The Use of a Y-Adapter

When using the same gas for backing as well as shielding, connect a Y-adapter with valves in each branch to the output of a flow regulator. Connect one branch to the backing gas device and use the other branch as the torch supply. With the branch valves turned off, open the cylinder valve slowly at first, and then open it all the way. At the torch, turn on the torch valve or purge control and, using the torch branch valve on the Y-adapter, adjust for the proper torch flow rate as indicated on the flow regulator. At the torch, turn off or deactivate the gas flow; then, open and adjust the backing gas branch valve on the Y-adapter for the proper backing gas flow as indicated on the flow regulator. Once the branch valves are adjusted for the proper flow rates, if welding must be interrupted for a time, use the cylinder valve to shut off the gas.

305SA02.EPS

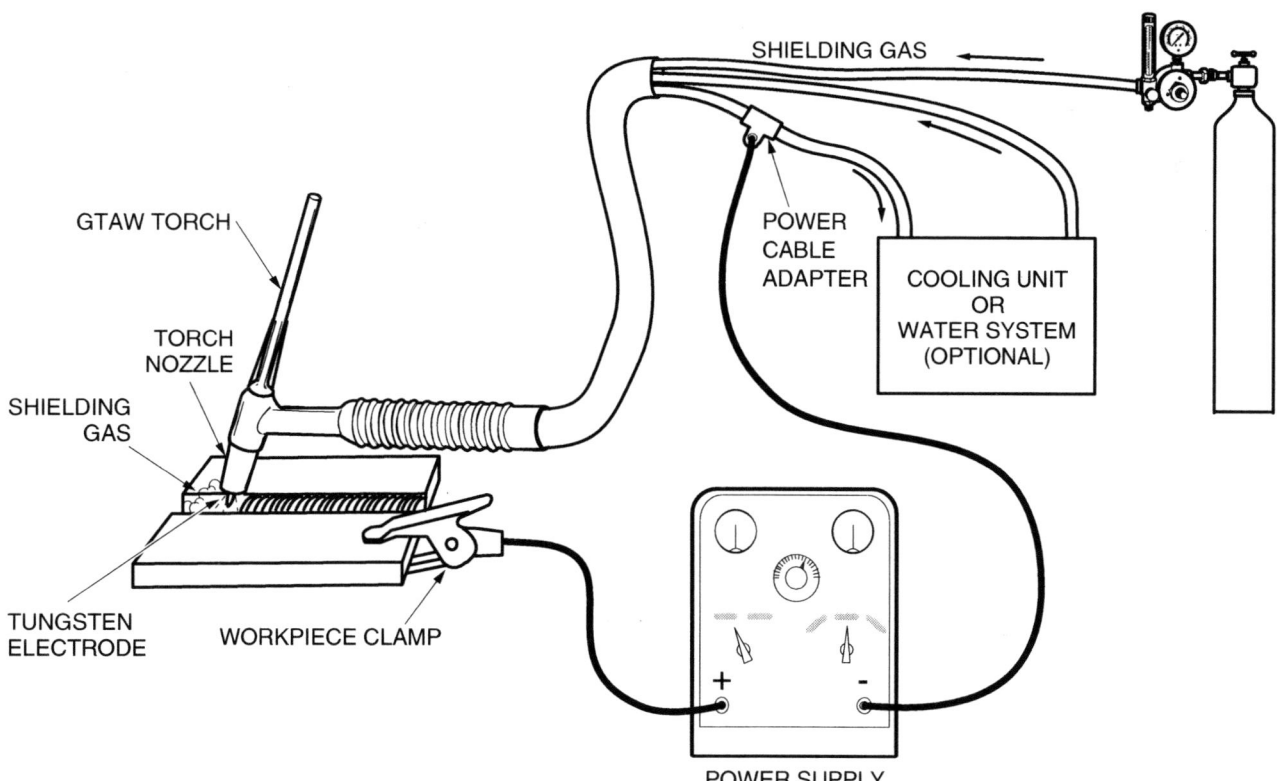

Figure 6 ◆ Configuration diagram of typical GTAW machine.

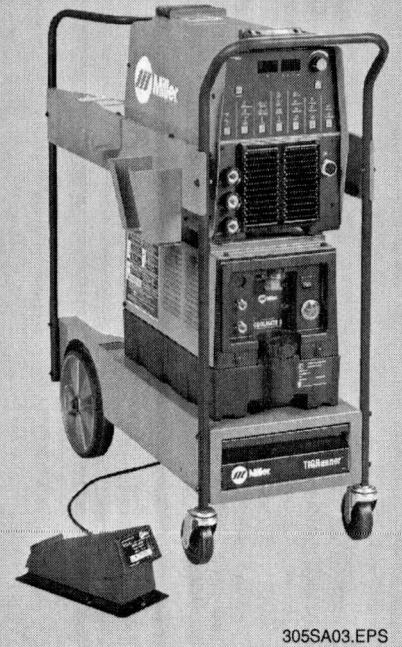

Advanced Inverter Power Sources

An advanced inverter power source, like the one shown here, can make welding pipe with GTAW easier. With features such as HF start only, current pulsing, and weld current sequencing (stop/start ramping), these machines provide much more control over some of the variables that affect the quality of a weld.

305SA03.EPS

5.0.0 ◆ GAS TUNGSTEN ARC WELDING TECHNIQUES

GTAW weld bead characteristics and quality are affected by several factors, each of which is influenced by the way you handle the torch. These factors include the following:

- Torch travel speed and arc length
- Torch angles
- Torch and filler metal handling techniques

5.1.0 Torch Travel Speed and Arc Length

Torch travel speed and arc length affect the GTAW weld puddle and weld penetration. A slow travel speed allows more heat to concentrate and forms a larger, more deeply penetrating puddle. Faster travel speeds prevent heat buildup and form smaller and shallower puddles. Arc length is the major control for bead width. As the torch is raised, arc length, voltage, and bead width increase.

5.2.0 Torch Angles

There are two basic torch angles you must control when performing GTAW: the work angle and the travel angle. The definition of these angles is the same as for all other methods of arc welding.

5.2.1 Work Angle

The torch work angle (*Figure 7*) is defined as a less-than-90° angle between a line perpendicular to the

major workpiece surface at the point of electrode contact and a plane determined by the electrode axis and the weld axis. For a T-joint or corner joint, the line is perpendicular to the nonbutting member. For pipe, the plane is determined by the electrode axis and a line tangent to the pipe surface at the same point.

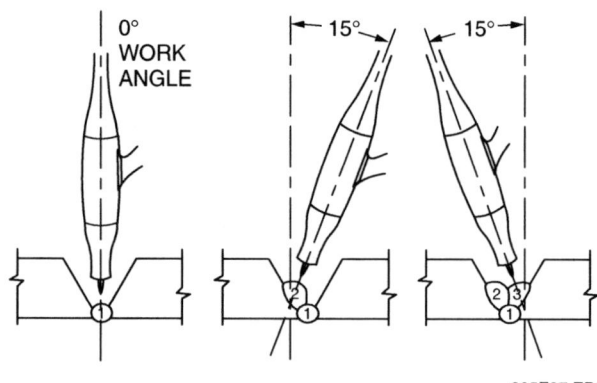

Figure 7 ◆ Typical torch work angles.

5.2.2 Travel Angle

The torch travel angle (*Figure 8*) is defined as a less-than-90° angle between the electrode axis and a line perpendicular to the weld axis at the point of electrode contact in a plane determined by the electrode axis and the weld axis. For pipe, the plane is determined by the electrode axis and a line tangent to the pipe surface at the same point. A push angle is used for GTAW and is created when the torch is tilted back so that the electrode tip precedes the torch in the direction of the weld.

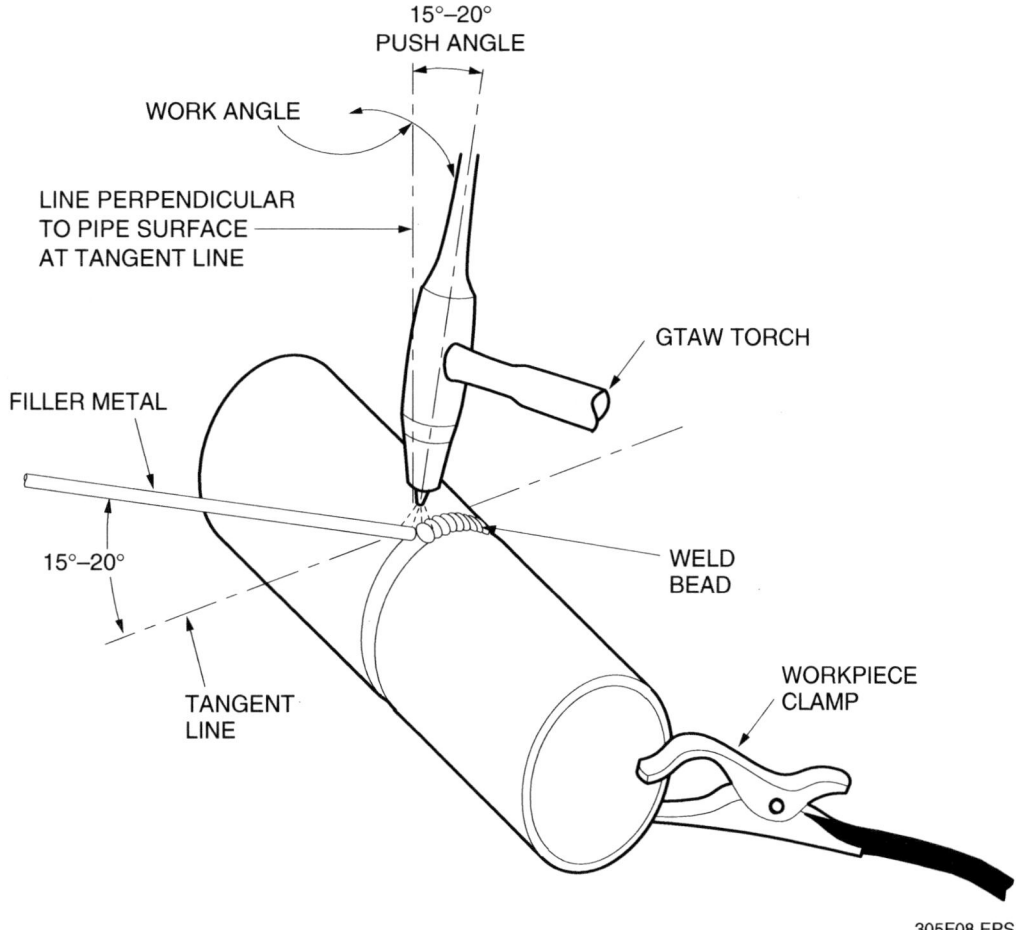

15°–20°
PUSH ANGLE

WORK ANGLE

LINE PERPENDICULAR
TO PIPE SURFACE
AT TANGENT LINE

FILLER METAL

GTAW TORCH

15°–20°

WELD
BEAD

TANGENT
LINE

WORKPIECE
CLAMP

305F08.EPS

Figure 8 ◆ Torch travel angle.

In this position, the electrode tip and shielding gas are directed ahead of the weld bead. Push angles of 15° to 20° are normally used for GTAW.

5.3.0 Torch and Filler Metal Handling Techniques

The two basic handling techniques used to perform GTAW are known as freehand and walking-the-cup. Try both techniques and use the one that gives the best results. Both techniques are explained in the following sections.

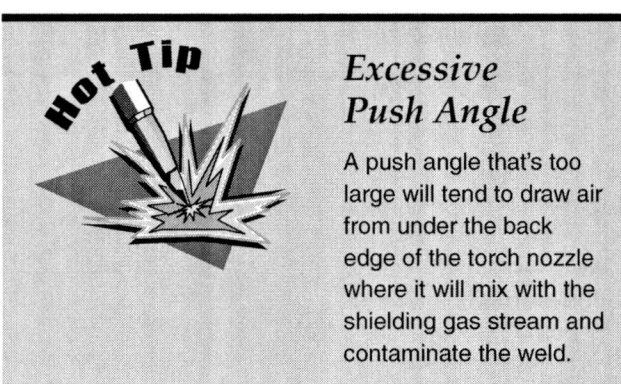

Excessive Push Angle

A push angle that's too large will tend to draw air from under the back edge of the torch nozzle where it will mix with the shielding gas stream and contaminate the weld.

5.3.1 Freehand Technique

In the freehand technique (*Figure 9*), hold the torch electrode tip just above the weld puddle or base metal. Support the torch with your hand. Steady your hand by resting some part of it on or against the base metal to maintain the proper arc length from the electrode tip to the puddle. If required, you can move the torch tip in a small circular motion within the molten puddle to maintain the puddle size and to advance the puddle. Add filler metal as needed.

Hold the GTAW filler metal in the hand not holding the torch. For all positions, hold the filler rod at an angle of about 20° above the base metal surface and in line with the weld. Always keep the tip of the filler rod within the shielding gas envelope to protect the filler rod from atmospheric contamination and to keep it preheated.

Dab the filler metal into the leading edge of the weld puddle, using extreme care not to touch the tungsten electrode with the end of the filler metal. If the tungsten electrode touches the filler metal or weld puddle, it will become contaminated with filler metal. The electrode must then be removed and cleaned by grinding (or chemical cleaning)

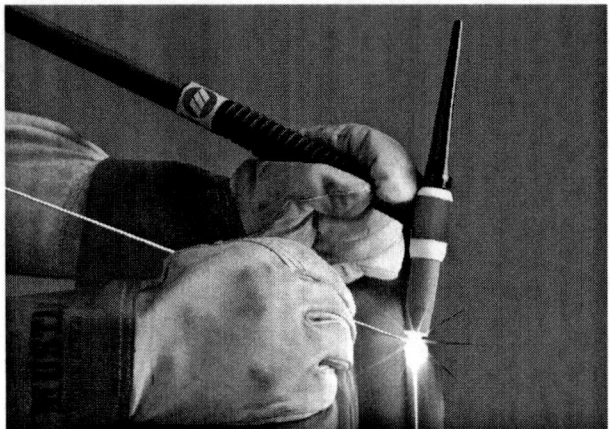

Figure 9 ◆ Freehand technique.

305F09.EPS

before proceeding. A technique for preventing contamination of the electrode is to move the electrode to the back edge of the weld puddle as the filler metal is dabbed into the leading edge of the weld puddle.

Figure 9 shows the use of the freehand technique for running stringer beads.

> **CAUTION**
>
> Do not insert the end of the filler metal rod into the molten puddle under the electrode and then attempt to melt it off. This can cause hard spots and weld defects.

5.3.2 Walking-the-Cup Technique

In the walking-the-cup technique (*Figure 10*), rest the edge of the torch nozzle (cup) against the base metal or groove edges to steady the torch and to maintain a constant arc length. Rock the torch from side to side on the edge of the cup as you advance it, to maintain the puddle size and to heat both sides of the groove. Add the filler metal in the same manner as with the freehand technique, using care not to contaminate the electrode.

6.0.0 ◆ OPEN-ROOT V-GROOVE PIPE WELDS

The open-root V-groove weld is the most common groove weld normally made on carbon, low-alloy, and stainless steel pipe. These welds are typically used for joining medium- and thick-walled pipe used in critical piping systems. Critical piping systems (including pipe, fittings, and welded joints) contain or carry material with the potential to cause long- or short-term catastrophic danger or injury to personnel and/or the environment should the piping system fail to contain or carry the material as designed. Welds in critical piping must meet the most stringent code requirements. Non-critical piping is low-pressure piping used for heating and air conditioning, simple water systems, and other service installations. Less-stringent code requirements are used to evaluate non-critical piping welds.

> **CAUTION**
>
> When welding pipe, do not make arc strikes outside the weld groove on the surface of the pipe. An arc strike can cause a hardened spot that can crack as the pipe expands and contracts. An arc strike on the pipe surface is considered a defect and will require repair or rework.

6.1.0 Root Pass

If a consumable insert or backing is not used, the most difficult part of making an open-root V-groove weld is the root pass. The root pass is made from the V-groove side of the joint and must have complete penetration but not an excessive amount of melt-through root reinforcement. The penetration is controlled with one of two techniques. One is called running a keyhole. A keyhole, shown in *Figure 11*, is a hole made when the root faces of two plates are melted away by the welding arc. The molten metal flows to a weld pool at the back of the keyhole, forming the weld.

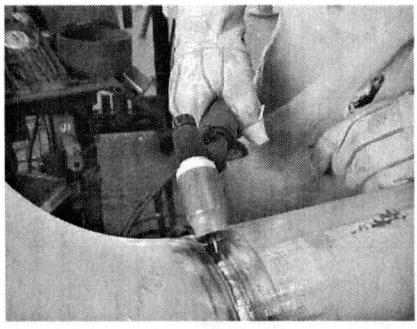

STEP 1

STEP 2

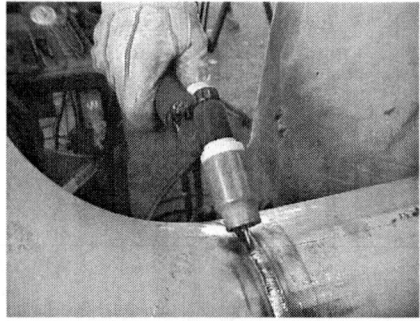

STEP 3

305F10.EPS

Figure 10 ◆ Walking-the-cup technique.

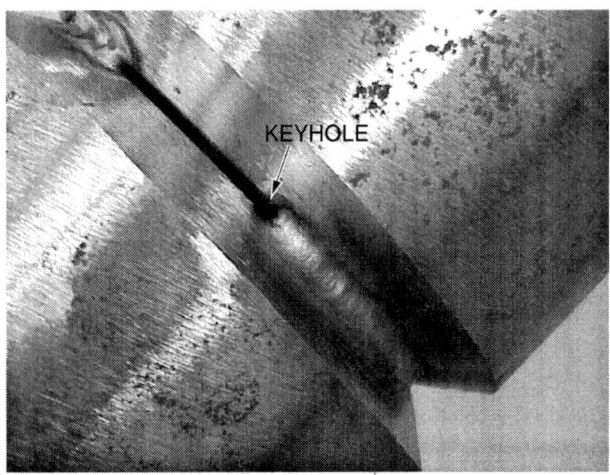

KEYHOLE

305F11.EPS

Figure 11 ◆ A root pass keyhole.

To run a keyhole, rest the filler metal rod on the joint root just ahead of the leading edge of the arc. When you require additional filler metal, move the arc back to the rear of the weld pool, and dab the rod into the weld pool to add more metal.

The other technique is similar to the on-the-wire method, which allows continuous feeding of the wire into the molten weld pool. The filler metal rod used is smaller than the root opening and is held between the root faces. The filler metal is melted as you pass the arc over it, which keeps the pool width to a minimum and achieves complete penetration.

When welding a consumable insert, hold the torch perpendicular to the pipe surface and point it radially toward the center of the pipe. Use an arc length of about ⅛" to melt the insert. Forward progression is governed by the melting rate of the insert and the characteristics of the weld pool. Increased fluidity and a rising weld pool indicate sufficient melting. When this occurs, advance the arc in a step-wise fashion.

When you do not use an insert, clean and inspect the root pass after running it. Remove any excess buildup or undercut with the edge of a narrow grinding wheel (not a cutoff wheel). This makes it easier to apply a hot pass. Use care not to widen the groove nor grind through the root pass. For most carbon steel pipe, the hot, fill, and cover passes are usually made with GMAW, FCAW, or SMAW. For high-pressure piping, it is essential that the hot pass is fused well with the root pass and pipe walls. The hot pass must melt out any discontinuities of the root pass weld. The hot pass is accomplished by running a stringer bead that covers both sides of the root pass using higher current or higher voltage, depending on the process used. After you complete and thoroughly clean the hot pass bead, apply the fill and cover passes. In this module, you will practice the root, hot, fill, and cover passes using GTAW.

6.2.0 Pipe Groove Weld Test Positions

Groove welds may be made in all positions on pipe. The weld position is determined by the axis of the pipe. Four standard weld test positions are used with pipe and are shown in *Figure 12*.

To destructively test 5G or 6G welds, the test specimens are cut from the four regions illustrated in *Figure 13*: midway between the 12-o'clock and 3-o'clock, 3-o'clock and 6-o'clock, 6-o'clock and 9-o'clock, and 9-o'clock and 12-o'clock positions. (The 12-o'clock position is the top of the coupon groove; 6-o'clock is the bottom.)

6.3.0 Acceptable and Unacceptable Pipe Weld Profiles

Pipe groove welds without backing should be made with slight reinforcement (not exceeding ⅛") and a gradual transition to the base metal at each toe. The root pass should have complete penetration. The root reinforcement on the inside of the pipe ranges from being flush to a maximum of ⅛". Pipe groove welds must not have excessive reinforcement or underfill at the face or root, excessive undercut, or overlap. Excessively large cover passes will reduce the pipe's strength. They cause the stresses in the pipe to be concentrated along

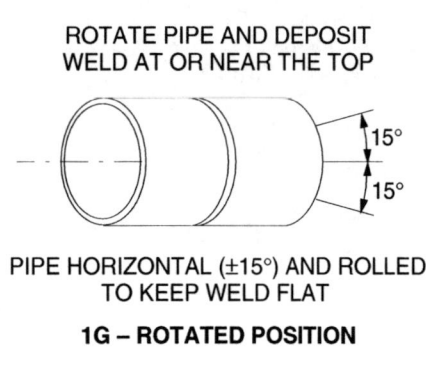

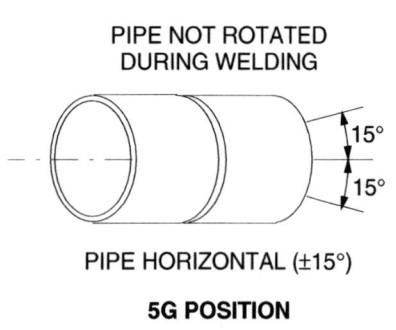

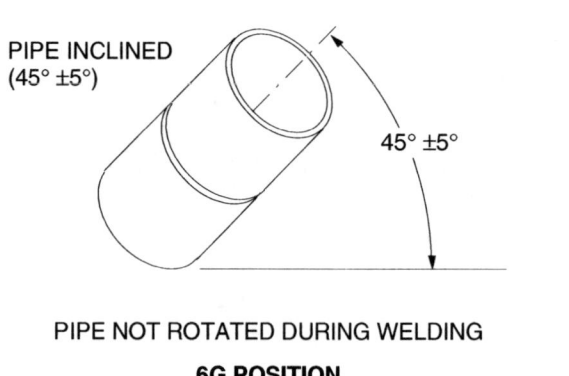

Figure 12 ◆ Four basic pipe groove weld test positions.

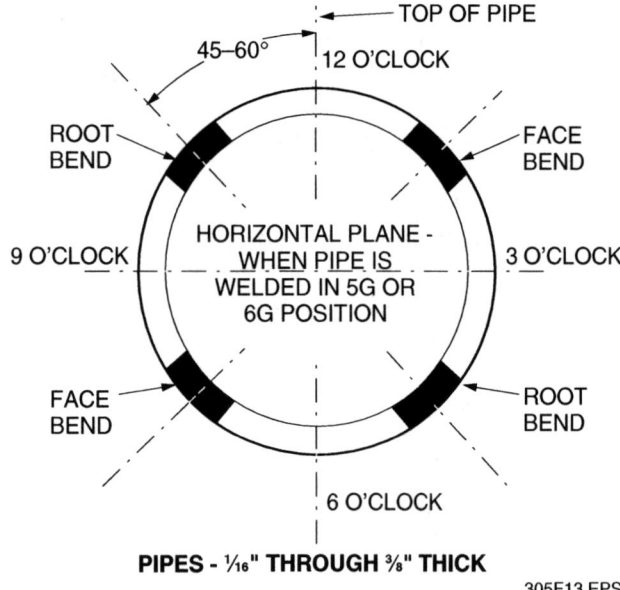

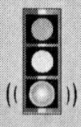

Figure 13 ◆ Test specimen regions of 5G or 6G position pipe.

the sides of the weld and do not permit the pipe to expand and contract in a uniform manner along its length. If a weld has any of these defects, as shown in *Figure 14*, it must be repaired.

7.0.0 ◆ PRACTICING OPEN-ROOT V-GROOVE WELDS

This section of the module explains how to perform the following open-root V-groove weld positions:

- Flat (1G-ROTATED)
- Horizontal (2G)
- Multiple (5G)
- Inclined multiple (6G)

> **NOTE**
>
> The following practice procedures assume that a consumable insert is not being used for the root pass.

7.1.0 Flat (1G-ROTATED) Position

Practice the 1G-ROTATED position open-root V-groove pipe welds using a shielding gas and a solid filler rod with a diameter specified by your instructor. Keep the torch angle at 90° (0° work angle) to the pipe axis with a 15° to 20° push angle. Use a slight circular motion to form the weld puddle for the root pass, paying particular attention to tie into the tack welds. Clean the face of the root pass with a brush.

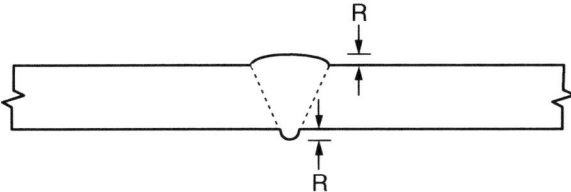

R = FACE AND ROOT REINFORCEMENT PER CODE
NOT TO EXCEED ⅛" MAX.

ACCEPTABLE WELD PROFILE

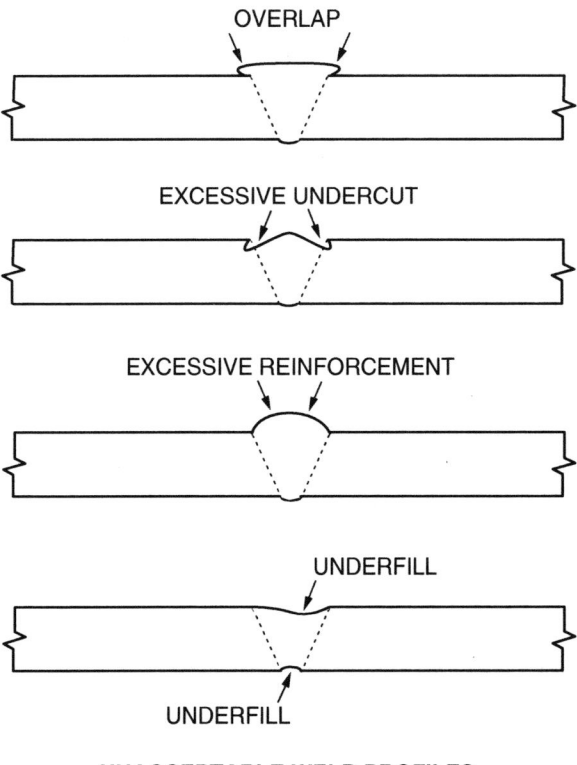

OVERLAP

EXCESSIVE UNDERCUT

EXCESSIVE REINFORCEMENT

UNDERFILL

UNDERFILL

UNACCEPTABLE WELD PROFILES

305F14.EPS

Figure 14 ◆ Acceptable and unacceptable pipe groove weld profiles.

Run the fill and cover passes as stringer beads using both the walking-the-cup and freehand torch handling techniques. Take particular care at the termination of the weld to fill the crater. Run all passes at or near the top of the pipe as the pipe is rotated.

Follow these steps to practice open-root V-groove pipe welds in the 1G-ROTATED position:

Step 1 Tack weld together the practice pipe weld coupon, as explained earlier.

Step 2 Position the pipe weld coupon horizontally on two sets of rollers at a comfortable welding height. *Figure 15* shows roller supports commonly found in pipe welding shops.

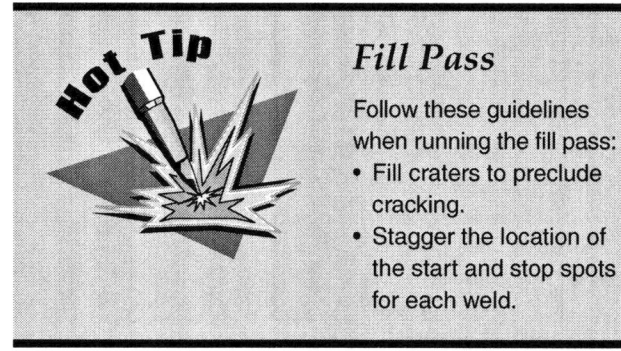

Fill Pass

Follow these guidelines when running the fill pass:
• Fill craters to preclude cracking.
• Stagger the location of the start and stop spots for each weld.

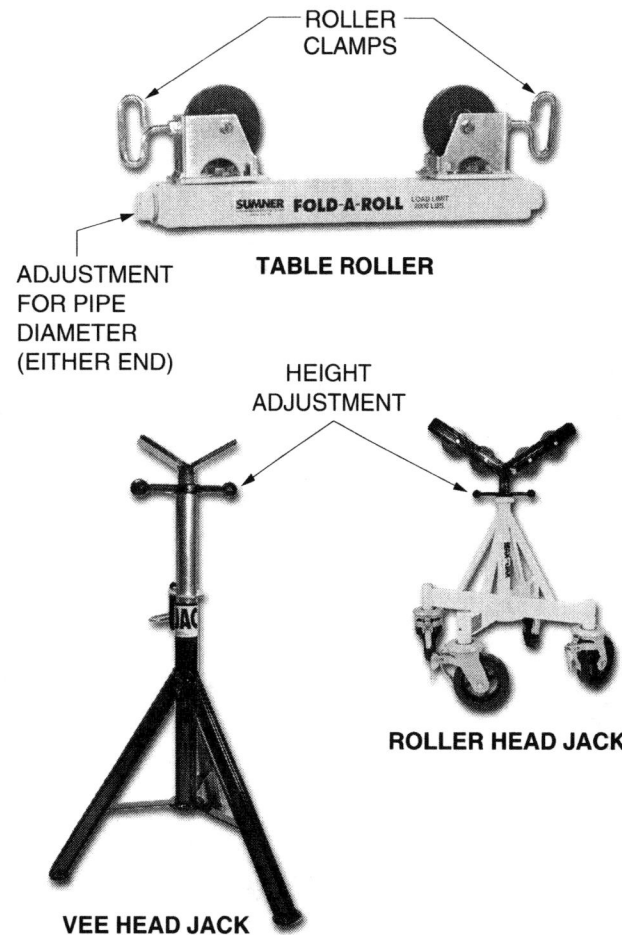

ROLLER CLAMPS

SUMNER FOLD-A-ROLL

TABLE ROLLER

ADJUSTMENT FOR PIPE DIAMETER (EITHER END)

HEIGHT ADJUSTMENT

ROLLER HEAD JACK

VEE HEAD JACK

FOLD-A-ROLL

FLOOR STAND ROLLER

305F15.EPS

Figure 15 ◆ Pipe roller supports.

Step 3 If gas backing is to be used, assemble the backing gas retaining devices to the pipe, and connect the backing gas hose. Set the backing gas flowmeter to the flow rate specified by the filler metal rod manufacturer, the WPS, or your instructor.

Step 4 Make sure the workpiece clamp is attached directly onto the pipe coupon. This will prevent the welding current from passing through the roller bearings or arcing between the rollers and the pipe coupon.

CAUTION
Failure to attach the workpiece clamp to the pipe coupon can result in variations in welding current, damage to the roller bearings, and arcing on the rollers and pipe coupon.

Step 5 Run the root pass using a circular motion. Position a tack weld at the 11 o'clock position, start the weld bead on the tack weld, and advance toward the 12 o'clock position.

Step 6 Roll the pipe as necessary to keep the weld in the flat position.

Step 7 Brush and, as necessary, grind the root pass to clean and shape the weld.

Step 8 Use the same rolling procedure to make the hot pass (bead 2 in *Figure 16*) and remaining passes. Use stringer beads. Pay particular attention at toes to tie in the welds. Overlap the passes, and clean the weld after each pass. Refer to *Figure 16* for the bead sequences and work angles.

Welding Thick-Walled Pipe

When welding a V-groove joint in thick-walled pipe, it is sometimes difficult to maintain an appropriate arc length because of the torch size or the depth/angle of the groove. In these cases, use a narrow nozzle so that the torch can be positioned closer to the root pass or use a gas lens that will allow a longer electrode stickout. In the latter case, you may have to use a larger electrode to prevent overheating the electrode.

Powered Pipe Roller

A pipe roller that is operated by an electric motor activated by a foot switch is shown here. These devices are primarily used for large, heavy pipe sections.

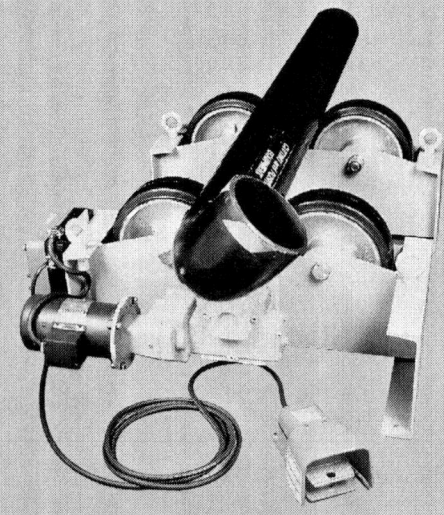

305SA04.EPS

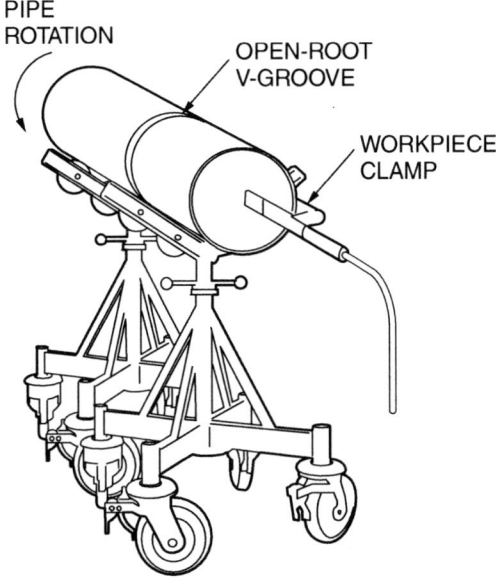

EXAMPLE OF PIPE ROTATION

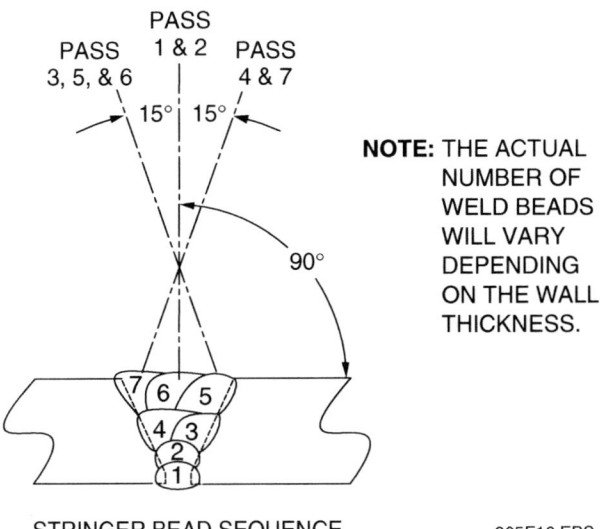

NOTE: THE ACTUAL NUMBER OF WELD BEADS WILL VARY DEPENDING ON THE WALL THICKNESS.

STRINGER BEAD SEQUENCE   305F16.EPS

Figure 16 ◆ Multiple-pass 1G-ROTATED bead sequence and work angles.

7.2.0 Horizontal (2G) Position

Practice 2G position open-root V-groove pipe welds using a shielding gas and a solid filler rod with a diameter specified by your instructor. The torch should be at a 15° to 20° push angle for the weld passes. (See *Figure 17*.) To prevent the weld puddle from sagging, you can drop the torch work angle slightly but not more than 10°. For the root pass, use a circular motion. Pay particular attention to tie into the tack welds.

When running the fill/cover passes, use stringer beads and both the walking-the-cup and freehand torch handling techniques. Ensure tie-in at the toes of the weld bead. Take particular care at the termination of the weld to fill the crater.

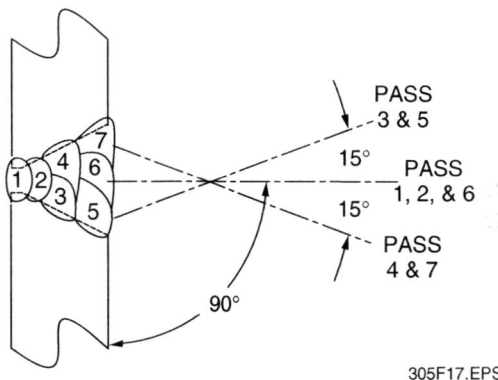

305F17.EPS

Figure 17 ◆ Multiple-pass 2G bead sequence and work angles.

Follow these steps to practice open-root V-groove pipe welds in the 2G position:

Step 1 Tack weld together the practice pipe weld coupon, as explained earlier.

Step 2 Clamp or tack weld the pipe coupon with the axis of the pipe vertical.

Step 3 If gas backing is to be used, assemble the backing gas retaining devices to the pipe, and connect the backing gas hose. Set the backing gas flowmeter to the flow rate specified by the filler metal rod manufacturer, the WPS, or your instructor.

Step 4 Run the root pass (pass 1) as shown in *Figure 17*. Start the root pass on a tack weld. Continue this procedure until the root pass is completed.

Step 5 Brush and, as necessary, grind the root pass to clean and shape the weld.

Step 6 Run the hot pass (bead 2) and remaining passes at the appropriate work angles. Overlap the passes, and clean the weld after each pass.

7.3.0 Multiple (5G) Position

Practice the multiple (5G) position open-root V-groove pipe welds using a shielding gas and a solid filler rod with a diameter specified by your instructor. Start on a tack weld positioned at the bottom of the pipe and weld uphill toward the top. Use a torch push angle of 15° to 20° to the pipe surface (*Figure 18*).

The torch should be at approximately a 15° to 20° push angle for the root, fill, and cover passes. (See *Figure 19*.) When running the fill and cover passes, use stringer beads or weave beads and both the walking-the-cup and freehand torch handling techniques. Ensure tie-in at the toes of the weld bead. Pay particular attention at the termination of the weld to fill the crater.

Follow these steps to practice open-root V-groove pipe welds in the 5G position:

Step 1 Tack weld together the practice pipe weld coupon, as explained earlier.

Step 2 Clamp or tack weld the pipe weld coupon into position with the pipe axis horizontal.

Step 3 If gas backing is to be used, assemble the backing gas retaining devices to the pipe, and connect the backing gas hose. Set the backing gas flowmeter to the flow rate specified by the filler metal rod manufacturer, the WPS, or your instructor.

Step 4 Run the root pass uphill with a stringer bead as shown in *Figure 18*. Repeat for the opposite side of the pipe.

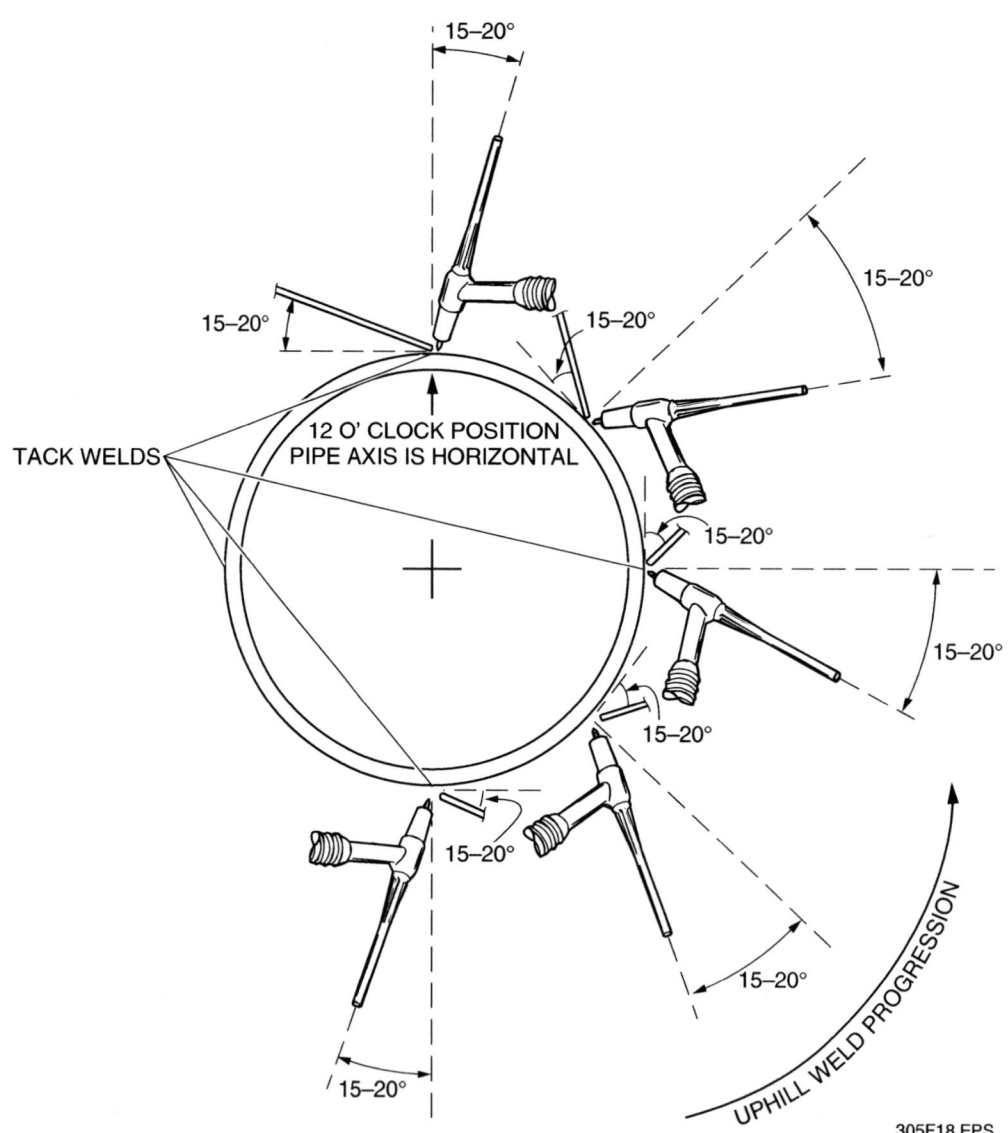

305F18.EPS

Figure 18 ◆ 5G pipe position torch and electrode rod angles.

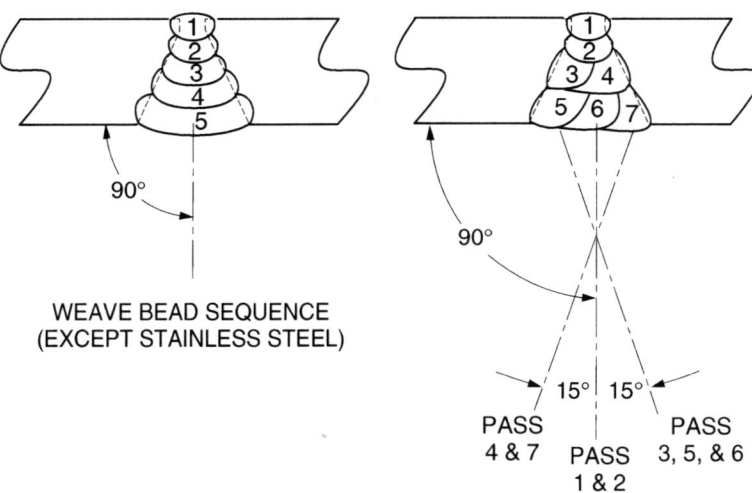

WEAVE BEAD SEQUENCE
(EXCEPT STAINLESS STEEL)

90°

15° 15°

PASS
4 & 7

PASS
1 & 2

PASS
3, 5, & 6

STRINGER BEAD SEQUENCE

305F19.EPS

Figure 19 ◆ Multiple-pass 5G bead sequences and work angles.

Step 5 Brush and, as necessary, grind the root pass to clean and shape the weld.

Step 6 Run the hot pass (bead 2) and remaining passes uphill at the appropriate work angles. Overlap the passes, and clean the weld after each pass.

7.4.0 Inclined Multiple (6G) Position

Practice the inclined multiple (6G) (45°) position open-root V-groove pipe welds using a shielding gas and a solid filler rod with a diameter specified by your instructor. Start at the bottom of the pipe and weld uphill toward the top. The torch should be at approximately a 15° to 20° push angle and at 90° (0° work angle) to the pipe for the root, fill, and cover passes. (See *Figure 20*.)

When running the fill and cover passes, use stringer beads and both the walking-the-cup and freehand torch handling techniques. Ensure tie-in at the toes of the weld bead. Pay particular attention at the termination of the weld to fill the crater.

Follow these steps to practice open-root V-groove pipe welds in the 6G position:

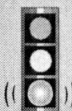 **NOTE**

If required for training purposes, a restricting ring may be added to the 6G position coupon to form a 6GR position coupon.

Step 1 Tack weld together the practice pipe weld coupon, as explained earlier.

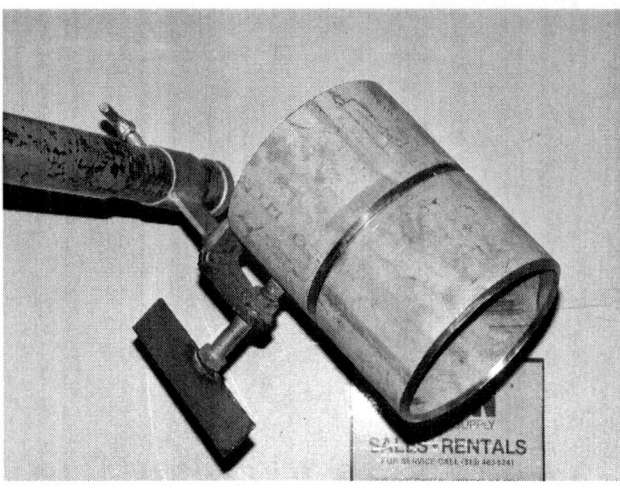

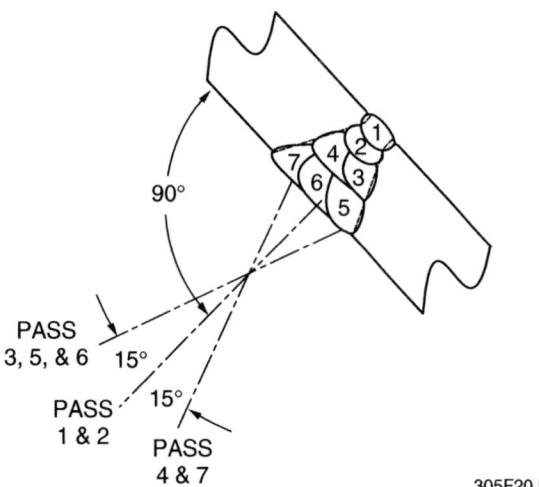

PASS
3, 5, & 6   15°

PASS
1 & 2   15°

PASS
4 & 7

305F20.EPS

Figure 20 ◆ Multiple-pass 6G bead sequence and work angles.

Step 2 Clamp or tack weld the pipe weld coupon into position with the pipe axis inclined 45° to the horizontal plane.

Step 3 If gas backing is to be used, assemble the backing gas retaining devices to the pipe, and connect the backing gas hose. Set the backing gas flowmeter to the flow rate specified by the filler metal rod manufacturer, the WPS, or your instructor.

Step 4 Run the root pass uphill with a stringer bead as shown in *Figure 18*. Repeat for the opposite side of the pipe.

Step 5 Brush and, as necessary, grind the root pass to clean and shape the weld.

Step 6 Run the hot pass (bead 2) and remaining passes uphill at the appropriate work angles. Overlap the passes, and clean the weld after each pass.

Summary

The ability to make open-root V-groove welds on pipe in all positions is one of the more difficult skills you must develop as a welder. The open-root V-groove weld is the most common weld joint used for joining medium- and thick-walled pipe. You must be able to set up the equipment, perform the welding, and recognize acceptable welds. Open-root V-groove pipe welds can be made in the 1G-ROTATED, 2G, 5G, and 6G positions. Practice these welds until you can consistently produce acceptable welds.

Review Questions

1. A major use of GTAW is to make manual high-quality welds in _____.
   a. ferrous and nonferrous piping used in critical and noncritical applications
   b. ferrous piping used in critical applications only
   c. nonferrous piping used in critical applications only
   d. ferrous and nonferrous piping used in noncritical applications only

2. The highest-quality protection for open-root joint welding is provided by _____.
   a. flux
   b. filler metal
   c. oxygen
   d. backing gas

3. Large piping can be plugged for welding using removable _____.
   a. dams
   b. backing gas
   c. flux
   d. caps

4. Flux prevents _____ from coming in contact with the molten metal.
   a. dirt
   b. slag
   c. water
   d. air

5. When welding, consumable inserts are used without backing gas protection _____.
   a. only when welding carbon steel pipe
   b. only when internal obstructions are acceptable
   c. when weld quality requirements are less critical
   d. wherever internal obstructions cannot be tolerated

6. For GTAW of carbon steel pipe, the bevel angle should be 30° or 37½° with no notches more than _____ deep.
   a. ½₂"
   b. ⅟₁₆"
   c. ⅛"
   d. ¼"

7. If the filler metal cannot be cleaned, _____.
   a. heat it with the torch
   b. turn up the purging gas
   c. do not use it
   d. put it in an electrode holder

8. If the tungsten electrode touches the filler metal or weld puddle, it will become contaminated with _____.
   a. shielding gas
   b. flux
   c. filler metal
   d. base metal

9. When using the freehand technique, always keep the tip of the filler rod _____.
   a. within the shielding gas envelope
   b. in contact with the tungsten electrode
   c. outside the shielding gas envelope
   d. within the molten puddle under the electrode

10. The edge of the torch nozzle (cup) is rested against the base metal in the _____ technique.
    a. stringer bead
    b. on-the-wire
    c. freehand
    d. walking-the-cup

11. When welding a consumable insert, use an arc length of about _____ to melt the insert.
    a. ½₂"
    b. ⅟₁₆"
    c. ⅛"
    d. ¼"

12. When welding a consumable insert, hold the torch _____ to the pipe surface.
    a. perpendicular
    b. parallel
    c. at a 10° angle
    d. at a 30° angle

13. There are _____ standard pipe-weld test positions.
    a. two
    b. three
    c. four
    d. five

14. When making flat (1G-ROTATED) position open-root V-groove carbon steel welds, use a _____ angle.
    a. 10° to 15° drag
    b. 10° to 15° push
    c. 15° to 20° drag
    d. 15° to 20° push

15. When making horizontal (2G) position open-root V-groove carbon steel pipe welds, use a _____ for the root pass.
    a. circular motion
    b. slight forward and backward motion
    c. slight side-to-side oscillation
    d. crescent-shaped motion

Trade Term Introduced in This Module

Backing gas (purge): An inert gas used on the back side of a joint to protect it from oxidation and atmospheric contamination during welding.

Performance Accreditation Tasks

The Performance Accreditation Tasks (PATCs) correspond to and support learning objectives in the *AWS Guide for the Training and Qualification of Welding Personnel.*

PATCs provide specific acceptable criteria for performance and help to ensure a true competency-based welding program for students.

The following tasks are designed to evaluate your ability to run open-root V-groove welds with GTAW equipment in three standard test positions using carbon steel filler rod of the appropriate diameter and argon shielding gas. Perform each task when you are instructed to do so by your instructor. As you complete each task, it to your instructor for evaluation. Do not proceed to the next task until told to do so by your instructor. For AWS 2G and 5G certifications, refer to *AWS EG3.0-96* for bend test requirements. For AWS 6G certifications, refer to *AWS EG4.0-96* for bend test requirements.

OPEN-ROOT V-GROOVE PIPE
WELD IN THE 2G POSITION

Using carbon steel filler rod of the appropriate diameter, argon shielding gas, and stringer beads, make an open-root V-groove weld on carbon steel pipe in the 2G position.

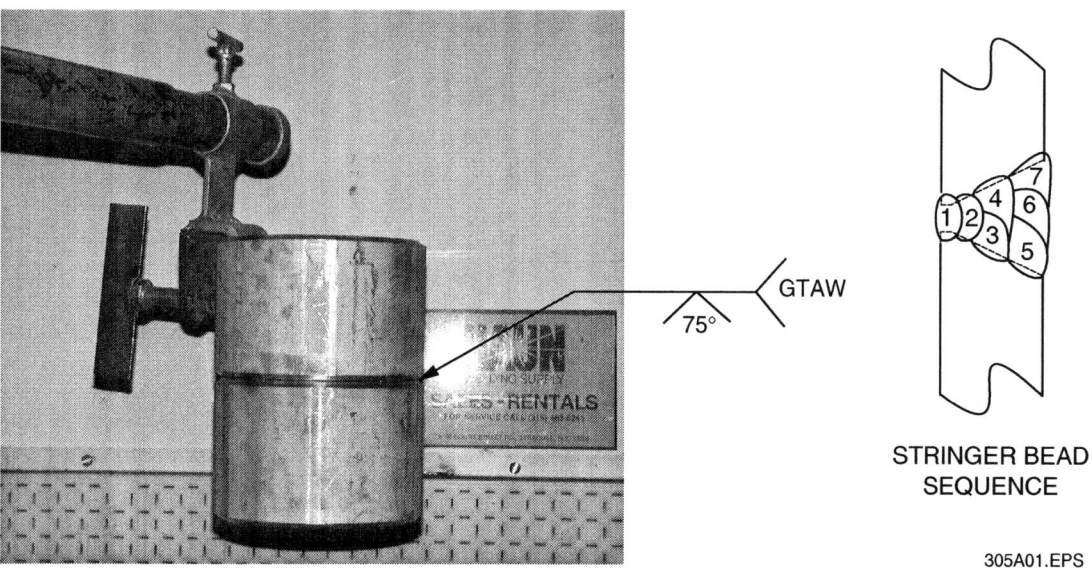

STRINGER BEAD
SEQUENCE

305A01.EPS

Criteria for Acceptance

- Uniform appearance on the bead face   _____

- Craters and restarts filled to the full cross section of the weld   _____

- Uniform weld width ±¹⁄₁₆"   _____

- Acceptable weld profile in accordance with the *ASME Boiler and Pressure Vessel Code, Section IX*   _____

- Smooth transition with complete fusion at the toes of the weld   _____

- Complete uniform root reinforcement at least flush with the inside of the pipe to a maximum of ¹⁄₈"   _____

- No porosity   _____

- No excessive undercut   _____

- No cracks   _____

- No overlap   _____

- No incomplete fusion   _____

OPEN-ROOT V-GROOVE PIPE WELD IN THE 5G POSITION

Using carbon steel filler rod of the appropriate diameter, argon shielding gas, and stringer or weave beads, make an open-root V-groove weld on carbon steel pipe in the 5G position.

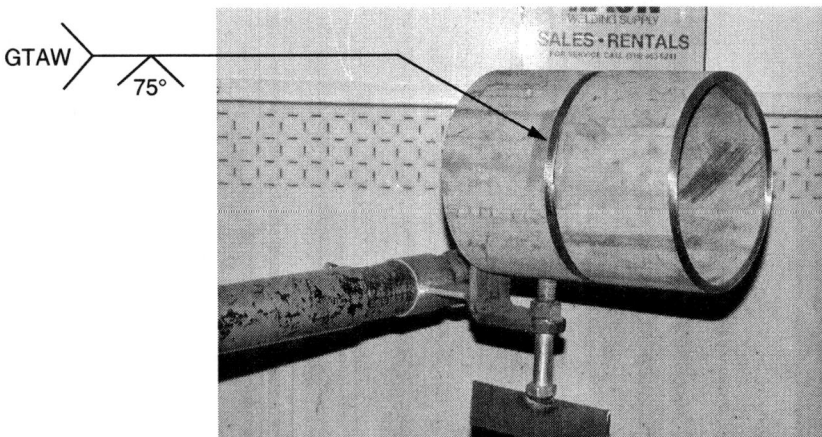

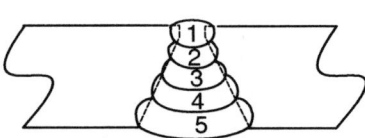

WEAVE BEAD SEQUENCE

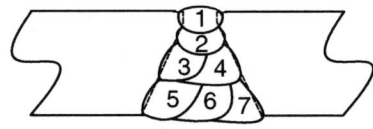

STRINGER BEAD SEQUENCE

305A02.EPS

Criteria for Acceptance

- Uniform appearance on the bead face
- Craters and restarts filled to the full cross section of the weld
- Uniform weld width ±1/16"
- Acceptable weld profile in accordance with the *ASME Boiler and Pressure Vessel Code, Section IX*
- Smooth transition with complete fusion at the toes of the weld
- Complete uniform root reinforcement at least flush with the inside of the pipe to a maximum of 1/8"
- No porosity
- No excessive undercut
- No cracks
- No overlap
- No incomplete fusion

OPEN-ROOT V-GROOVE PIPE WELD IN THE 6G (OR 6GR) POSITION

Using carbon steel filler rod of the appropriate diameter, argon shielding gas, and stringer beads, make an open-root V-groove weld on carbon steel pipe in the 6G (or 6GR) position.

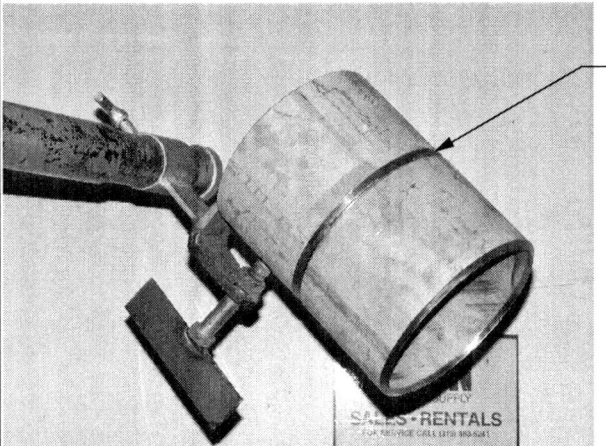

GTAW

75°

NOTE: IF REQUIRED FOR QUALIFICATION PURPOSES, A RESTRICTING RING MAY BE ADDED TO THE 6G POSITION COUPON TO FORM A 6GR POSITION COUPON.

STRINGER BEAD SEQUENCE

305A03.EPS

Criteria for Acceptance

- Uniform appearance on the bead face   _____

- Craters and restarts filled to the full cross section of the weld   _____

- Uniform weld width ±$\frac{1}{16}$"   _____

- Acceptable weld profile and guided bend test in accordance with the *ASME Boiler and Pressure Vessel Code, Section IX*   _____

- Smooth transition with complete fusion at the toes of the weld   _____

- Complete uniform root reinforcement at least flush with the inside of the pipe to a maximum of $\frac{1}{8}$"   _____

- No porosity   _____

- No excessive undercut   _____

- No cracks   _____

- No overlap   _____

- No incomplete fusion   _____

Additional Resources

This module is intended to be a thorough resource for task training. The following reference works are suggested for further study. These are optional materials for continued education rather than for task training.

ANSI/AWS C5.5, Recommended Practices for Tungsten Arc Welding. Miami, FL: American Welding Society.

ASME – Boiler and Pressure Vessel Code: Section 9, Welding and Brazing Qualifications, Current Edition. New York, NY: ASME International.

AWS D10.11, Recommended Practices for Root Pass Welding of Pipe Without Backing. Miami, FL: American Welding Society.

AWS D10.12M/D10.12, Guide for Welding Mild Carbon Steel Pipe. Miami, FL: American Welding Society.

MIG Welding Handbook. Florence, SC: L-TEC Welding & Cutting Systems.

Welding of Pipelines and Related Facilities, API Standard 1104, Latest Edition. New York, NY: ASME International.

Welding Pressure Pipe Lines and Piping Systems. Cleveland, OH: The Lincoln Electric Company.

Welding Skills. R.T. Miller. Homewood, IL: American Technical Publishers.

Figure Credits

Gerald Shannon	305F03, 305F04, 305F05, 305F17, 305F19, 305F20
G.A.L. Gage Co.	305SA01
Bill Cherry	305SA02
Miller Electric Mfg. Co.	305SA03 305F09
Tom Atkinson	305F10
Terry Lowe	305F11
Sumner Manufacturing Co., Inc.	305F15
Koike Aronson, Inc.	305SA04

The NCCER makes every effort to keep these textbooks up-to-date and free of technical errors. We appreciate your help in this process. If you have an idea for improving this textbook, or if you find an error, a typographical mistake, or an inaccuracy in NCCER's Contren® textbooks, please write us, using this form or a photocopy. Be sure to include the exact module number, page number, a detailed description, and the correction, if applicable. Your input will be brought to the attention of the Technical Review Committee. Thank you for your assistance.

Instructors – If you found that additional materials were necessary in order to teach this module effectively, please let us know so that we may include them in the Equipment/Materials list in the Instructor's Guide.

Write: Curriculum Revision and Development Department
National Center for Construction Education and Research
P.O. Box 141104, Gainesville, FL 32614-1104

Fax: 352-334-0932

E-mail: curriculum@nccer.org

Craft _____ Module Name _____

Copyright Date _____ Module Number _____ Page Number(s) _____

Description _____

_____

_____

_____

(Optional) Correction _____

_____

_____

(Optional) Your Name and Address _____

_____

_____

Gas Tungsten Arc Welding (GTAW) – Low-Alloy and Stainless Steel Pipe

COURSE MAP

This course map shows all of the modules in the third level of the Welding curriculum. The suggested training order begins at the bottom and proceeds up. Skill levels increase as you advance on the course map. The local Training Program Sponsor may adjust the training order.

WELDING LEVEL THREE

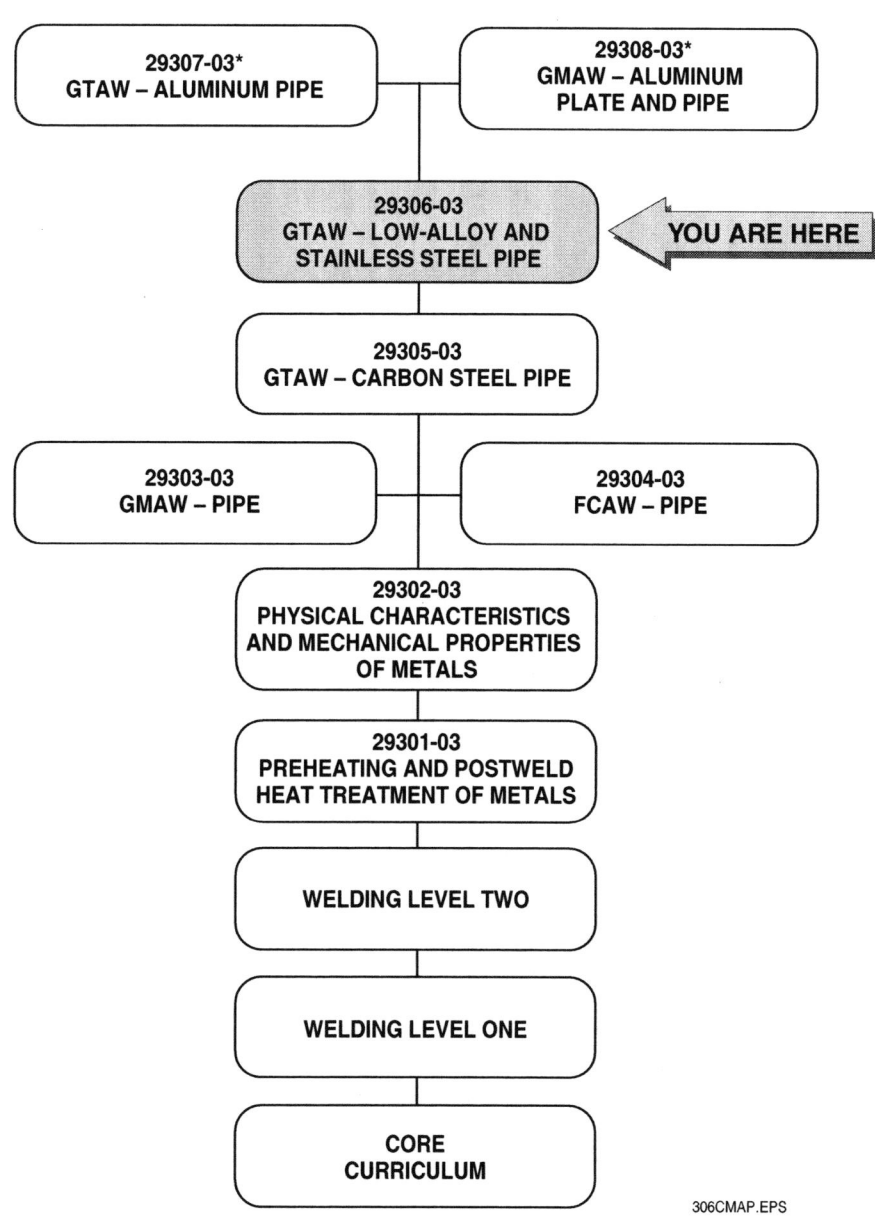

29307-03*
GTAW – ALUMINUM PIPE

29308-03*
GMAW – ALUMINUM PLATE AND PIPE

29306-03
GTAW – LOW-ALLOY AND STAINLESS STEEL PIPE

YOU ARE HERE

29305-03
GTAW – CARBON STEEL PIPE

29303-03
GMAW – PIPE

29304-03
FCAW – PIPE

29302-03
PHYSICAL CHARACTERISTICS AND MECHANICAL PROPERTIES OF METALS

29301-03
PREHEATING AND POSTWELD HEAT TREATMENT OF METALS

WELDING LEVEL TWO

WELDING LEVEL ONE

CORE CURRICULUM

306CMAP.EPS

***Please note that Modules 29307-03 and 29308-03 are electives for those progressing through the *Welding Level Three* program.**

Figures

Gas Tungsten Arc Welding (GTAW) – Low-Alloy and Stainless Steel Pipe

Objectives

When you have completed this module, you will be able to do the following:

1. Set up GTAW equipment to perform stainless and/or low-alloy steel pipe welding.
2. Identify and explain open-root V-groove pipe weld techniques.
3. Perform open-root V-groove pipe welds using GTAW in the following positions:
   - 1G-ROTATED
   - 2G
   - 5G
   - 6G

Prerequisites

Before you begin this module, it is recommended that you successfully complete the following: Core Curriculum; Welding Levels One and Two; Welding Level Three, Modules 29301-03 through 29305-03.

Required Trainee Materials

1. Pencil and paper
2. Appropriate personal protective equipment

1.0.0 ◆ INTRODUCTION

Gas tungsten arc welding (GTAW) is an arc welding process that uses an arc between a tungsten electrode and the base metal to melt the base metal. The electrode, the arc, and the molten base metal are shielded from atmospheric contamination by a flow of inert gas from the torch nozzle. (See *Figure 1*.) The filler metal is a rod of similar composition to

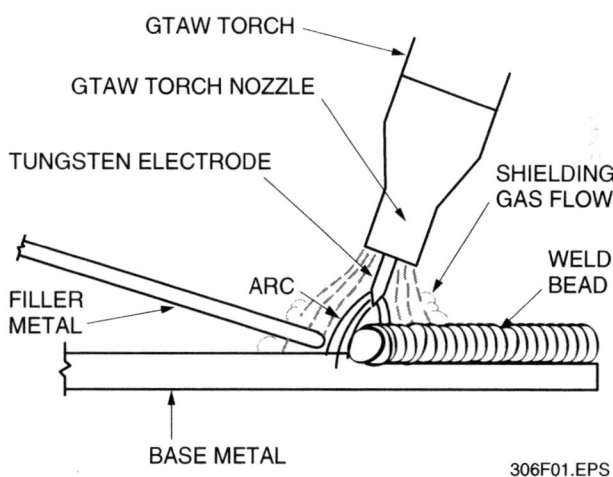

Figure 1 ◆ GTAW process.

the base metal and is usually hand-held and manually fed into the leading edge of the weld puddle. GTAW produces high-quality welds without slag or oxidation. Because there is no flux, there will be no corrosion due to slag entrapment; therefore, no postweld cleaning is necessary. An exception is the slag that may be left by some flux cored rods, which must be removed.

A major use of GTAW is to make manual high-quality welds in stainless steel piping used in both critical and non-critical applications. The GTAW process allows greater control of root penetration and fill than almost any other process. For this reason, GTAW is often used to make the root pass on pipe, even when the fill and cover passes are made with SMAW, GMAW, or FCAW, which all three have higher deposition rates.

This module explains how to set up GTAW equipment and perform open-root V-groove welds on stainless and low-alloy (or carbon) steel pipe with stainless steel and/or low-alloy filler metal in the 2G, 5G, and 6G welding positions. The dimensions and specifications in this module are designed to be representative of codes in general and are not specific to any certain code. Always follow the proper codes for your site.

2.0.0 ◆ SAFETY SUMMARY

The following is a summary of safety procedures and practices you must observe while cutting or welding. Keep in mind that this is just a summary. Complete safety coverage is provided in the Level One module *Welding Safety*. If you have not completed that module, do so before continuing. Above all, be sure to wear appropriate protective clothing and equipment when welding or cutting.

2.1.0 Protective Clothing and Equipment

- Always use safety goggles with a full face shield or a helmet. The goggles, face shield, or helmet lens must have the proper light-reducing tint for the type of welding or cutting to be performed. Never directly or indirectly view an electric arc without using a properly tinted lens.

- Wear proper protective leather and/or flame retardant clothing along with welding gloves to protect from flying sparks, molten metal, and heat.

- Wear 8" or taller high-top safety shoes or boots. Make sure that the tongue and lace area of the footwear is covered by a pant leg. If the tongue and lace area is exposed or the footwear must be protected from burn marks, wear leather spats under the pants or chaps and over the front top of the footwear.

- Wear a solid material (non-mesh) hat with a bill pointing to the rear or, if much overhead cutting or welding is required, a full leather hood with a welding faceplate and the correct tinted lens. If a hard hat is required, use one that allows attachment of both rear deflector material and a face shield.

- If a full leather hood is not worn, wear a face shield and snug fitting welding goggles over safety glasses for gas welding or cutting. Either the face shield or the lenses of the welding goggles must be an approved shade 5 or 6 filter. For electric arc welding or cutting, wear safety goggles and a welding hood with the correct tinted lens (shade 9 to 14).

- If a full leather hood is not worn, wear earplugs to protect your ear canals from sparks.

2.2.0 Fire/Explosion Prevention

- Never carry matches or gas-filled lighters in your pockets. Sparks can cause the matches to ignite or the lighter to explode, causing serious injury.

- Never perform any type of heating, cutting, or welding until a hot-work permit is obtained and an approved fire watch is established. Most work-site fires caused by these types of operations are started by cutting torches.

- Never use oxygen to blow off clothing. The oxygen can remain trapped in the fabric for a time. If a spark hits the clothing during this period, the clothing can burn rapidly and violently out of control.

- Make sure that any flammable material in the work area is moved or is shielded by a fire-resistant covering. Approved fire extinguishers must be available before attempting any heating, welding, or cutting operations.

- Always comply with any site requirement for a hot-work permit and/or a fire watch.

- Never release a large amount of oxygen nor use oxygen as compressed air. Its presence around flammable materials or sparks can cause rapid and uncontrolled combustion. Keep oxygen away from oil, grease, and other petroleum products.

- Never release a large amount of fuel gas, especially acetylene. Methane and propane tend to concentrate in and along low areas and can ignite at a considerable distance from the release point. Acetylene is lighter than air but is even more dangerous than methane. When mixed with air or oxygen, it will explode at much lower concentrations than any other fuel.

- To prevent fires, maintain a neat and clean work area, and make sure that any metal scrap or slag is cold before disposal.

- Before cutting or welding containers such as tanks or barrels, check to see if they have contained any explosive, hazardous, or flammable materials, including petroleum products, citrus products, or chemicals that decompose into toxic fumes when heated. As a standard practice, always clean and then fill any tanks or barrels with water, or purge them with a flow of inert gas to displace any oxygen.

2.3.0 Work Area Ventilation

- Make sure to follow confined space procedures before conducting any welding or cutting in the confined space.

- Never use oxygen to ventilate confined spaces.
- Always perform cutting or welding operations in a well-ventilated area. Such operations involving zinc or cadmium materials or coatings result in toxic fumes. For long-term cutting or welding of these materials, always wear an approved, full-face, supplied-air respirator (SAR) that uses breathing air supplied from outside of the work area. For occasional, very short-term exposure, you may use a high-efficiency particulate arresting (HEPA)-rated or metal-fume filter on a standard respirator.
- Make sure confined spaces are ventilated properly for cutting or welding operations.

3.0.0 ◆ ROOT BACK SIDE PROTECTION

Welding processes that use inert gas to protect the molten puddle on the face of the weld from the atmosphere do not provide protection for the back side of the weld. Although back side protection is usually not required for carbon steel and some low alloys, it is essential for stainless and other low-alloy steels. This is because alloying elements are more susceptible to combining with the oxygen in the atmosphere to form oxides. The oxides are objectionable because they can decrease the corrosion resistance to an unacceptable level and can be pulled into the weld puddle during subsequent welding, causing defects. The most common methods of providing back side protection are the following:

- Backing gas (purge)
- Backup flux
- Consumable insert

3.1.0 Backing Gas

Backing gas provides the best and highest-quality protection for welding open-root joints. Argon provides better protection than nitrogen, but nitrogen typically costs one-fourth to one-third less. Because it costs less, nitrogen is used where it will not adversely affect the weld. However, you must use the gas specified by the WPS. To be effective, the backing gas has to do the following:

- Replace the atmospheric gases to an acceptable level.
- Be contained to provide protection where needed.
- Have a continuous flow to compensate for gas lost during welding.

For short and small-diameter piping, you can plug or cap ends and any openings to provide containment and to reduce the gas amount and expense. This method of containment is not cost-effective for large piping or long sections. To decrease the volume of piping requiring backing gas replacement, plug the piping near the weld, on both sides, with removable devices called dams. Dams may be inflatable bladders, inflated plastic bags, or foam rubber, or they may be water-soluble dams, which are typically made of water-soluble paper. (See *Figure 2*.) When you use dams, they must be removed after welding is completed. When argon is required as the backing gas, another procedure to reduce the cost is to purge with nitrogen before purging with the argon.

When setting up backing gas, establish the containment, and then seal the weld joint. This is typically done with masking tape. Introduce the backing gas through a tube or hose that passes through or past one of the containment devices or through a nozzle (inlet) that directs the gas through the root opening of the weld. Argon backing gas should always enter at the bottom, and because the gas is typically heavier than the air it is displacing, there should be a vent located near the top as an exit for the atmospheric gases. If the vent hole is in the bottom, the gas will vent out without displacing the atmospheric air in the top section of the pipe. However, if nitrogen is used as the backing gas, the vent should be at the bottom, because nitrogen is lighter than air and will rise to the top.

Check the WPS to determine the backing gas flow rate range. This range is for maintaining the backing gas atmosphere during welding after the pipe has been purged. Higher flow rates can be

Maintaining Proper Backing Gas Flow

As the weld progresses around a pipe, you may have to reduce the backing gas flow to the minimum to prevent blowing the molten weld pool out of the joint and causing holes in the root pass. The backing gas flow should be maintained until at least the next passes are placed over the root pass and the root pass is completely covered from view.

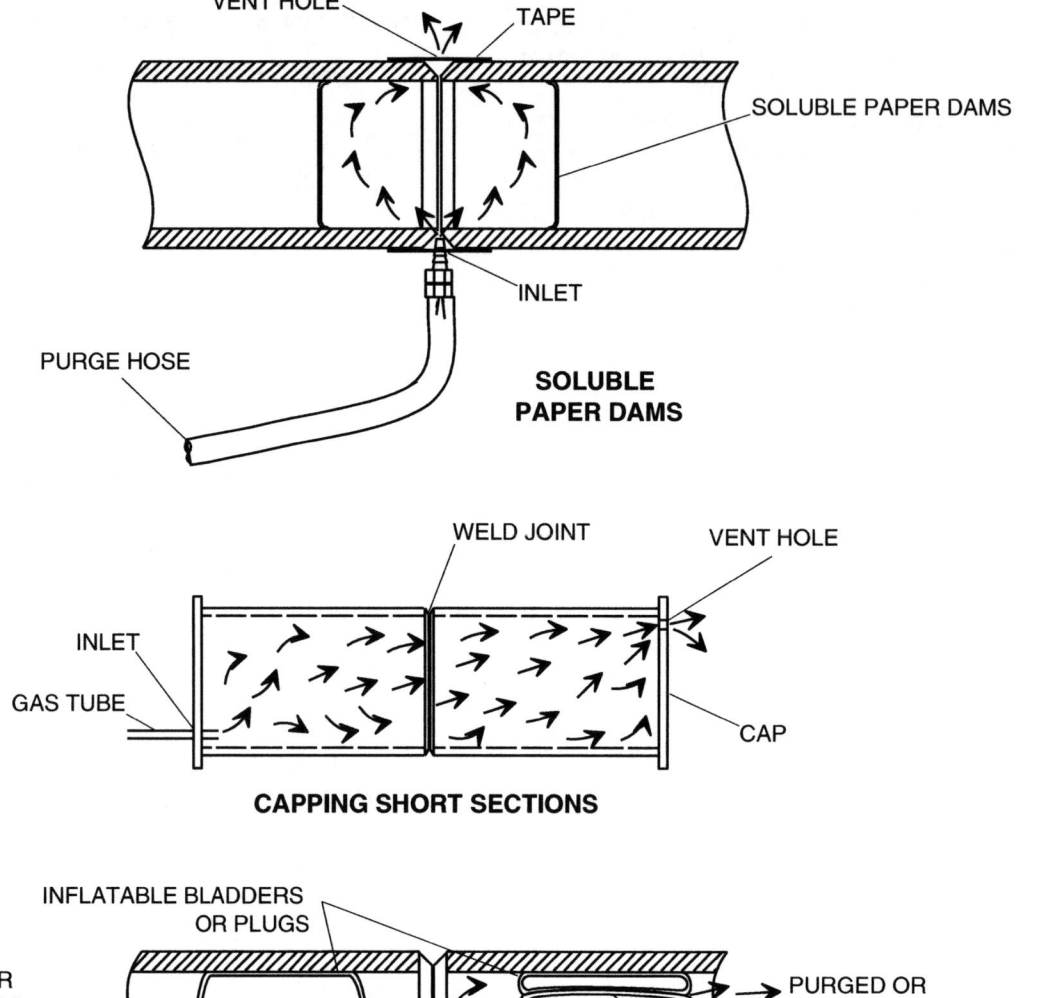

VENT HOLE
TAPE
SOLUBLE PAPER DAMS
INLET
PURGE HOSE

**SOLUBLE
PAPER DAMS**

WELD JOINT
VENT HOLE
INLET
GAS TUBE
CAP

CAPPING SHORT SECTIONS

INFLATABLE BLADDERS
OR PLUGS
PURGING OR
BACKING GAS
PURGED OR
VENTED ATMOSPHERE
GAS TUBE OR HOSE
VENT IN PLUG
OPENINGS IN
GAS TUBES OR INLET

PLUGS FOR LONG SECTIONS

306F02.EPS

Figure 2 ◆ Typical pipe-purging methods.

used for the initial purge to reduce the time it takes to purge the pipe at the onset. An oxygen analyzer is often used to verify that the atmosphere inside the pipe is correct for the welding being performed. When the oxygen analyzer is used, 1% maximum oxygen is the typical criterion for an acceptable welding atmosphere.

3.2.0 Backup Flux

Using backup flux is another method of providing root protection for open-root welds when weld quality requirements are less critical. Special flux

cored rods are available to provide root back side protection during the GTAW process, but typically, you will use a dry powder flux that you mix with alcohol (methanol) or acetone to a consistency of thick cream. After the mixture has set for several minutes to allow the alcohol or acetone to react with the flux, you brush an even coat of the mixture onto the back side of the joint. The mixture dries rapidly and leaves a coating of flux. During welding, the flux powder melts, forming a barrier that prevents air from coming in contact with the molten metal. To remove the flux after welding, use a wire brush and hot water.

WARNING!

Backup flux may contain silica, fluorides, or other toxic materials. Weld only in well-ventilated spaces. Failure to provide proper ventilation could result in personal injury or death.

CAUTION

Backup fluxes are formulated for welding specific alloy types. Check the manufacturer's recommendations to be sure you use the correct flux for the alloy being welded. Using the wrong flux could result in weld defects.

3.3.0 Consumable Inserts

Consumable inserts are preplaced filler metal rings that completely fuse into the root of the joint and become part of the weld. Use consumable inserts when welding critical piping systems that cannot tolerate internal obstructions and also when welding pipe, tube, and other components that require uniformity and a smooth root bead. The insert becomes an integral part of the initial root-weld bead and ensures complete fusion and smooth blending of the insert and adjacent pipe, tube, or fitting ends. Inserts reduce flow restriction within the bore to a minimum and help eliminate root-bead cracking. Backing gas is used with inserts to prevent oxidation of the weld root when welding stainless or low-alloy steel pipe.

4.0.0 ◆ WELDING PREPARATION

Before welding can begin, the area has to be readied, the welding equipment must be set up, and the metal to be welded must be prepared. The following sections explain how to set up the equipment for welding.

To practice welding, you need a welding table, bench, or stand. The welding surface must be steel, and provisions must be available for placing weld coupons out of position. (See *Figure 3*.)

To set up the area for welding, follow these steps:

Step 1 Check to be sure the area is properly ventilated. Make use of doors, windows, and fans.

Step 2 Check the area for fire hazards. Remove any flammable materials before proceeding.

Step 3 Locate the nearest fire extinguisher. Do not proceed unless the extinguisher is charged and you know how to use it.

306F03.EPS

Figure 3 ◆ Welding station.

Step 4 Position a welding table near the welding machine.

Step 5 Set up flash shields around the welding area.

4.1.0 Practice Pipe Weld Coupons

Pipe weld coupons should be cut from 3" to 12" diameter Schedule 40 or Schedule 80 stainless, low-alloy, or carbon steel pipe. Each welded joint will require two coupons of the same size and schedule pipe.

NOTE

Most codes allow you to substitute carbon steel pipe for stainless or low-alloy steel pipe as base metal for welder qualification. However, you must use applicable stainless or low-alloy steel filler metals. Always refer to the WPS or site quality standards for the proper qualification information.

Figure 4 shows typical ASME International, American Petroleum Institute (API), and AWS pipe bevel specifications for bevel angles, root faces, and root openings.

To prepare carbon steel practice weld coupons for open-root V-groove weld joints, follow these steps:

Step 1 Clean all rust or mill scale from the inside and outside of the carbon steel pipe with a grinder or wire brush. Clean for a distance of ½" from the weld joint.

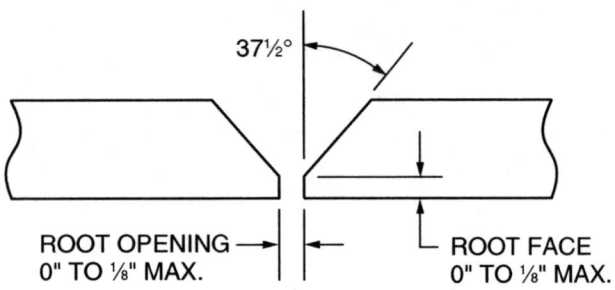

ROOT OPENING
0" TO ⅛" MAX.

ROOT FACE
0" TO ⅛" MAX.

TYPICAL ASME AND AWS SPECIFICATION

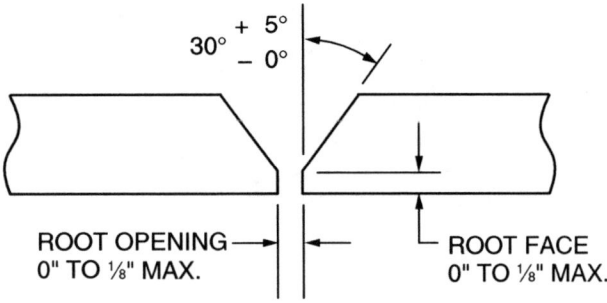

$30° \begin{array}{c} + 5° \\ - 0° \end{array}$

ROOT OPENING
0" TO ⅛" MAX.

ROOT FACE
0" TO ⅛" MAX.

TYPICAL API SPECIFICATION

306F04.EPS

Figure 4 ◆ Pipe bevel specifications.

NOTE

If you use stainless steel or low-alloy pipe and a grinding wheel, make sure the grinding wheel is approved for use on stainless or low-alloy steel. Ensure that it has not been used on any other metal or for removing hydrocarbon contaminants.

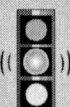

CAUTION

There is little or no surface cleaning action when using a GTAW torch for stainless or low-allow steel welding; therefore, pipe surfaces must be thoroughly cleaned to prevent weld discontinuities.

Step 2 Bevel the end of the pipe to 30° or 37½° by any acceptable beveling method, such as by using a mechanical cutter or by thermal cutting or grinding.

Step 3 Cut off a section of the beveled pipe end (1½" minimum).

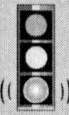

NOTE

For 1G-ROTATED welding, you may have to cut longer coupons to fit the rollers, if used.

Step 4 Check the bevel. There should be no dross, and the bevel angle should be 30° or 37½° with no notches more than ¹⁄₁₆" deep.

Step 5 Grind or file a 0" to ⅛" root face on the bevel as specified by your instructor.

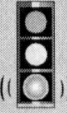

NOTE

Welding codes allow both the root opening and the root face on open root welds to be from 0" to ⅛" wide. Select and adjust the opening and root face as needed when you start the welding practices.

Conserve Materials

Pipe for practice welding is expensive and difficult to obtain. Make every effort to conserve and not waste available material. Completely use weld coupons until all surfaces have been welded by cutting the coupon apart and reusing the pieces. Check with the instructor for the appropriate size coupon.

Welding Large-Diameter Pipe

The practice coupons on large-diameter pipe must be greater than 1½" in length. They need more metal to help absorb the increased heat used to make these welds.

Step 6 Provide for root backing using backup flux or gas backing.

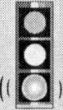
Step 7 Align the two pipe sections so that the inside diameter (ID) surfaces are even all around. Align small-diameter pipe by clamping both pieces to a piece of angle iron. Align large-diameter pipe with the aid of a pipe alignment jig or by holding a straightedge across the joint, parallel to the pipe axis. The straightedge must be used all around the pipe in case one or both sections are distorted.

Step 8 Gap the root opening at 0" to ⅛" with pieces of filler wire, metal shims, or pieces of bare welding electrode of the correct diameter, as directed by your instructor.

Step 9 When the root opening is correct and the pipe ends are aligned, make the first of four tack welds of no greater than 1".

Step 10 After the first tack weld, on the opposite side, check the root opening, adjust the gap if necessary, and make the second tack weld.

Step 11 Check the root opening again and weld the third tack midway between the first two tacks.

Step 12 Weld the fourth tack opposite the third tack and midway between the first and second tacks. There should now be four tack welds evenly spaced (90°) around the pipe coupon. This is illustrated by *Figure 5*.

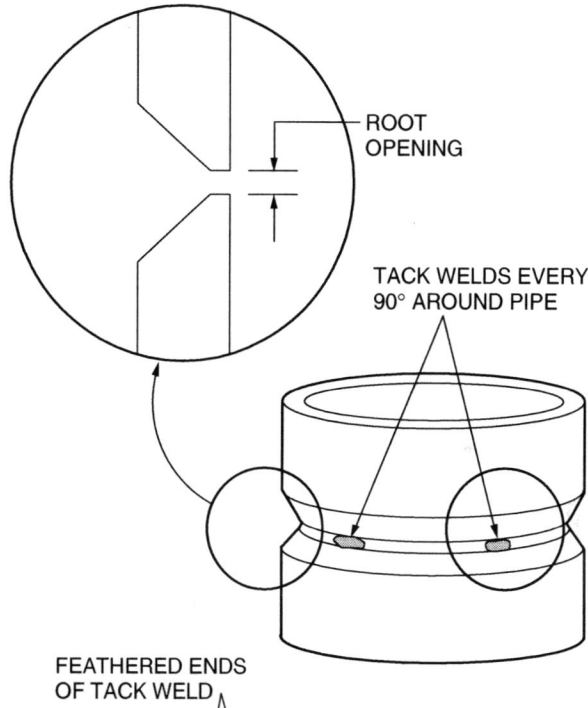

306F05.EPS

Figure 5 ◆ Tacked open-root V-groove weld coupon.

Step 13 Feather the tack welds with the edge of a grinding wheel. Feathering the ends of the tack welds with a grinder helps fuse the tack welds into the root pass.

4.2.0 The Welding Machine

Identify a proper constant-current welding machine for GTAW and follow these steps to set it up for use.

Step 1 Verify that the welding machine can be used for GTAW with or without internally controlled gas shielding.

Step 2 Identify an air-cooled or water-cooled GTAW torch. Make sure it is compatible with the welding machine and any cooling unit.

Step 3 Check to be sure that the welding machine is properly grounded through the primary current receptacle.

Step 4 Verify the location of the primary disconnect.

Step 5 Configure the welding machine for GTAW parameters as directed by your instructor (*Figure 6*). Configure the torch polarity and equip the torch with the correct diameter and type of tungsten (³⁄₃₂" or ⅛" EWTh-2 or EWTh-1) for the filler metal used for the application.

Step 6 Connect the proper shielding gas for the application as described in the previous level and as specified by the filler metal manufacturer, WPS, site quality standards, or your instructor.

Step 7 Connect the clamp of the workpiece lead to the workpiece.

Step 8 Turn on the welding machine, and purge the torch as directed by the manufacturer's instructions.

4.3.0 Filler Metals

Filler metals are selected to be compatible with the base metal to be welded. The WPS or site standards specify the filler metal type and size to use.

For the welding exercises in this module, you should use ³⁄₃₂" or ⅛" stainless and/or low-alloy steel filler metals, as specified by your instructor, to weld the pipe coupons. Remove only a small number of filler metals at a time. Keep the remainder in the package to keep them clean. Before using the filler metal, check it for burned ends or contamination such as corrosion, dirt, oil, or grease, which can all cause weld discontinuities. Clean the filler metal with a clean, oil-free rag and chemcal cleaner, and snip any burned ends. If the filler metal cannot be cleaned, do not use it.

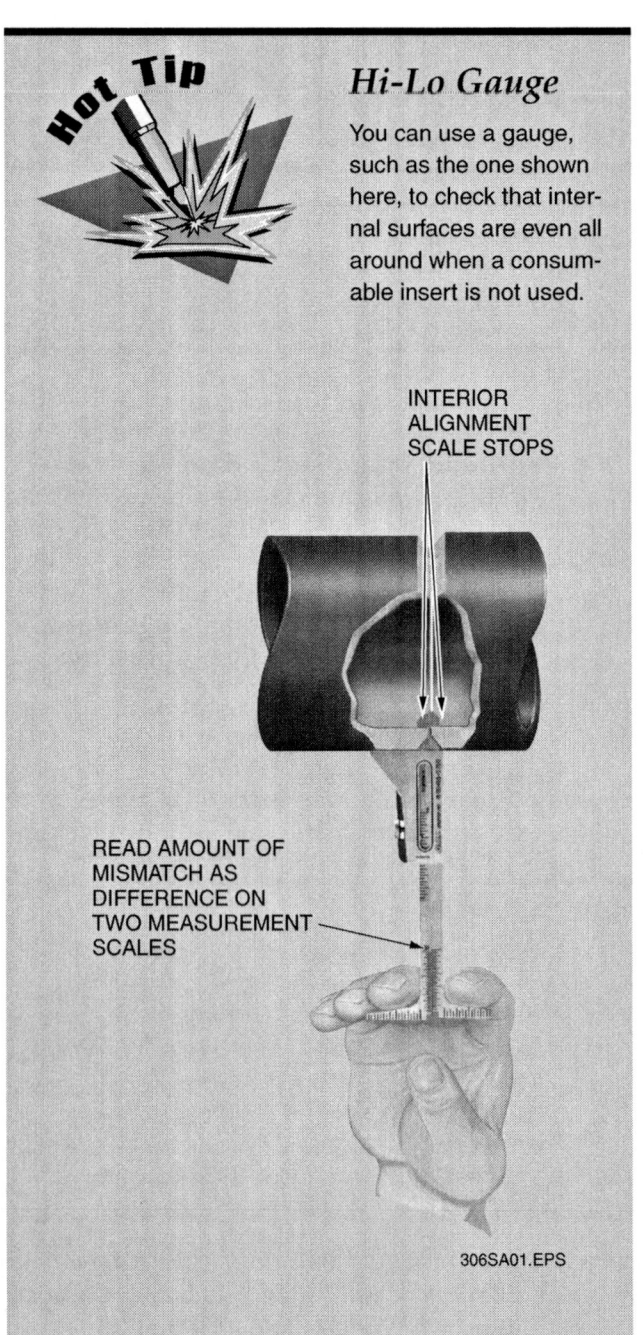

Hot Tip

Hi-Lo Gauge

You can use a gauge, such as the one shown here, to check that internal surfaces are even all around when a consumable insert is not used.

INTERIOR ALIGNMENT SCALE STOPS

READ AMOUNT OF MISMATCH AS DIFFERENCE ON TWO MEASUREMENT SCALES

306SA01.EPS

CAUTION

When the WPS or site quality standards specify a filler metal, it must be used to prevent defective welds.

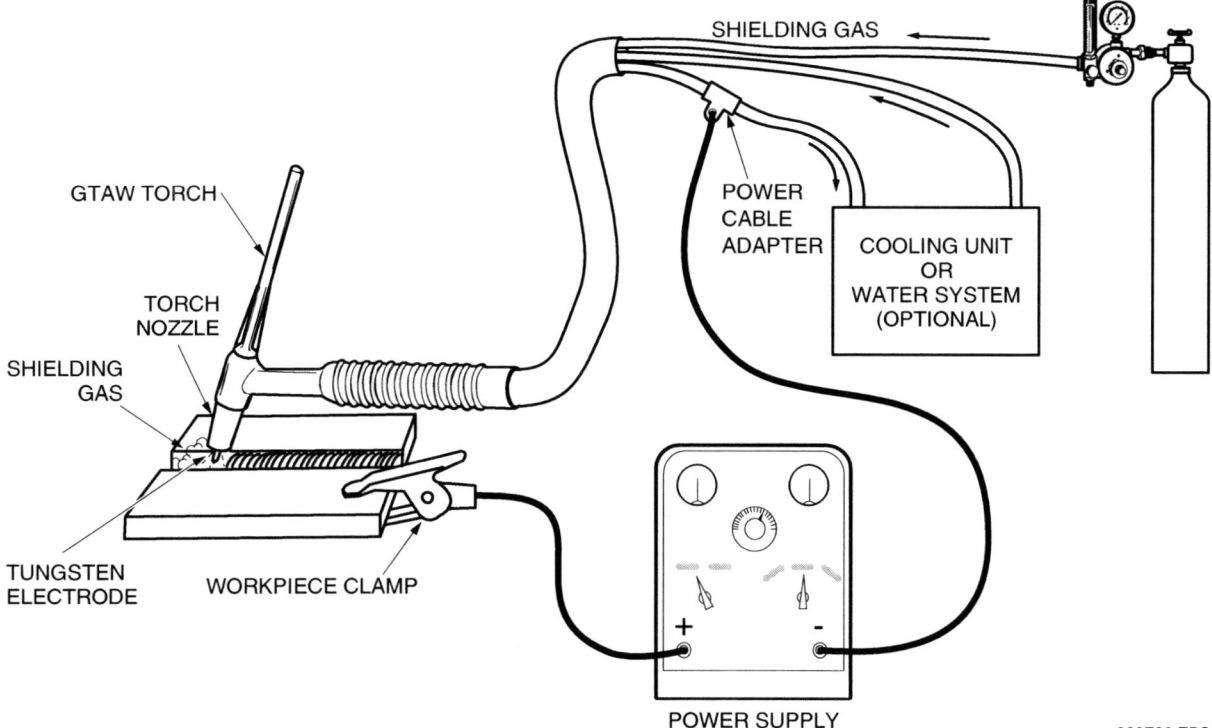

SHIELDING GAS

GTAW TORCH

TORCH
NOZZLE

SHIELDING
GAS

TUNGSTEN
ELECTRODE

WORKPIECE CLAMP

POWER
CABLE
ADAPTER

COOLING UNIT
OR
WATER SYSTEM
(OPTIONAL)

+   −

POWER SUPPLY

306F06.EPS

Figure 6 ◆ Configuration diagram of typical GTAW machine.

5.0.0 ◆ GAS TUNGSTEN ARC WELDING TECHNIQUES

GTAW weld bead characteristics and quality are affected by several factors, each of which is influenced by the way you handle the torch. These factors include the following:

- Torch travel speed and arc length
- Torch angles
- Torch and filler metal handling techniques

5.1.0 Torch Travel Speed and Arc Length

Torch travel speed and/or arc length may affect the GTAW weld puddle and weld penetration. A slow travel speed allows more heat to concentrate and forms a larger, more deeply penetrating puddle. Faster travel speeds prevent heat buildup and form smaller and shallower puddles. Torch arc length is the major control for bead width. As the torch is raised, arc length, voltage, and bead width increase.

5.2.0 Torch Angles

There are two basic torch angles you must control when performing GTAW: the work angle and the travel angle. The definition of these angles is the same as for all other methods of arc welding.

5.2.1 Work Angle

The torch work angle (*Figure 7*) is defined as a less-than-90° angle between a line perpendicular to the major workpiece surface at the point of electrode contact and a plane determined by the electrode axis and the weld axis. For a T-joint or corner joint, the line is perpendicular to the nonbutting member. For pipe, the plane is determined by the electrode axis and a line tangent to the pipe surface at the same point.

5.2.2 Travel Angle

The torch travel angle (*Figure 8*) is defined as an less-than-90° angle between the electrode axis and

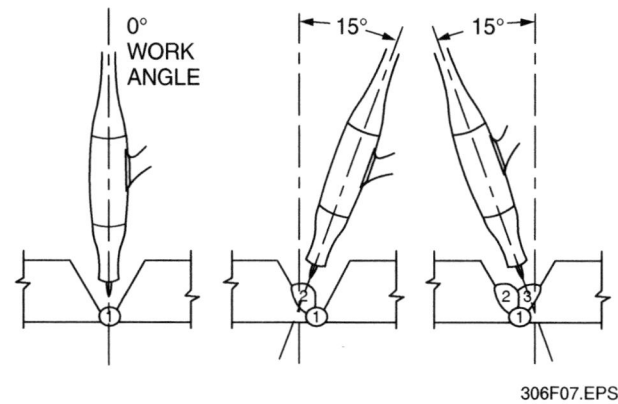

0°
WORK
ANGLE

15°   15°

306F07.EPS

Figure 7 ◆ Typical torch work angles.

a line perpendicular to the weld axis at the point of electrode contact in a plane determined by the electrode axis and the weld axis. For pipe, the plane is determined by the electrode axis and a line tangent to the pipe surface at the same point. A push angle is used for GTAW and is created when the torch is tilted back so that the electrode tip precedes the torch in the direction of the weld. In this position, the electrode tip and shielding gas are directed ahead of the weld bead. Push angles of 15° to 20° are normally used for GTAW.

5.3.0 Torch and Filler Metal Handling Techniques

The two basic handling techniques used to perform GTAW are known as freehand and walking-the-cup. Try both techniques and use the one that gives the best results. Both techniques are explained in the following sections.

5.3.1 Freehand Technique

In the freehand technique (*Figure 9*), hold the torch electrode tip just above the weld puddle or base metal. Support the torch with your hand. Steady

your hand by resting some part of it on or against the base metal to maintain the proper arc length from the electrode tip to the puddle. If required, you can move the torch tip in a small circular motion within the molten puddle to maintain the puddle size and to advance the puddle. Add filler metal as needed.

Hold the GTAW filler metal in the hand not holding the torch. For all positions, hold the filler

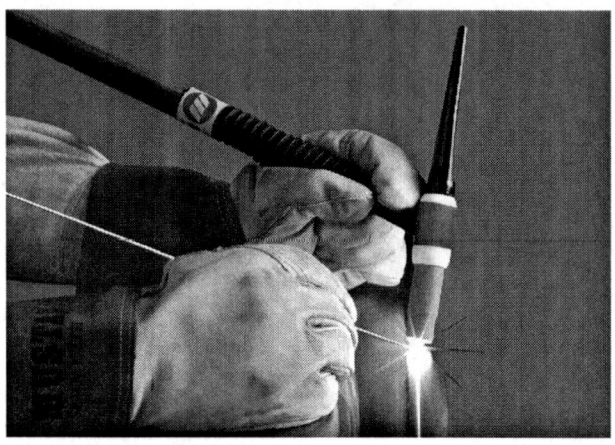

306F09.EPS

Figure 9 ◆ Freehand technique.

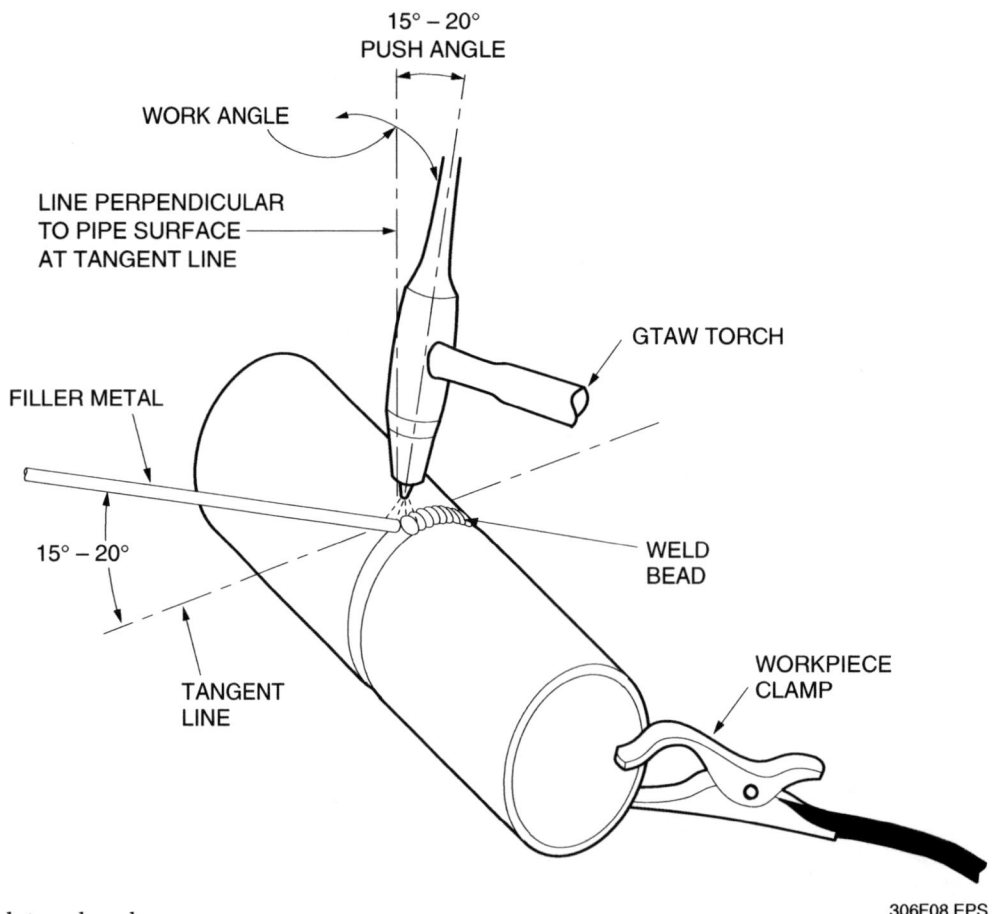

Figure 8 ◆ Torch travel angle.

306F08.EPS

The Use of a Y-Adapter

When using the same gas for backing as well as shielding, connect a Y-adapter with valves in each branch to the output of a flow regulator. Connect one branch to the backing gas device and use the other branch as the torch supply. With the branch valves turned off, open the cylinder valve slowly at first, and then open it all the way. At the torch, turn on the torch valve or purge control and, using the torch branch valve on the Y-adapter, adjust for the proper torch flow rate as indicated on the flow regulator. At the torch, turn off or deactivate the gas flow; then, open and adjust the backing gas branch valve on the Y-adapter for the proper backing gas flow as indicated on the flow regulator. Once the branch valves are adjusted for the proper flow rates, if welding must be interrupted for a time, use the cylinder valve to shut off the gas.

306SA02.EPS

rod at an angle of about 20° above the base metal surface and in line with the weld. Always keep the tip of the filler rod within the shielding gas envelope to protect the filler rod from atmospheric contamination and to keep it preheated.

Dab the filler metal into the leading edge of the weld puddle, using extreme care not to touch the tungsten electrode with the end of the filler metal. If the tungsten electrode touches the filler metal or weld puddle, it will become contaminated with filler metal. The electrode must then be removed and cleaned by grinding (or chemical cleaning) before proceeding. A technique for preventing contamination of the electrode is to move the electrode to the back edge of the weld puddle as the filler metal is dabbed into the leading edge of the weld puddle.

 **CAUTION**

Do not insert the end of the filler metal rod into the molten puddle under the electrode and then attempt to melt it off. This can cause hard spots and weld defects.

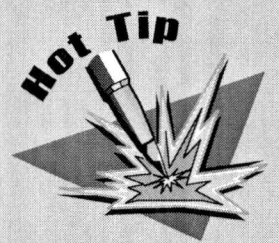
Grinding Wheels

Some WPS or site quality standards may prohibit grinding stainless steel welds. Always check before using a grinding wheel. When using a grinding wheel, ensure that it is a type approved for use on stainless steel. Also, ensure that it has not been used to grind a hydrocarbon-contaminated surface or any other metals.

Excessive Push Angle

A push angle that's too large will tend to draw air from under the back edge of the torch nozzle where it will mix with the shielding gas stream and contaminate the weld.

5.3.2 Walking-the-Cup Technique

In the walking-the-cup technique (*Figure 10*), rest the edge of the torch nozzle (cup) against the base metal or groove edges to steady the torch and to maintain a constant arc length. Rock the torch from side to side on the edge of the cup as you advance it, to maintain the puddle size and to heat both sides of the groove. Add the filler metal in the same manner as with the freehand technique, using care not to contaminate the electrode.

5.4.0 Welding Stainless Steel

Welding stainless steel is slightly more difficult than welding carbon steel. When compared to carbon steel, stainless steel has the following characteristics:

- Conducts heat more slowly
- Expands more when heated
- Has a higher resistance to electrical current flow

These traits are similar for all stainless steels but vary greatly with the metallurgical structure. Those in the same metallurgical group (austenitic, martensitic, or ferritic) have similar welding characteristics.

NOTE

Information in this module is provided as only a general guideline. Refer to your WPS or site quality standards for specific requirements for a weld. Failure to follow WPS or site quality standards can result in quality issues.

5.4.1 Austenitic Stainless Steels

Austenitic stainless steels are generally easier to weld than the other types of stainless steels. It normally does not require preheating or postheating, but the interpass temperature must be controlled to prevent carbide precipitation.

Advanced Inverter Power Sources

An advanced inverter power source, like the one shown here, can make welding pipe with GTAW easier. With features such as HF start only, current pulsing, and weld current sequencing (stop/start ramping), these machines provide much more control over some of the variables that affect the quality of a weld.

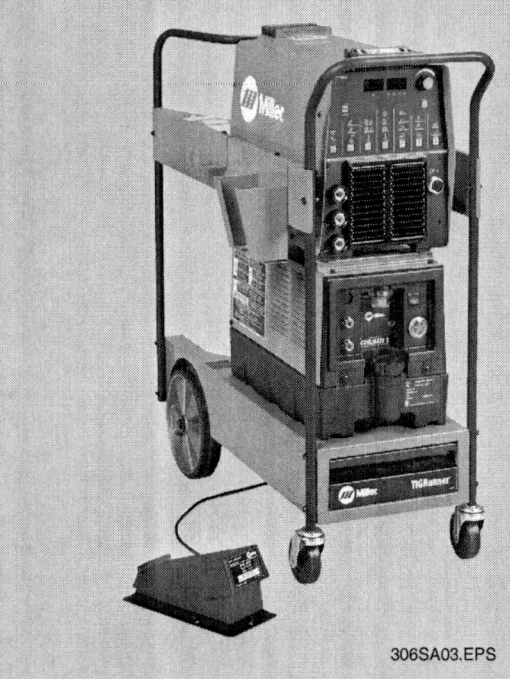

306SA03.EPS

NOTE

Interpass temperature should not exceed 350°F.

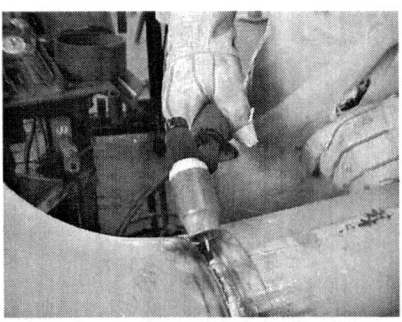

STEP 1

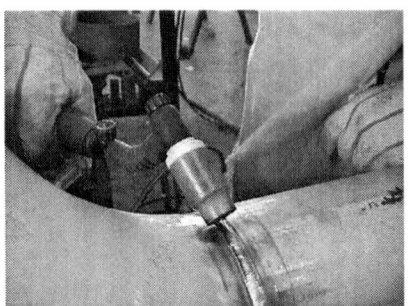

STEP 2

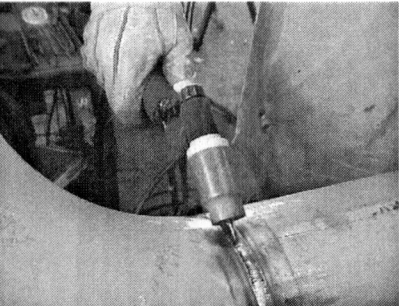

STEP 3

306F10.EPS

Figure 10 ◆ Walking-the-cup technique.

5.4.2 Martensitic Stainless Steels

Martensitic stainless steels air-harden in proportion to the chromium and carbon content. This makes the weld and weld zone very hard and brittle unless the piece is heat-treated.

Martensitic heating requirements are based on carbon content. Approximate heating requirements are as follows:

- *0–0.1% carbon* – No preheat or postheat required.
- *0.1–0.2% carbon* – Preheat to 500°F and cool slowly.
- *0.2–0.5% carbon* – Preheat to 500°F, and then anneal after welding (anneal to 1,500°F– 1,600°F, furnace-cool to 1,100°F, and then air-cool).

If preheating and postheating cannot be used, weld martensitic stainless steels with a ductile filler metal, such as ER309 or ER310, if allowed by the procedure. The heat-affected zone will still be hard and brittle, but the filler metal will be ductile and able to absorb some stresses to protect the heat-affected zones.

5.4.3 Ferritic Stainless Steels

Ferritic stainless steels have the poorest weldability of the stainless types. The types with higher chromium and carbon are susceptible to carbide precipitation and intergranular corrosion, and they may require annealing after welding to redissolve the carbides and restore corrosion resistance.

Welding ferritic stainless steels, such as 405, 430, 430F, 430FSe, and 446, causes embrittlement because of severe grain growth. Weld them with one of the recommended electrode rods described in a previous level.

When heat treatment is not an option, weld ferritic stainless steel with ER308, ER309, or ER310 electrodes. The ductile filler metal is better at resisting cracking.

5.5.0 Welding Low-Alloy Steel

Alloy steel is more difficult to weld than carbon steel. The molten weld puddle tends to be more fluid, so it is slightly more difficult to control. Problems will also occur if you use the wrong filler metal or if you do not closely follow the proper preheat, interpass temperature control, or postheat recommendations. When compared to carbon steel, alloy steel has the following characteristics:

- Is more likely to have hydrogen-induced underbead cracking
- Requires a closer match of filler metal and base metal
- Has higher hardenability (causes cracks during cooling)
- Requires the heating cycle to be more closely managed (preheat, interpass temperature, postheat, and welding heat input)
- Requires greater care during joint fit-up to ensure a more uniform root opening and root face

These characteristics are similar for all alloy steels but vary greatly with the metallurgical structure. Those in the same group (HSLA, Q&T, HTLA, or Cr-Mo) have similar welding characteristics.

5.5.1 Controlling the Heat Cycle

Hardness caused by either improper preheat, interpass temperature control, or postheat (when required) is one of the major causes of cracking during low-alloy steel welding. As the alloy content or carbon content of the alloy steel increases, so does the steel's hardenability. Therefore, you must carefully control the heating cycle with every increase in the alloy and/or carbon content.

Because controlling the heating cycle is so important in welding alloy steel, heat treatment specialists are often brought in to work directly with the welder. In addition, instruments are often connected to the weldment to provide a continuous record of the time and temperature of the weldment during the entire heating (welding) cycle.

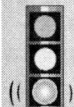

NOTE

If possible, interpass temperature should not exceed 350°F.

Weld Pool Size

When using GTAW in horizontal, vertical, and overhead positions, you must keep the weld pool smaller than when in the flat position. The weld pool is generally held in the joint by capillary action. If the pool is too large, gravitational force overcomes the capillary action and the liquid metal runs out of the joint.

5.5.2 Protecting the Root

Low-alloy steel has a higher alloy content than carbon steel, so back side protection may be required during tacking and welding. To determine if back side protection is required, check the WPS or site quality standards.

Because low-alloy steel piping is susceptible to weld discontinuities, welding it requires a more refined technique. You must use greater control over arc length and electrode manipulation. When running a pass, do not oscillate the tungsten wider than the filler metal rod. The filler metal is typically of a much higher quality (purer) than the base metal. Staying on the rod minimizes the melting of the base metal, which has more impurities than the filler metal. If the puddle overheats and falls to the inside during welding, stop and file or grind the area to remove any carburized metal impurities. If a gas root backing is not being used, the metal is more likely to carburize at a fall-through.

6.0.0 ◆ OPEN-ROOT V-GROOVE PIPE WELDS

The open-root V-groove weld is the most common groove weld normally made on carbon, low-alloy, and stainless steel pipe. These welds are typically used for joining medium- and thick-walled pipe used in critical piping systems. Critical piping systems (including pipe, fittings, and welded joints) contain or carry material with the potential to cause long- or short-term catastrophic danger and injury to personnel and/or the environment should the piping system fail to contain or carry the material as designed. Welds in critical piping must meet the most stringent code requirements. Non-critical piping is low-pressure piping used for heating and air conditioning, simple water systems, and other service installations. Less-stringent code requirements are used to evaluate non-critical piping welds.

6.1.0 Root Pass

If a consumable insert or backing is not used, the most difficult part of making an open-root V-groove weld is the root pass. The root pass is made from the V-groove side of the joint and must have complete penetration but not an excessive amount of root reinforcement. The penetration is controlled with a technique similar to the on-the-wire method, which allows continuous feeding of the wire into the molten weld pool. The filler metal rod used is smaller than the root opening and is held between the root faces. The

filler metal is melted as you pass the arc over it, which keeps the pool width to a minimum and achieves complete penetration.

When welding a consumable insert, hold the torch perpendicular to the pipe surface and point it radially toward the center of the pipe. Use an arc length of about ⅛" to melt the insert. Forward progression is governed by the melting rate of the insert and the characteristics of the weld pool. Increased fluidity and a rising weld pool indicate sufficient melting.

6.2.0 Pipe Groove Weld Test Positions

Groove welds may be made in all positions on pipe. The weld position is determined by the axis of the pipe. Four standard weld test positions are used with pipe and are shown in *Figure 11.*

To destructively test 5G or 6G welds, the test specimens are cut from the four regions illustrated in *Figure 12*: midway between the 12-o'clock and 3-o'clock, 3-o'clock and 6-o'clock, 6-o'clock and 9-o'clock, and 9-o'clock and 12-o'clock positions. (The 12-o'clock position is the top of the coupon groove; 6-o'clock is the bottom.)

6.3.0 Acceptable and Unacceptable Pipe Weld Profiles

Pipe groove welds without backing should be made with slight reinforcement (not exceeding ⅛") and a gradual transition to the base metal at each toe. The root pass should have complete penetration. The root reinforcement on the inside of the ⅛". Pipe groove welds must not have excessive reinforcement or underfill at the face or root, excessive undercut, or overlap. Excessively large cover passes will reduce the pipe's strength. They cause the stresses in the pipe to be concentrated along the sides of the weld and will not permit the pipe to expand and contract in a uniform manner along its length. If a weld has any of these defects, as shown in *Figure 13*, it must be repaired.

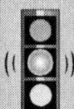

CAUTION

When welding pipe, do not make arc strikes outside the weld groove on the surface of the pipe. An arc strike can cause a hardened spot that can crack as the pipe expands and contracts. An arc strike on the pipe surface is considered a defect and will require repair or rework.

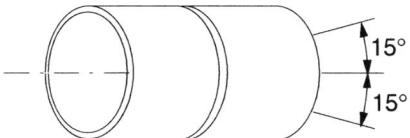

ROTATE PIPE AND DEPOSIT
WELD AT OR NEAR THE TOP

15°
15°

PIPE HORIZONTAL (±15°) AND ROLLED
TO KEEP WELD FLAT

1G – ROTATED POSITION

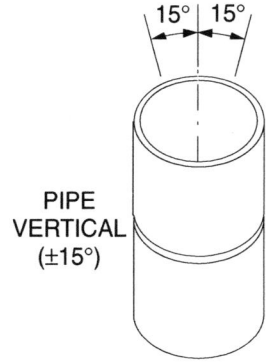

15° | 15°

PIPE
VERTICAL
(±15°)

PIPE NOT ROTATED DURING WELDING

2G POSITION

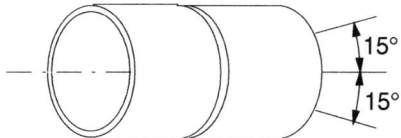

PIPE NOT ROTATED
DURING WELDING

15°
15°

PIPE HORIZONTAL (±15°)

5G POSITION

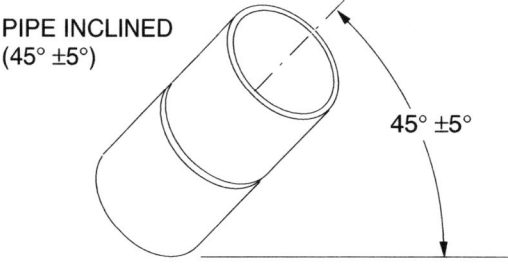

PIPE INCLINED
(45° ±5°)

45° ±5°

PIPE NOT ROTATED DURING WELDING

6G POSITION

306F11.EPS

Figure 11 ◆ Four basic pipe groove weld test positions.

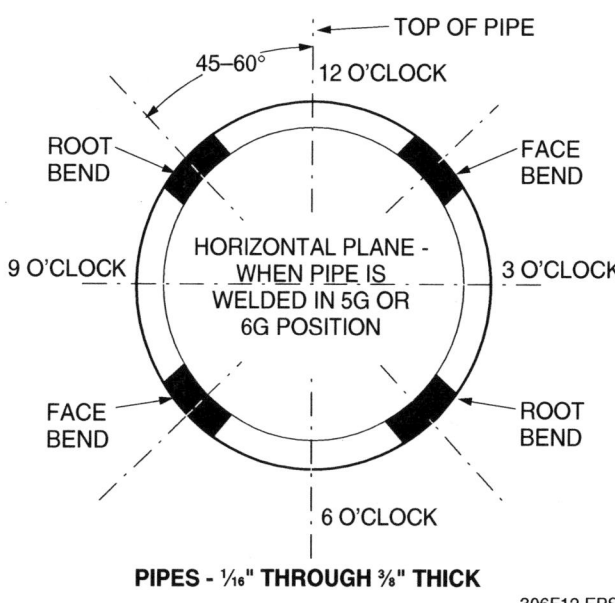

TOP OF PIPE

45–60° 12 O'CLOCK

ROOT
BEND

FACE
BEND

9 O'CLOCK

HORIZONTAL PLANE -
WHEN PIPE IS
WELDED IN 5G OR
6G POSITION

3 O'CLOCK

FACE
BEND

ROOT
BEND

6 O'CLOCK

PIPES - ¹⁄₁₆" THROUGH ⅜" THICK

306F12.EPS

Figure 12 ◆ Test specimen regions of 5G or 6G position pipe.

7.0.0 ◆ PRACTICING OPEN-ROOT V-GROOVE WELDS

This section of the module explains how to perform the following open-root V-groove weld positions:

- Horizontal (2G)
- Multiple (5G)
- Inclined multiple (6G)

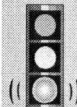

NOTE

The following practice procedures assume that a consumable insert is not being used for the root pass.

7.1.0 Horizontal (2G) Position

Practice 2G position open-root V-groove pipe welds using a shielding gas and a solid filler rod with a diameter specified by your instructor. The torch should be at a 15° to 20° push angle for the weld passes. (See *Figure 14*.) To prevent the weld puddle from sagging, you can drop the torch work angle slightly but not more than 10°. For the root pass, use a circular motion to form the weld puddle. Pay particular attention to tie into the tack welds.

When running the fill/cover passes, use stringer beads and both the walking-the-cup and freehand torch handling techniques. Ensure tie-in at the toes of the weld bead. Take particular care at the termination of the weld to fill the crater.

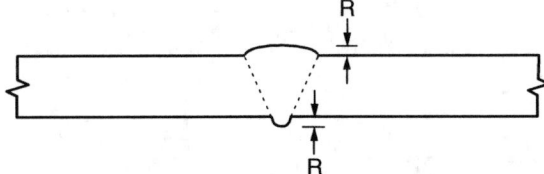

R = FACE AND ROOT REINFORCEMENT PER CODE
NOT TO EXCEED ⅛" MAX.

ACCEPTABLE WELD PROFILE

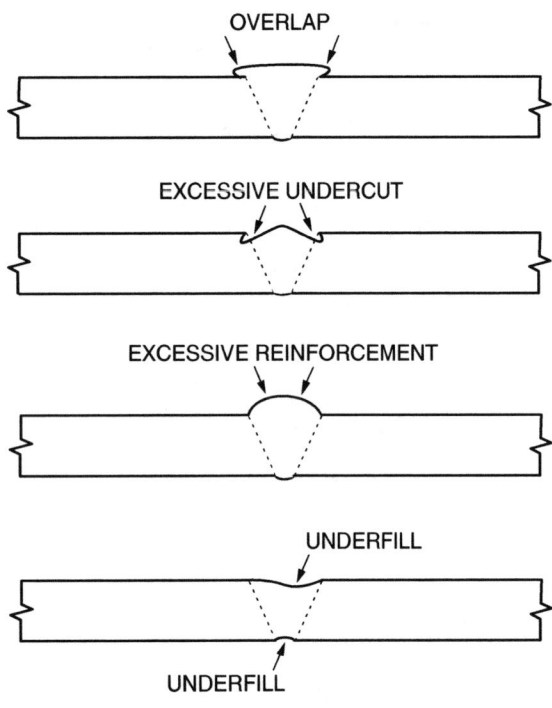

OVERLAP

EXCESSIVE UNDERCUT

EXCESSIVE REINFORCEMENT

UNDERFILL

UNDERFILL

UNACCEPTABLE WELD PROFILES

306F13.EPS

Figure 13 ◆ Acceptable and unacceptable pipe groove weld profiles.

Follow these steps to practice open-root V-groove pipe welds in the 2G position:

Step 1 Purge and tack weld together the practice pipe weld coupon as explained earlier.

Step 2 Clamp or tack weld the pipe coupon with the axis of the pipe vertical.

Step 3 Assemble the backing gas retaining devices to the pipe, and connect the backing gas hose. Set the backing gas flowmeter to the flow rate specified by the filler metal rod manufacturer, the WPS, or your instructor.

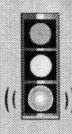

NOTE

If required by a WPS, site quality standard, or your instructor, observe any preheat/interpass temperatures or PWHT requirements.

Step 4 Run the root pass (pass 1) as shown in *Figure 14*. Continue this procedure until the root pass is completed.

Step 5 Brush the root pass to clean the weld.

Step 6 Run the remaining passes at the appropriate work angles. Overlap the passes, and clean the weld after each pass.

7.2.0 Multiple (5G) Position

Practice the multiple (5G) position open-root V-groove pipe welds using a shielding gas and a solid filler rod with a diameter specified by your instructor. Start at the bottom of the pipe and weld uphill toward the top. Use a torch push angle of 15° to 20° to the pipe surface (*Figure 15*).

The torch should be at approximately a 15° to 20° push angle for the root, fill, and cover passes. (See *Figure 16*.) When running the fill and cover passes, use stringer beads and both the walking-the-cup and freehand torch handling techniques. Ensure tie-in at the toes of the weld bead. Pay particular attention at the termination of the weld to fill the crater.

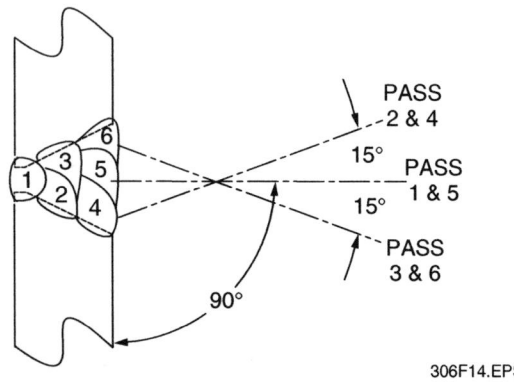

PASS 2 & 4

15°

PASS 1 & 5

15°

PASS 3 & 6

90°

306F14.EPS

Figure 14 ◆ Multiple-pass 2G bead sequence and work angles.

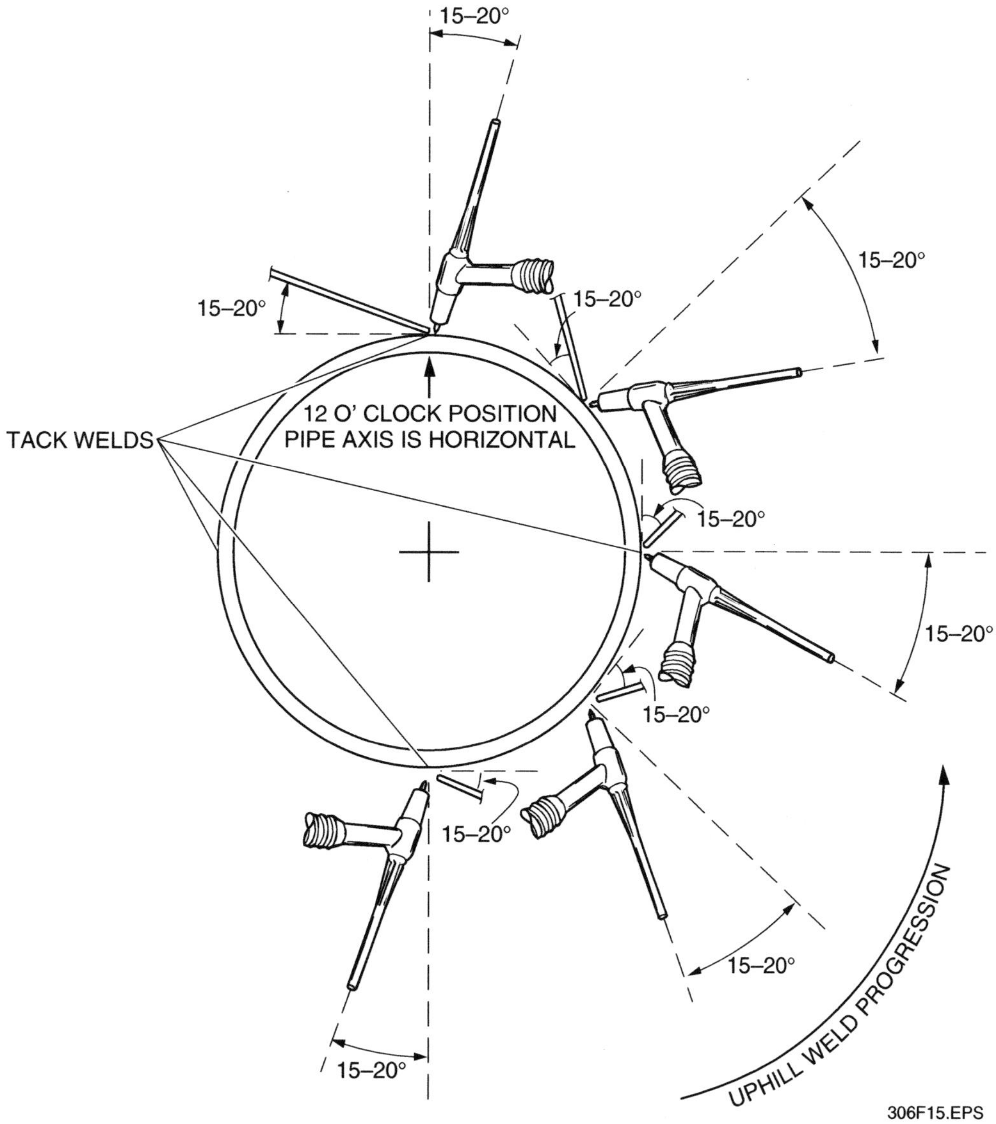

15–20°

15–20°

15–20°

TACK WELDS

12 O' CLOCK POSITION
PIPE AXIS IS HORIZONTAL

15–20°

15–20°

15–20°

15–20°

15–20°

UPHILL WELD PROGRESSION

15–20°

306F15.EPS

Figure 15 ◆ 5G pipe position torch and electrode rod angles.

Follow these steps to practice open-root V-groove pipe welds in the 5G position:

Step 1 Purge and tack weld together the practice pipe weld coupon as explained earlier.

Step 2 Clamp or tack weld the pipe weld coupon into position with the pipe axis horizontal.

Step 3 Assemble the backing gas retaining devices to the pipe, and connect the backing gas hose. Set the backing gas flowmeter

On-the-Wire Method

The on-the-wire method, used when welding open-root V-grooves for plate, can also be used for open-root V-groove welds of pipe. It is similar to using an insert. The filler rod diameter is selected to be the same size as the root opening, and the rod is bent in a circle to match the pipe diameter between the root faces. The rod can be manually held in this position to start, or it can be clamped between the two root faces. As the torch is advanced to melt the rod into the root faces, the position of the wire is adjusted slightly, as necessary, to keep it centered between the root faces. Like an insert, this method has the advantage of not having to use tape to prevent excessive backing gas flow out of the open-root V-groove.

to the flow rate specified by the filler metal rod manufacturer, the WPS, or your instructor.

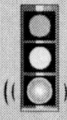

NOTE

If required by a WPS, a site quality standard, or your instructor, observe any preheat/interpass temperatures or PWHT requirements.

Step 4 Run the root pass uphill with a stringer bead as shown in *Figure 15*. Repeat for the opposite side of the pipe.

Step 5 Brush the root pass to clean the weld.

Step 6 Run the remaining passes uphill at the appropriate work angles. Overlap the passes, and clean the weld after each pass.

7.3.0 Inclined Multiple (6G) Position

Practice the inclined multiple (6G) (45°) position open-root V-groove pipe welds using a shielding gas and a solid filler rod with a diameter specified by your instructor. Start at the bottom of the pipe and weld uphill toward the top. The torch should be at approximately a 15° to 20° push angle and at 90° (0° work angle) to the pipe for the root, fill, and cover passes. (See *Figure 17*.)

When running the fill and cover passes, use stringer beads and both the walking-the-cup and freehand torch handling techniques. Ensure tie-in at the toes of the weld bead. Pay particular attention at the termination of the weld to fill the crater.

Follow these steps to practice open-root V-groove pipe welds in the 6G position:

NOTE

If required for training purposes, a restricting ring may be added to the 6G position coupon to form a 6GR position coupon.

Test Position

Welding pipe in the 6G position is a common welding test. A ring may be added to test for restricted accessibility (6GR).

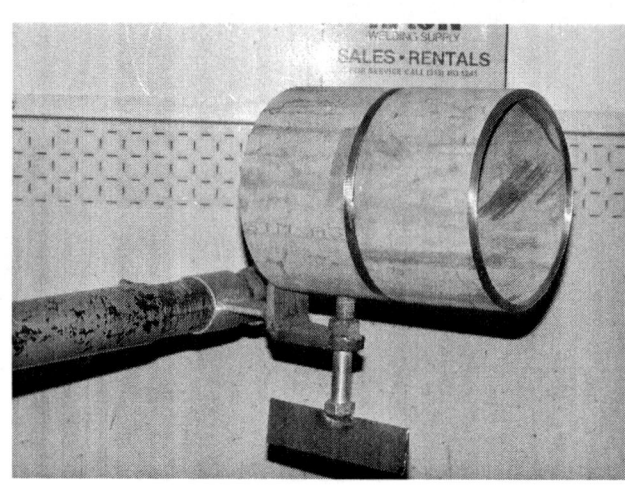

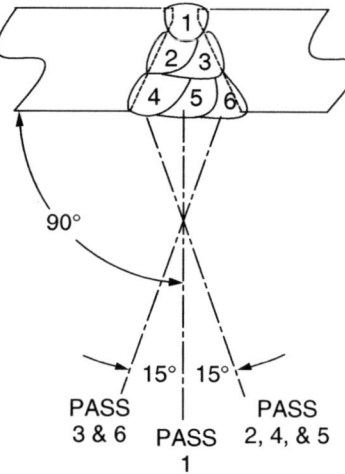

STRINGER BEAD SEQUENCE   306F16.EPS

Figure 16 ◆ Multiple-pass 5G bead sequence and work angles.

Step 1 Purge and tack weld together the practice pipe weld coupon as explained earlier.

Step 2 Clamp or tack weld the pipe weld coupon into position with the pipe axis inclined 45° to the horizontal plane.

Step 3 Assemble the backing gas retaining devices to the pipe, and connect the backing gas hose. Set the backing gas flowmeter to the flow rate specified by the filler metal rod manufacturer, the WPS, or your instructor.

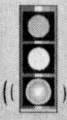

NOTE

If required by a WPS, a site quality standard, or your instructor, observe any preheat/interpass temperatures or PWHT requirements.

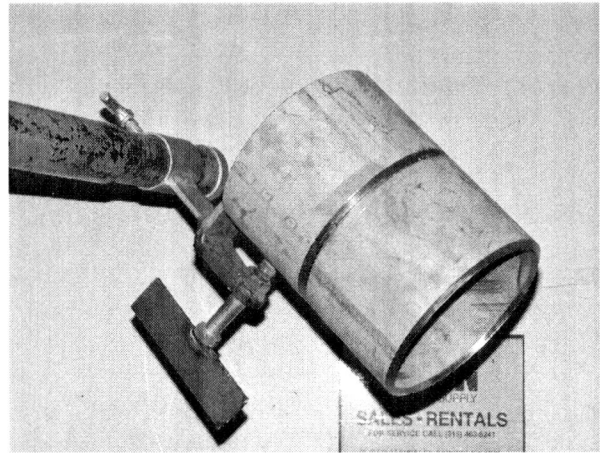

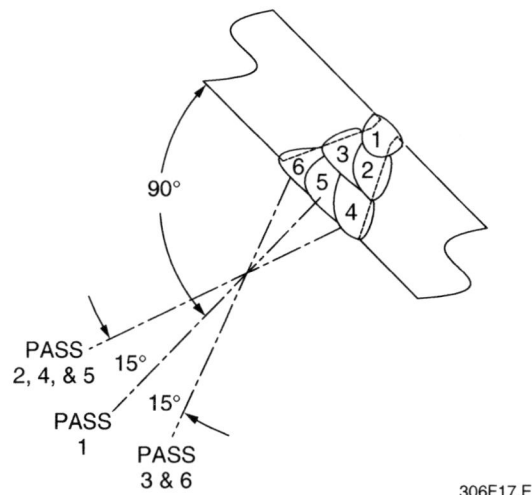

Figure 17 ◆ Multiple-pass 6G bead sequence and work angles.

Step 4 Run the root pass uphill with a stringer bead as shown in *Figure 15*. Repeat for the opposite side of the pipe.

Step 5 Brush the root pass to clean the weld.

Step 6 Run the remaining passes uphill at the appropriate work angles. Overlap the passes, and clean the weld after each pass.

Summary

The ability to make open-root V-groove welds on pipe in all positions is one of the more difficult skills you must develop as a welder. The open-root V-groove weld is the most common weld joint used for joining medium- and thick-walled pipe. You must be able to set up the equipment, perform the welding, and recognize acceptable welds. Open-root V-groove pipe welds can be made in the 1G-ROTATED, 2G, 5G, and 6G positions. Practice the 2G, 5G, and 6G position welds until you can consistently produce acceptable welds.

Review Questions

1. The welding process that provides the greatest control of root penetration and fill is _____.
   a. GTAW
   b. SMAW
   c. GMAW
   d. FCAW

2. GTAW produces high-quality welds _____.
   a. without slag or oxidation
   b. with slag and oxidation
   c. with a low degree of flux entrapment
   d. with slag but without oxidation

3. In GTAW, a flow of _____ shields the electrode, arc, and molten base metal from atmospheric contamination.
   a. inert gas
   b. oxygen
   c. flux
   d. water

4. Alloy elements in steel combine with _____.
   a. nitrogen in the atmosphere to form nitrates
   b. oxygen in the atmosphere to form oxides
   c. carbon dioxide in the atmosphere to form carbides
   d. moisture in the atmosphere to form hydrogen

5. Back side protection is essential when welding _____ and most low-alloy steels.
   a. tungsten
   b. high-carbon
   c. low-carbon
   d. stainless

6. The best method of providing back side protection for welding open-root joints is _____.
   a. backup flux
   b. consumable inserts
   c. backing gas
   d. dry powder flux

7. During welding, the maximum acceptable oxygen percentage inside a pipe is _____.
   a. 1%
   b. 5%
   c. 10%
   d. 20%

8. Argon backing gas is _____ than air.
   a. lighter
   b. more toxic
   c. heavier
   d. less toxic

9. When welding critical stainless steel piping systems where internal obstructions cannot be tolerated, use _____ as back side protection.
   a. powder flux
   b. backing gas
   c. flux cored rods
   d. consumable inserts

10. Tack welds become part of the _____.
    a. root pass
    b. fill passes
    c. cover passes
    d. temporary backing

11. Bead width in GTAW depends mainly upon _____.
    a. travel speed
    b. torch arc length
    c. torch angle
    d. filler metal handling techniques

12. _____ angles of _____ are normally used for GTAW.
    a. Push; 10° to 15°
    b. Push; 15° to 20°
    c. Pull; 10° to 15°
    d. Pull; 15° to 20°

13. Compared to carbon steel, stainless steel _____.
    a. conducts heat more quickly
    b. expands more when heated
    c. has a lower resistance to electrical current flow
    d. is much less difficult to weld

14. Compared to carbon steel, alloy steel _____.
    a. requires greater care during joint fit-up
    b. has lower hardenability
    c. is less likely to have hydrogen-induced underbead cracking
    d. requires the heating cycle to be less closely managed

15. With alloy steel, the molten weld puddle _____ than with carbon steel.
    a. tends to be more fluid
    b. tends to be less fluid
    c. will have more slag
    d. will have less slag

Performance Accreditation Tasks

The Performance Accreditation Tasks (PATCs) correspond to and support learning objectives in the *AWS Guide for the Training and Qualification of Welding Personnel*.

PATCs provide specific acceptable criteria for performance and help to ensure a true competency-based welding program for students.

The following tasks are designed to evaluate your ability to run open-root V-groove welds with GTAW equipment in three standard test positions using stainless and/or low-alloy steel filler rod of the appropriate diameter and argon shielding gas. Perform each task when you are instructed to do so by your instructor. As you complete each task, show it to your instructor for evaluation. Do not proceed to the next task until told to do so by your instructor. For AWS 2G and 5G certifications, refer to *AWS EG3.0-96* for bend test requirements. For AWS 6G certifications, refer to *AWS EG4.0-96* for bend test requirements.

OPEN-ROOT V-GROOVE PIPE WELD IN THE 2G POSITION

Using stainless and/or low-alloy steel filler rod of the appropriate diameter, argon shielding gas, and stringer beads, make an open-root V-groove weld on carbon steel pipe in the 2G position.

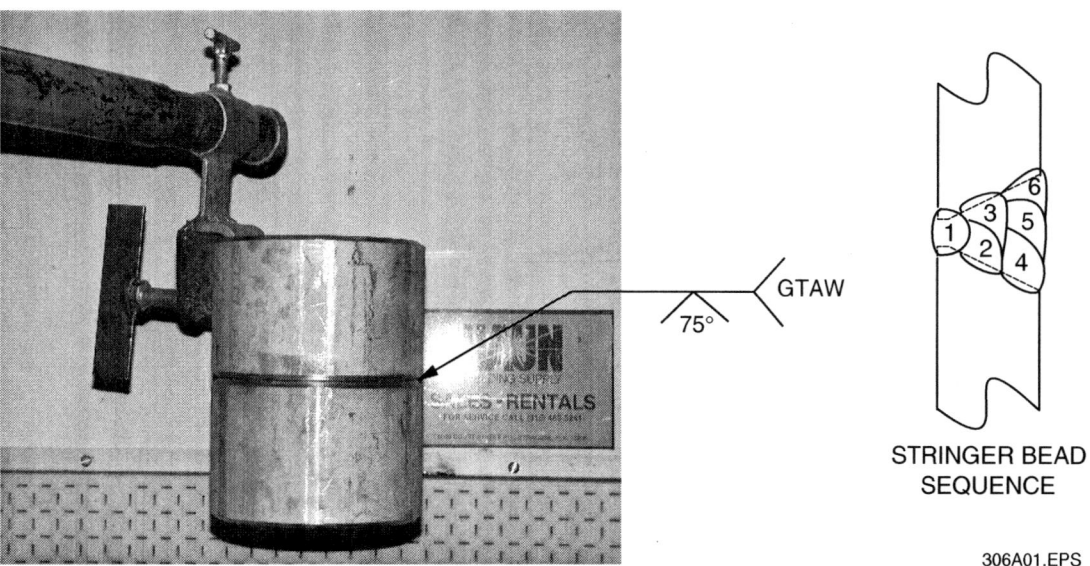

STRINGER BEAD
SEQUENCE

306A01.EPS

Criteria for Acceptance

- Uniform appearance on the bead face   _____

- Craters and restarts filled to the full cross section of the weld   _____

- Uniform weld width ±1/16"   _____

- Acceptable weld profile in accordance with the
  ASME Boiler and Pressure Vessel Code, Section IX   _____

- Smooth transition with complete fusion at the toes of the weld   _____

- Complete uniform root reinforcement at least flush with the inside of the
  pipe to a maximum of 1/8"   _____

- No porosity   _____

- No excessive undercut   _____

- No cracks   _____

- No overlap   _____

- No incomplete fusion   _____

OPEN-ROOT V-GROOVE PIPE WELD IN THE 5G POSITION

Using stainless and/or low-alloy steel filler rod of the appropriate diameter, argon shielding gas, and stringer beads, make an open-root V-groove weld on carbon steel pipe in the 5G position.

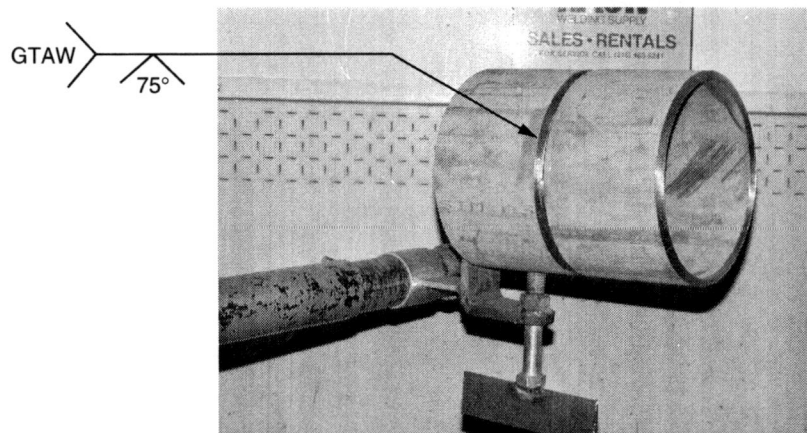

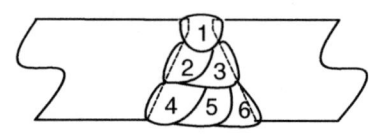

STRINGER BEAD SEQUENCE

306A02.EPS

Criteria for Acceptance

- Uniform appearance on the bead face   _____

- Craters and restarts filled to the full cross section of the weld   _____

- Uniform weld width ±¹⁄₁₆"   _____

- Acceptable weld profile in accordance with the
  ASME Boiler and Pressure Vessel Code, Section IX   _____

- Smooth transition with complete fusion at the toes of the weld   _____

- Complete uniform root reinforcement at least flush with the inside of the
  pipe to a maximum of ¹⁄₈"   _____

- No porosity   _____

- No excessive undercut   _____

- No cracks   _____

- No overlap   _____

- No incomplete fusion   _____

OPEN-ROOT V-GROOVE PIPE
WELD IN THE 6G (OR 6GR) POSITION

Using stainless and/or low-alloy steel filler rod of the appropriate diameter, argon shielding gas, and stringer beads, make an open-root V-groove weld on carbon steel pipe in the 6G (or 6GR) position.

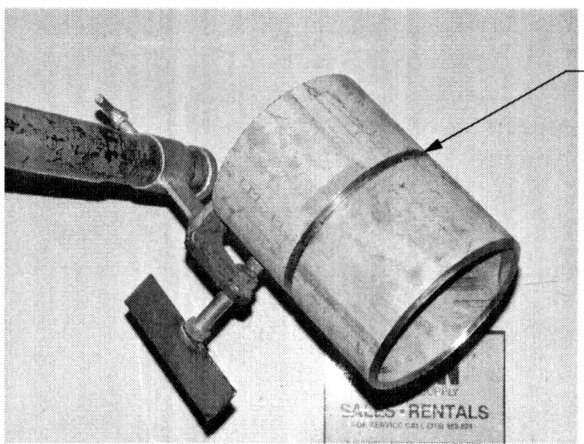

NOTE: IF REQUIRED FOR QUALIFICATION PURPOSES, A RESTRICTING RING MAY BE ADDED TO THE 6G POSITION COUPON TO FORM A 6 GR POSITION COUPON.

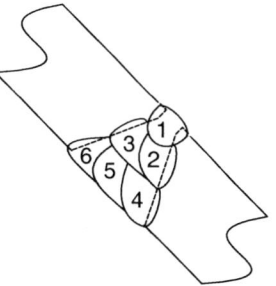

306A03.EPS

Criteria for Acceptance

- Uniform appearance on the bead face   _____

- Craters and restarts filled to the full cross section of the weld   _____

- Uniform weld width ±¹⁄₁₆"   _____

- Acceptable weld profile and guided bend test in accordance with the *ASME Boiler and Pressure Vessel Code, Section IX*   _____

- Smooth transition with complete fusion at the toes of the weld   _____

- Complete uniform root reinforcement at least flush with the inside of the pipe to a maximum of ¹⁄₈"   _____

- No porosity   _____

- No excessive undercut   _____

- No cracks   _____

- No overlap   _____

- No incomplete fusion   _____

Additional Resources

This module is intended to be a thorough resource for task training. The following reference works are suggested for further study. These are optional materials for continued education rather than for task training.

ANSI/AWS C5.5, Recommended Practices for Tungsten Arc Welding. Miami, FL: American Welding Society.

ASME – Boiler and Pressure Vessel Code: Section 9, Welding and Brazing Qualifications, Current Edition. New York, NY: ASME International.

AWS D10.11, Recommended Practices for Root Pass Welding of Pipe Without Backing. Miami, FL: American Welding Society.

MIG Welding Handbook. Florence, SC: L-TEC Welding & Cutting Systems.

Welding of Pipelines and Related Facilities, API Standard 1104, Latest Edition. New York, NY: ASME International.

Welding Pressure Pipe Lines and Piping Systems. Cleveland, OH: The Lincoln Electric Company.

Welding Skills. R.T. Miller. Homewood, IL: American Technical Publishers.

Figure Credits

Gerald Shannon	306F03, 306F04, 306F05, 306F14, 306F16, 306F17
G.A.L. Gage	306SA01
Bill Cherry	306SA02
Miller Electric Mfg. Co.	306SA03, 306F09
Tom Atkinson	306F10

The NCCER makes every effort to keep these textbooks up-to-date and free of technical errors. We appreciate your help in this process. If you have an idea for improving this textbook, or if you find an error, a typographical mistake, or an inaccuracy in NCCER's Contren® textbooks, please write us, using this form or a photocopy. Be sure to include the exact module number, page number, a detailed description, and the correction, if applicable. Your input will be brought to the attention of the Technical Review Committee. Thank you for your assistance.

Instructors – If you found that additional materials were necessary in order to teach this module effectively, please let us know so that we may include them in the Equipment/Materials list in the Instructor's Guide.

Write: Curriculum Revision and Development Department
National Center for Construction Education and Research
P.O. Box 141104, Gainesville, FL 32614-1104

Fax: 352-334-0932

E-mail: curriculum@nccer.org

Craft _____ Module Name _____

Copyright Date _____ Module Number _____ Page Number(s) _____

Description _____

_____

_____

_____

(Optional) Correction _____

_____

_____

(Optional) Your Name and Address _____

_____

_____

Gas Tungsten Arc Welding (GTAW) – Aluminum Pipe

COURSE MAP

This course map shows all of the modules in the third level of the Welding curriculum. The suggested training order begins at the bottom and proceeds up. Skill levels increase as you advance on the course map. The local Training Program Sponsor may adjust the training order.

WELDING LEVEL THREE

```
┌─────────────────────────┐      ┌─────────────────────────┐
│      29307-03*          │      │      29308-03*          │
│  GTAW – ALUMINUM PIPE   │      │   GMAW – ALUMINUM       │
│                         │      │   PLATE AND PIPE        │
└─────────────────────────┘      └─────────────────────────┘

        YOU ARE HERE

            ┌─────────────────────────┐
            │       29306-03          │
            │  GTAW – LOW-ALLOY AND    │
            │  STAINLESS STEEL PIPE    │
            └─────────────────────────┘

            ┌─────────────────────────┐
            │       29305-03          │
            │  GTAW – CARBON STEEL PIPE│
            └─────────────────────────┘

┌─────────────────┐              ┌─────────────────┐
│    29303-03     │              │    29304-03     │
│  GMAW – PIPE    │              │  FCAW – PIPE    │
└─────────────────┘              └─────────────────┘

        ┌───────────────────────────────┐
        │          29302-03             │
        │  PHYSICAL CHARACTERISTICS      │
        │  AND MECHANICAL PROPERTIES     │
        │         OF METALS             │
        └───────────────────────────────┘

        ┌───────────────────────────────┐
        │          29301-03             │
        │  PREHEATING AND POSTWELD       │
        │  HEAT TREATMENT OF METALS      │
        └───────────────────────────────┘

        ┌───────────────────────────────┐
        │      WELDING LEVEL TWO         │
        └───────────────────────────────┘

        ┌───────────────────────────────┐
        │      WELDING LEVEL ONE         │
        └───────────────────────────────┘

        ┌───────────────────────────────┐
        │          CORE                  │
        │        CURRICULUM              │
        └───────────────────────────────┘
```

307CMAP.EPS

*Please note that Modules 29307-03 and 29308-03 are electives for those progressing through the *Welding Level Three* program.

Figures

Gas Tungsten Arc Welding (GTAW) – Aluminum Pipe

Objectives

When you have completed this module, you will be able to do the following:

1. Set up GTAW equipment to perform aluminum pipe welding.
2. Identify and explain V-groove and modified U-groove pipe weld techniques.
3. Perform V-groove or modified U-groove pipe welds using GTAW in the following positions:

   - 2G
   - 5G
   - 6G

Prerequisites

Before you begin this module, it is recommended that you successfully complete the following: Core Curriculum; Welding Levels One and Two; Welding Level Three, Modules 29301-03 through 29306-03.

Required Trainee Materials

1. Pencil and paper
2. Appropriate personal protective equipment

1.0.0 ◆ INTRODUCTION

Gas tungsten arc welding (GTAW) is an arc welding process that uses an arc between a tungsten electrode and the base metal to melt the base metal. The electrode, the arc, and the molten base metal are shielded from atmospheric contamination by a flow of inert gas from the torch nozzle. (See *Figure 1*.) The filler metal is a rod that is compatible with the base metal, and it is usually hand-held and

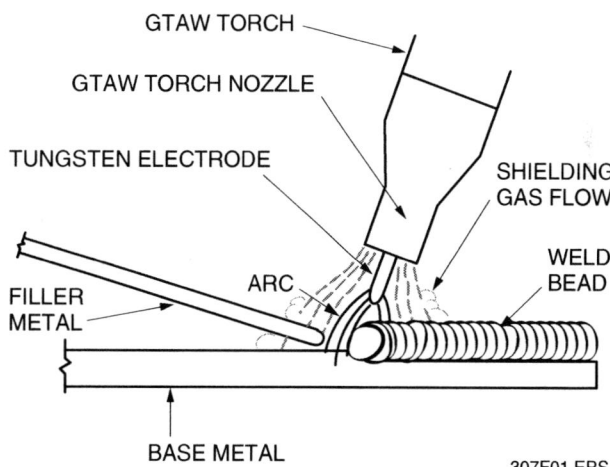

Figure 1 ◆ GTAW process.

manually fed into the leading edge of the weld puddle. GTAW produces high-quality welds without slag or oxidation. Because there is no flux, there will be no corrosion due to slag entrapment; therefore, no postweld cleaning is necessary.

A major use of GTAW is to make manual high-quality welds in aluminum piping used in both critical and non-critical applications. The GTAW process allows greater control of root penetration and fill than almost any other process. For this reason, GTAW is often used to make the root pass on pipe, even when the fill and cover passes are made with GMAW, which has a higher deposition rate.

This module explains how to set up GTAW equipment and perform open-root V-groove welds on aluminum pipe with aluminum filler metal in the 2G, 5G, and 6G welding positions. The dimensions and specifications in this module are designed to be representative of codes in general and are not specific to any certain code. Always follow the proper codes for your site.

2.0.0 ◆ SAFETY SUMMARY

The following is a summary of safety procedures and practices you must observe while cutting or welding. Keep in mind that this is just a summary. Complete safety coverage is provided in the Level One module *Welding Safety*. If you have not completed that module, do so before continuing. Above all, be sure to wear appropriate protective clothing and equipment when welding or cutting.

2.1.0 Protective Clothing and Equipment

- Always use safety goggles with a full face shield or a helmet. The goggles, face shield, or helmet lens must have the proper light-reducing tint for the type of welding or cutting to be performed. Never directly or indirectly view an electric arc without using a properly tinted lens.

- Wear proper protective leather and/or flame retardant clothing along with welding gloves to protect from flying sparks, molten metal, and heat.

- Wear 8" or taller high-top safety shoes or boots. Make sure that the tongue and lace area of the footwear is covered by a pant leg. If the tongue and lace area is exposed or the footwear must be protected from burn marks, wear leather spats under the pants or chaps and over the front top of the footwear.

- Wear a solid material (non-mesh) hat with a bill pointing to the rear or, if much overhead cutting or welding is required, a full leather hood with a welding faceplate and the correct tinted lens. If a hard hat is required, use one that allows attachment of both rear deflector material and a face shield.

- If a full leather hood is not worn, wear a face shield and snug fitting welding goggles over safety glasses for gas welding or cutting. Either the face shield or the lenses of the welding goggles must be an approved shade 5 or 6 filter. For electric arc welding or cutting, wear safety goggles and a welding hood with the correct tinted lens (shade 10 to 14).

- If a full leather hood is not worn, wear earplugs to protect your ear canals from sparks.

2.2.0 Fire/Explosion Prevention

- Never carry matches or gas-filled lighters in your pockets. Sparks can cause the matches to ignite or the lighter to explode, causing serious injury.

- Never perform any type of heating, cutting, or welding until a hot-work permit is obtained and an approved fire watch is established. Most work-site fires caused by these types of operations are started by cutting torches.

- Never use oxygen to blow off clothing. The oxygen can remain trapped in the fabric for a time. If a spark hits the clothing during this period, the clothing can burn rapidly and violently out of control.

- Make sure that any flammable material in the work area is moved or is shielded by a fire-resistant covering. Approved fire extinguishers must be available before attempting any heating, welding, or cutting operations.

- Always comply with any site requirement for a hot-work permit and/or a fire watch.

- Never release a large amount of oxygen nor use oxygen as compressed air. Its presence around flammable materials or sparks can cause rapid and uncontrolled combustion. Keep oxygen away from oil, grease, and other petroleum products.

- Never release a large amount of fuel gas, especially acetylene. Methane and propane tend to concentrate in and along low areas and can ignite at a considerable distance from the release point. Acetylene is lighter than air but is even more dangerous than methane. When mixed with air or oxygen, it will explode at much lower concentrations than any other fuel.

- To prevent fires, maintain a neat and clean work area, and make sure that any metal scrap or slag is cold before disposal.

- Before cutting or welding containers such as tanks or barrels, check to see if they have contained any explosive, hazardous, or flammable materials, including petroleum products, citrus products, or chemicals that decompose into toxic fumes when heated. As a standard practice, always clean and then fill any tanks or barrels with water, or purge them with a flow of inert gas to displace any oxygen.

2.3.0 Work Area Ventilation

- Make sure to follow confined space procedures before conducting any welding or cutting in the confined space.

- Never use oxygen to ventilate confined spaces.

- Always perform cutting or welding operations in a well-ventilated area. Such operations involving zinc or cadmium materials or coatings result in toxic fumes. For long-term cutting or welding of these materials, always wear an approved, full-face, supplied-air respirator (SAR) that uses breathing air supplied from outside of the work area. For occasional, very short-term exposure, you may use a high-efficiency particulate arresting (HEPA)-rated or metal-fume filter on a standard respirator.

- Make sure confined spaces are ventilated properly for cutting or welding operations.

3.0.0 ◆ WELDING PREPARATION

Before welding can begin, the area has to be readied, the welding equipment must be set up, and the metal to be welded must be prepared. The following sections explain how to set up the equipment for welding.

To practice welding, you need a welding table, bench, or stand. The welding surface can be steel, but an aluminum surface is preferred. Provisions must be available for placing weld coupons out of position.

To set up the area for welding, follow these steps:

Step 1 Check to be sure the area is properly ventilated. Make use of doors, windows, and fans.

Step 2 Check the area for fire hazards. Remove any flammable materials before proceeding.

Step 3 Locate the nearest fire extinguisher. Do not proceed unless the extinguisher is charged and you know how to use it.

Step 4 Position a welding table near the welding machine.

Step 5 Set up flash shields around the welding area.

3.1.0 Practice Pipe Weld Coupons

Pipe weld coupons should be cut from 3" to 12" diameter Schedule 40 aluminum pipe, and each welded joint will require two coupons of the same size. *Figure 2* shows typical AWS aluminum pipe specifications for bevel angles, modified U-grooves, root faces, and root openings. For this module, the practice coupons will be tacked together with no root opening or backing rings and will be used without any backing gas.

To prepare weld coupons for groove weld joints, follow these steps:

Step 1 Use an approved solvent to remove any grease or oil within a minimum of 1" from the weld joint. Clean the inside and outside of the aluminum pipe with a grinder or stainless steel brush.

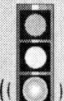
NOTE

If you use a grinding wheel, make sure it is a special high-speed grinding wheel approved to use on aluminum. Ensure that it has not been used on any other metal or used to remove hydrocarbon contaminates.

WARNING!

To prevent injury, always comply with the MSDS guidelines for the cleaning solvent(s) being used.

Step 2 Bevel or groove the end of the pipe as shown in *Figure 2* by any acceptable method, such as by using a mechanical cutter or by thermal cutting or grinding. The preferred GTAW joint for aluminum pipe without backing is the modified U-groove.

Step 3 Cut off a section of the beveled pipe end (4" minimum).

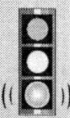
NOTE

For 1G-ROTATED welding, you may have to cut longer coupons to fit the rollers, if used.

Step 4 Check the groove. There should be no dross, and the groove angle should be as shown in *Figure 2* with no notches more than 1/16" deep.

Step 5 Grind or file a root face on the bevel as specified by your instructor or as shown in *Figure 2*.

NOTE

If the groove bevel has been cut thermally, machine the face 1/8" back from the face of the bevel.

Conserve Materials

Pipe for practice welding is expensive and difficult to obtain. Make every effort to conserve and not waste available material. Completely use weld coupons until all surfaces have been welded by cutting the coupon apart and reusing the pieces. Check with the instructor for the appropriate size coupon.

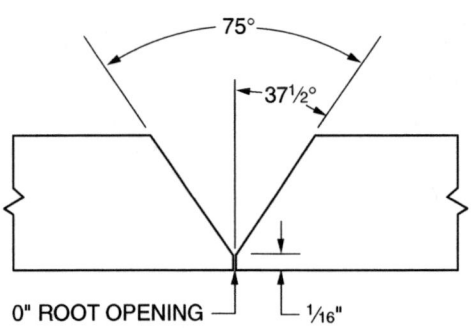

STANDARD V-GROOVE
(WALL THICKNESS ¾" OR LESS)

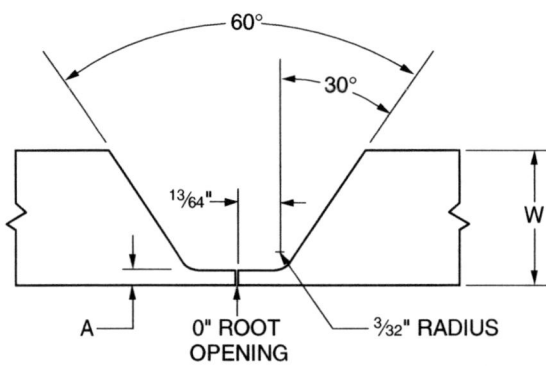

PIPING DIMENSIONS			
NOMINAL PIPE DIAMETER SIZE NUMBER		OUTSIDE DIAMETER (OD): EXTENDED ROOT FACE THICKNESS	WALL THICKNESS
DN	NPS	A	W (Max.)
6 – 65	⅛ – 2½"	¹³⁄₃₂ – 2⅞"	¹⁄₁₆"	⁹⁄₃₂"
80 – 300	3 – 12"	3½ – 12¾"	³⁄₃₂"	½"

MODIFIED U-GROOVE (PREFERRED) FOR AC GTAW

307F02.EPS

Figure 2 ◆ AWS aluminum pipe groove specifications for use without backing.

Machined Grooves and Root Faces

Machining the grooves and root faces of aluminum pipe is the cleanest and most uniform method of preparing pipe for GTAW. This is especially true for aluminum where you must remove all contaminants from thermal cutting to eliminate weld discontinuities. Most aluminum pipe can be purchased with the ends pre-machined for the desired groove.

Welding Large-Diameter Pipe

The practice coupons on large-diameter pipe should be greater than 6" in length. They need more metal to help absorb the increased heat used to make these welds.

Step 6 Align the two pipe sections so that the ID surfaces are even all around. Align small-diameter pipe by clamping both pieces to a piece of angle iron. Align large-diameter pipe with the aid of a pipe alignment jig or by holding a straightedge across the joint, parallel to the pipe axis. The straightedge must be used all around the pipe in case one or both sections are distorted.

Step 7 When the pipe ends are aligned with no root opening, make the first of four tack welds of no greater than 1".

NOTE

Heavy-wall or large-diameter pipe may require longer tacks or more than four tacks.

Step 8   After the first tack weld, on the opposite side, make sure there is no root opening, eliminate any gap, and make the second tack weld.

Step 9   Check the root again and weld the third tack midway between the first two tacks.

Step 10   Weld the fourth tack opposite the third tack and midway between the first and second tacks. There should now be four tack welds evenly spaced (90°) around the pipe coupon. This is illustrated by *Figure 3*.

Step 11   Feather the tack welds with the edge of a grinding wheel. Feathering the ends of the tack welds with a grinder helps fuse the tack welds into the root pass.

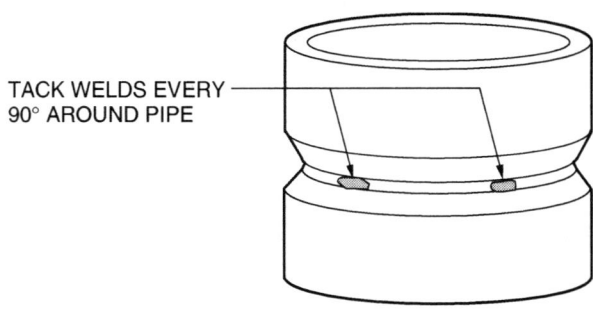

TACK WELDS EVERY 90° AROUND PIPE

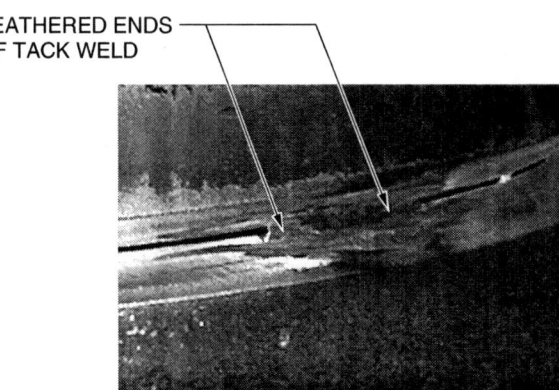

FEATHERED ENDS OF TACK WELD

307F03.EPS

Figure 3 ◆ Tacked V-groove weld coupon.

3.2.0 The Welding Machine

Identify a proper constant-current welding machine for GTAW and follow these steps to set it up for use.

Step 1   Verify that the welding machine can be used for GTAW with or without internally controlled gas shielding.

Step 2   Identify an air-cooled or water-cooled GTAW torch. Make sure it is compatible with the welding machine and any cooling unit.

Step 3   Check to be sure that the welding machine is properly grounded through the primary current receptacle.

Step 4   Verify the location of the primary disconnect.

Step 5   Configure the welding machine for GTAW parameters as directed by your instructor (*Figure 4*). Configure the torch polarity to AC and equip the torch with a properly prepared ½2" or ⅛" EWP or EWZr tungsten.

Step 6   Connect the proper shielding gas for the application as described in the previous level and as specified by the filler metal manufacturer, WPS, site quality standards, or your instructor.

Step 7   Connect the clamp of the workpiece lead to the workpiece.

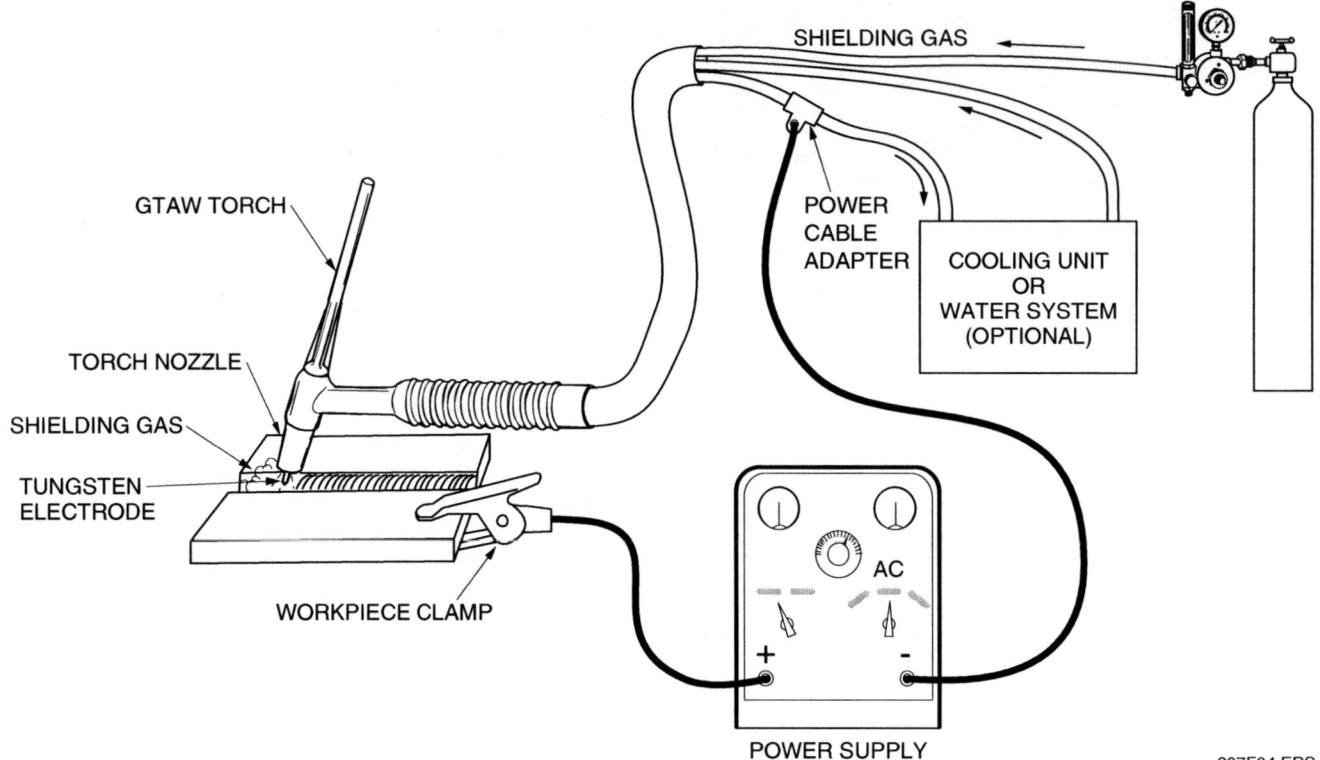

SHIELDING GAS

GTAW TORCH

TORCH NOZZLE

SHIELDING GAS

TUNGSTEN ELECTRODE

WORKPIECE CLAMP

POWER CABLE ADAPTER

COOLING UNIT OR WATER SYSTEM (OPTIONAL)

AC

+ −

POWER SUPPLY

307F04.EPS

Figure 4 ◆ Configuration diagram of typical GTAW machine.

Step 8 Turn on the welding machine, and purge the torch as directed by the manufacturer's instructions.

3.3.0 Filler Metals

Filler metals are selected to be compatible with the base metal to be welded. The WPS or site standards specify the filler metal type and size to use.

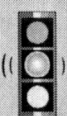

> **CAUTION**
>
> When the WPS or site quality standards specify a filler metal, it must be used to prevent weld discontinuities.

For the welding exercises in this module, you should use ³⁄₃₂" and/or ⅛" aluminum filler metals, as specified by your instructor, to weld the aluminum pipe coupons. Remove only a small number of filler metals at a time. Keep the remainder in the package to keep them clean. Before using the filler metal, check it for burned ends or contamination such as corrosion, dirt, oil, or grease, which can all cause weld discontinuities. Chemically clean the filler metal with a clean rag, use emery cloth or stainless steel wool to remove oxides, and snip any burned ends. If the filler metal cannot be cleaned, do not use it.

4.0.0 ◆ GAS TUNGSTEN ARC WELDING TECHNIQUES

GTAW weld bead characteristics and quality are affected by several factors, each of which is influenced by the way you handle the torch. These factors include the following:

- Torch travel speed and arc length
- Torch angles
- Torch and filler metal handling techniques

4.1.0 Torch Travel Speed and Arc Length

Torch travel speed and/or arc length may affect the GTAW weld puddle and weld penetration. A slow travel speed allows more heat to concentrate and forms a larger, more deeply penetrating puddle. Faster travel speeds prevent heat buildup and form smaller and shallower puddles. Arc length is the major control for bead width. As the torch is raised, voltage, arc length, and bead width increase.

4.2.0 Torch Angles

There are two basic torch angles you must control when performing GTAW: the work angle and the travel angle. The definition of these angles is the same as for all other methods of arc welding.

4.2.1 Work Angle

The torch work angle (*Figure 5*) is defined as a less-than-90° angle between a line perpendicular to the major workpiece surface at the point of electrode contact and a plane determined by the electrode axis and the weld axis. For a T-joint or corner joint, the line is perpendicular to the nonbutting member. For pipe, the plane is determined by the electrode axis and a line tangent to the pipe surface at the same point.

4.2.2 Travel Angle

The torch travel angle (*Figure 5*) is defined as a less-than-90° angle between the electrode axis and a line perpendicular to the weld axis at the point of electrode contact in a plane determined by the electrode axis and the weld axis. For pipe, the plane is determined by the electrode axis and a line tangent to the pipe surface at the same point. A push angle is used for GTAW and is created

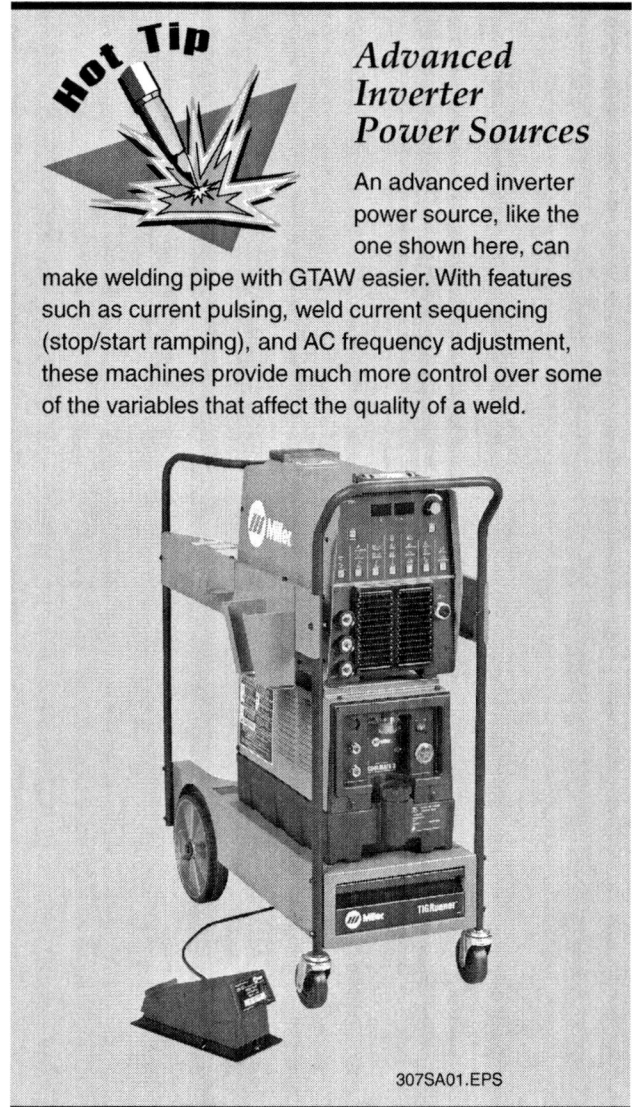

Hot Tip

Advanced Inverter Power Sources

An advanced inverter power source, like the one shown here, can make welding pipe with GTAW easier. With features such as current pulsing, weld current sequencing (stop/start ramping), and AC frequency adjustment, these machines provide much more control over some of the variables that affect the quality of a weld.

307SA01.EPS

Hot Tip

Advanced Excessive Push Angle

A push angle that's too large will tend to draw air from under the back edge of the torch nozzle where it will mix with the shielding gas stream and contaminate the weld.

when the torch is tilted back so that the electrode tip precedes the torch in the direction of the weld. In this position, the electrode tip and shielding gas are directed ahead of the weld bead to provide base metal cleaning on the positive portion of the AC voltage cycle. Push angles of 10° to 15° are normally used for GTAW of aluminum.

4.3.0 Torch and Filler Metal Handling Techniques

The two basic handling techniques used to perform GTAW are the freehand and walking-the-cup techniques, which were explained in previous modules. Try both techniques and use the one that gives the best results.

5.0.0 ◆ V-GROOVE AND MODIFIED U-GROOVE PIPE WELDS

The modified U-groove and V-groove weld with backing are the most common groove welds normally made on aluminum pipe. These welds are typically used for joining medium- and thick-walled pipe used in most non-critical piping systems. Critical piping systems (including pipe, fittings, and welded joints) contain or carry material with the potential to cause long- or short-term catastrophic danger or damage to personnel and/or the environment, should the components fail to contain or carry the material as designed. Non-critical piping is low-pressure piping used for heating and air conditioning, simple water systems, and other service installations. Less-stringent code requirements are used to evaluate non-critical piping welds.

CAUTION

When welding pipe, do not make arc strikes outside the weld groove on the surface of the pipe. An arc strike can cause a hardened spot that can crack as the pipe expands and contracts. An arc strike on the pipe surface is considered a defect and will require repair or rework.

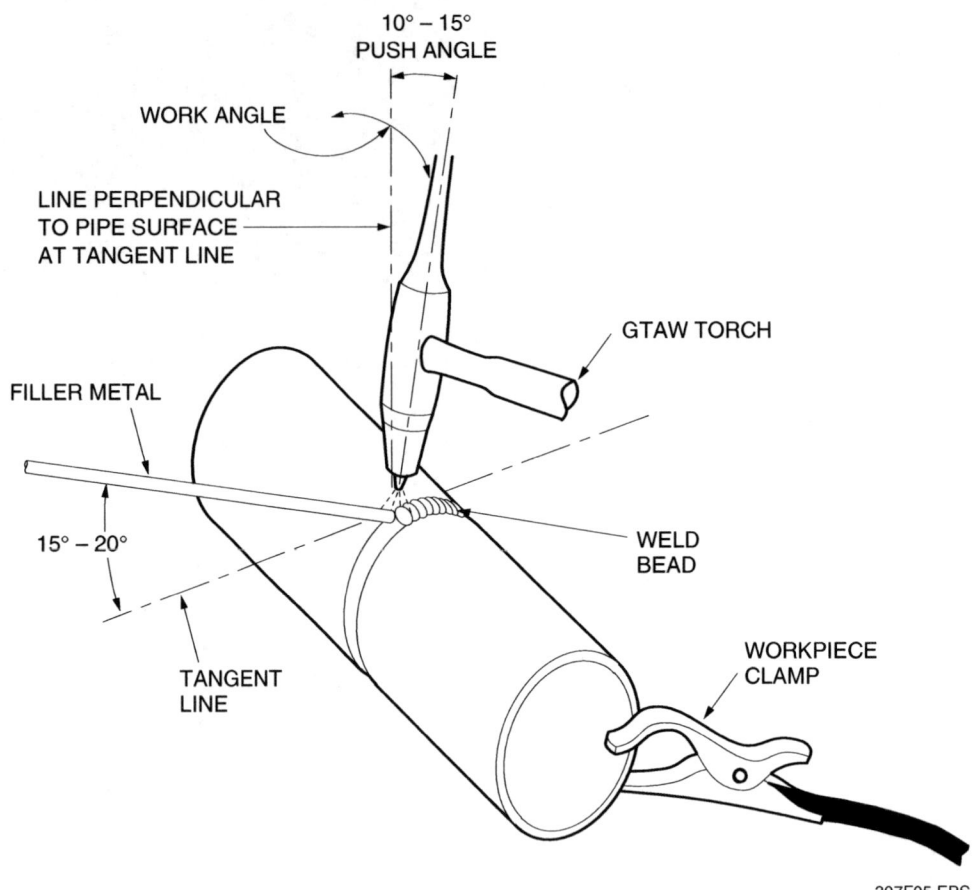

10° – 15°
PUSH ANGLE

WORK ANGLE

LINE PERPENDICULAR
TO PIPE SURFACE
AT TANGENT LINE

GTAW TORCH

FILLER METAL

15° – 20°

WELD
BEAD

WORKPIECE
CLAMP

TANGENT
LINE

307F05.EPS

Figure 5 ◆ Typical torch work and travel angles.

5.1.0 Techniques for Aluminum Pipe GTAW

Aluminum pipe is commonly welded with alternating current (AC) with argon as a shielding gas. In some cases, an argon/helium mix is used for increased penetration. Higher currents are used when welding aluminum because of high thermal conductivity in the base metal. The modified U-groove, covered in a previous section, is preferred for non-backed joints because it decreases the heat conducted away from the weld joint and allows complete root penetration and fusion with a smaller weld pool. This smaller weld pool is more easily controlled in both vertical and overhead positions by capillary action and the shielding gas flow. Without backing material, obtaining proper penetration and fusion of the root pass is the most

Weld Pool Size

When using GTAW in horizontal, vertical, and overhead positions, you must keep the weld pool smaller than when in the flat position. The weld pool is generally held in the joint by capillary action. If the pool is too large, gravitational force overcomes the capillary action and the liquid metal runs out of the joint.

Test Position

Welding pipe in the 6G position is a common welding test. A ring may be added to test for restricted accessibility (6GR).

Filler Metal

Clean the filler metal rods with an approved chemical to remove hydrocarbon contaminants and with steel wool to remove any oxides.

difficult part of the welding. The challenge lies in establishing a weld pool that penetrates and fuses with the pipe without falling through.

5.2.0 Pipe Groove Weld Test Positions

Groove welds may be made in all positions on pipe. The weld position is determined by the axis of the pipe. Four standard weld test positions are used with pipe and are shown in *Figure 6*.

To destructively test 5G or 6G welds, the test specimens are cut from the four regions illustrated in *Figure 7*: midway between the 12-o'clock and 3-o'clock, 3-o'clock and 6-o'clock, 6-o'clock and 9-o'clock, and 9-o'clock and 12-o'clock positions. (The 12-o'clock position is the top of the coupon groove; 6-o'clock is the bottom.)

5.3.0 Acceptable and Unacceptable Pipe Weld Profiles

Pipe groove welds without backing should be made with slight reinforcement (not exceeding ⅛") and a gradual transition to the base metal at each toe. The root pass should have complete penetration. The root reinforcement on the inside of the pipe ranges from being flush to a maximum of ⅛". Pipe groove welds must not have excessive reinforcement or underfill at the face or root, excessive

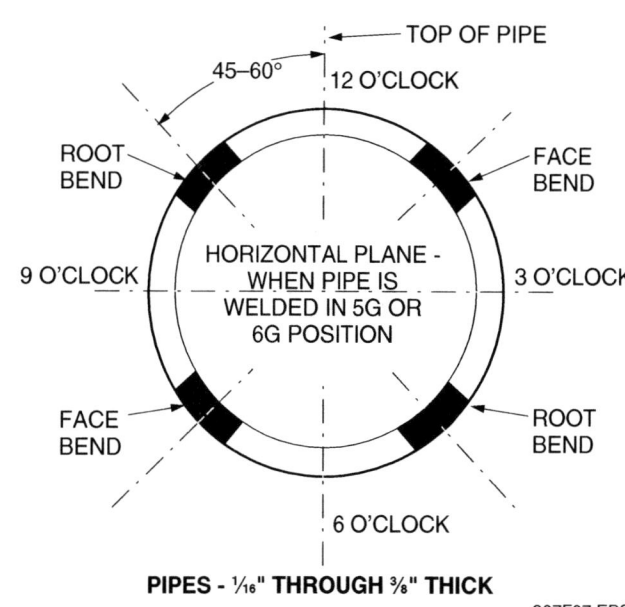

PIPES - ¹⁄₁₆" THROUGH ⅜" THICK

307F07.EPS

Figure 7 ◆ Test specimen regions of 5G or 6G position pipe.

undercut, or overlap. Excessively large cover passes will reduce the pipe's strength. They cause the stresses in the pipe to be concentrated along the sides of the weld and will not permit the pipe to expand and contract in a uniform manner along its length. If a weld has any of these defects, as shown in *Figure 8*, it must be repaired.

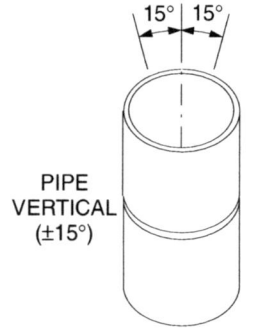

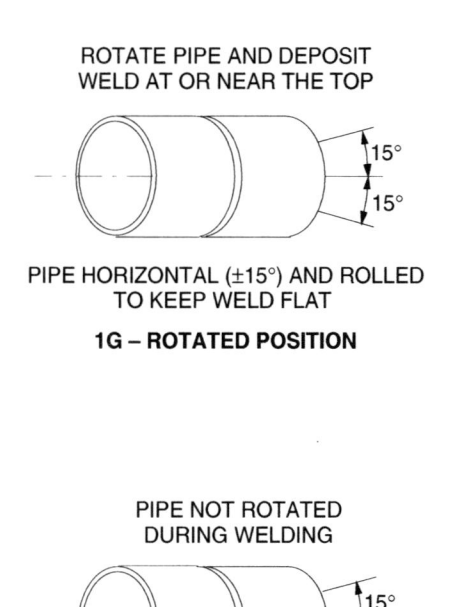

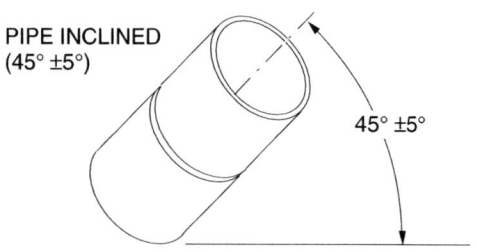

307F06.EPS

Figure 6 ◆ Four basic pipe groove weld test positions.

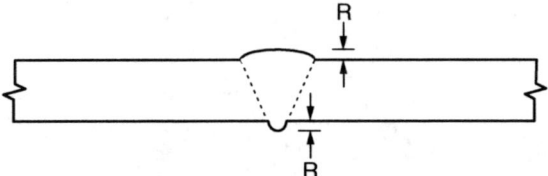

R = FACE AND ROOT REINFORCEMENT PER CODE
NOT TO EXCEED ⅛" MAX.

ACCEPTABLE WELD PROFILE

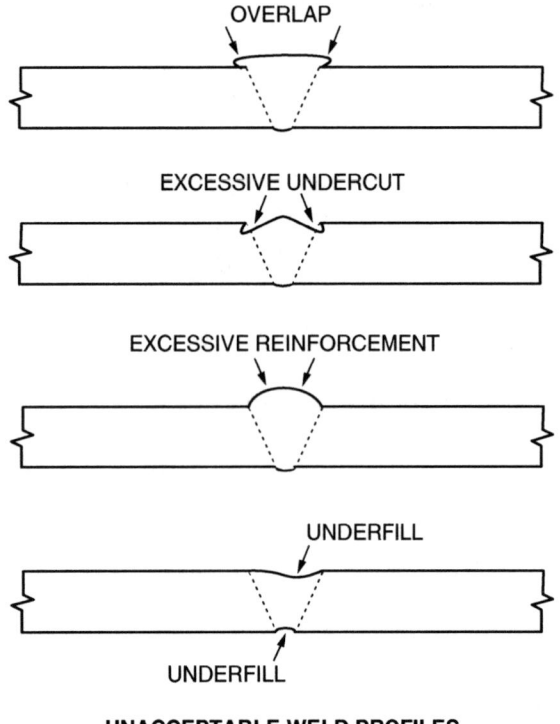

OVERLAP

EXCESSIVE UNDERCUT

EXCESSIVE REINFORCEMENT

UNDERFILL

UNDERFILL

UNACCEPTABLE WELD PROFILES

307F08.EPS

Figure 8 ◆ Acceptable and unacceptable pipe groove weld profiles.

6.0.0 ◆ PRACTICING V-GROOVE OR MODIFIED U-GROOVE WELDS

This section of the module explains how to perform the following V-groove or modified U-groove weld positions:

- Horizontal (2G)
- Multiple (5G)
- Inclined multiple (6G)

6.1.0 Horizontal (2G) Position

Practice 2G position groove pipe welds using a shielding gas and a solid filler rod with a diameter specified by your instructor. The torch should be at a 10° to 15° push angle for the weld passes. (See *Figure 9*.) To prevent the weld puddle from

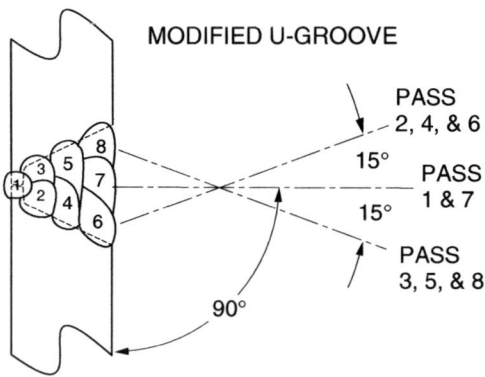

MODIFIED U-GROOVE

PASS
2, 4, & 6

15°

PASS
1 & 7

15°

PASS
3, 5, & 8

90°

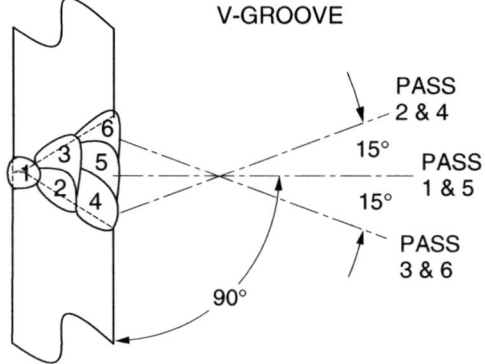

V-GROOVE

PASS
2 & 4

15°

PASS
1 & 5

15°

PASS
3 & 6

90°

STRINGER BEAD SEQUENCES

307F09.EPS

Figure 9 ◆ Multiple-pass 2G bead sequences and work angles.

sagging, the torch work angle can be dropped slightly but not more than 10°. For the root pass, use a circular motion to form the weld puddle. Pay particular attention to tie into the tack welds.

When running the fill/cover passes, use stringer beads and both the walking-the-cup and freehand torch handling techniques. Ensure tie-in at the toes of the weld bead. Take particular care at the termination of the weld to fill the crater.

Follow these steps to practice groove pipe welds in the 2G position:

Step 1 Tack weld together the practice pipe weld coupon, as explained earlier.

Step 2 Clamp or tack weld the pipe coupon with the axis of the pipe vertical.

Step 3 Run the root pass (pass 1) as shown in *Figure 9*, starting on a tack weld. Continue this procedure until the root pass is completed.

Step 4 Brush the root pass to clean the weld.

Step 5 Run the remaining passes at the appropriate work angles. Overlap the passes, and clean the weld after each pass.

6.2.0 Multiple (5G) Position

Practice the multiple (5G) position groove pipe welds using a shielding gas and a solid filler rod with a diameter specified by your instructor. Start on a tack weld positioned at the bottom of the pipe and weld uphill toward the top. Use a torch push angle of 10° to 15° to the pipe surface (*Figure 10*).

The torch should be at approximately a 10° to 15° push angle for the root, fill, and cover passes. (See *Figure 11*.) When running the fill and cover passes, use stringer beads and both the walking-the-cup and freehand torch handling techniques. Ensure tie-in at the toes of the weld bead. Pay particular attention at the termination of the weld to fill the crater.

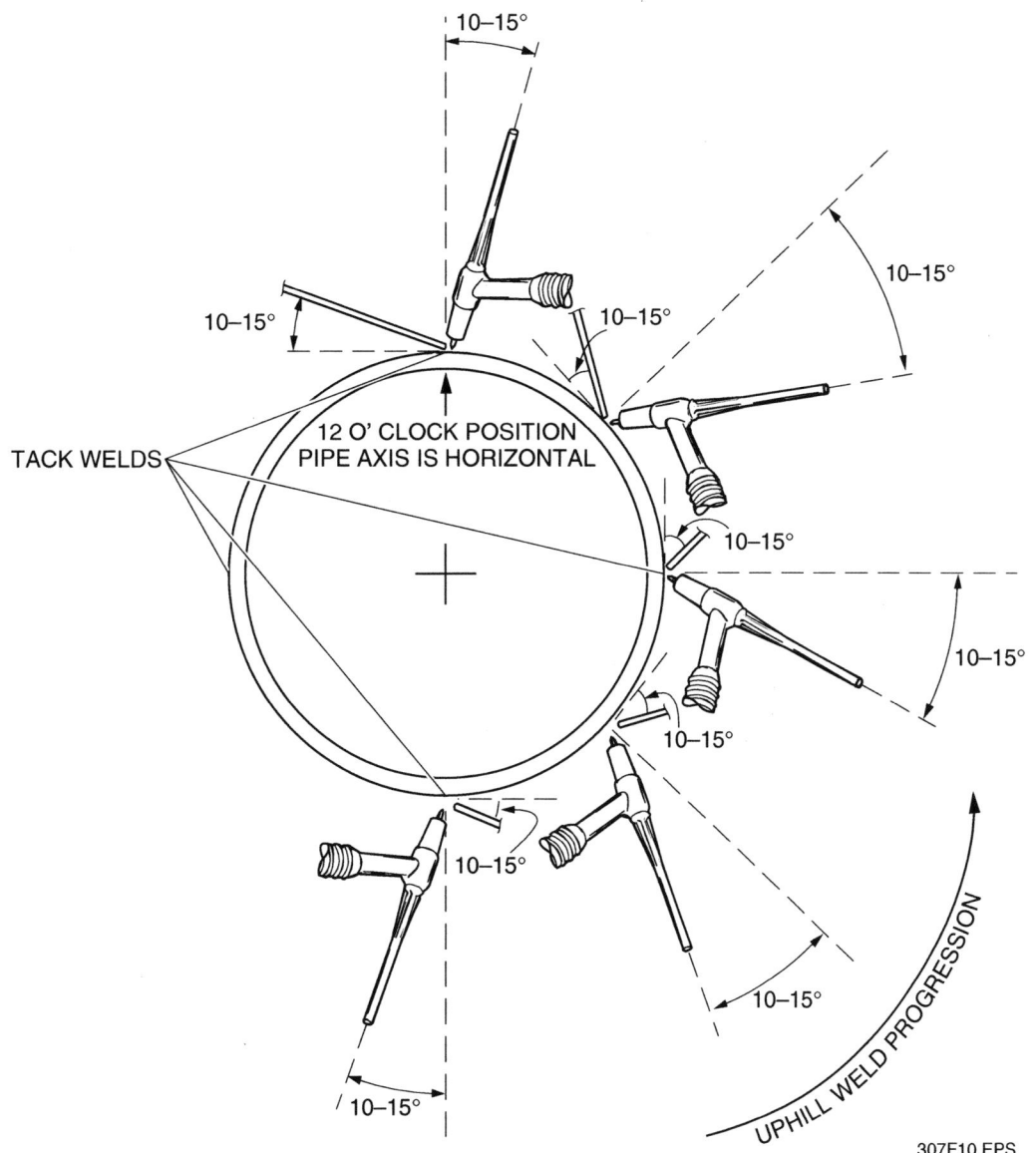

307F10.EPS

Figure 10 ◆ 5G pipe position torch and electrode rod angles.

MODIFIED U-GROOVE

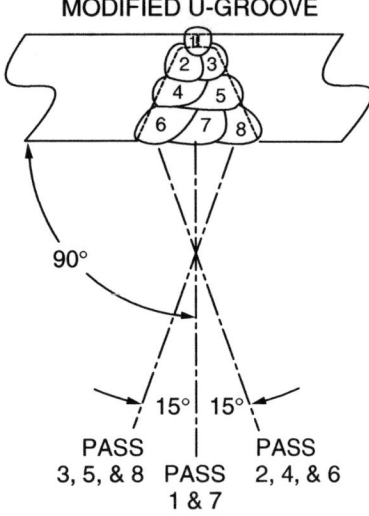

V-GROOVE

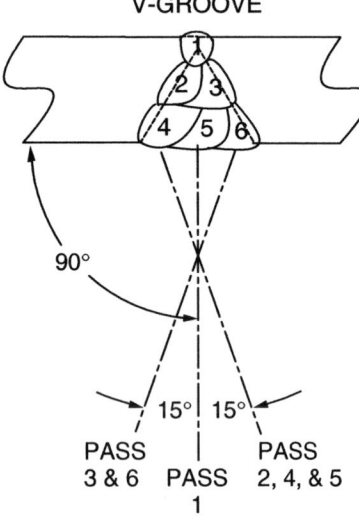

STRINGER BEAD SEQUENCES

307F11.EPS

Figure 11 ◆ Multiple-pass 5G bead sequences and work angles.

Follow these steps to practice groove pipe welds in the 5G position:

Step 1 Tack weld together the practice pipe weld coupon, as explained earlier.

Step 2 Clamp or tack weld the pipe weld coupon into position with the pipe axis horizontal.

Step 3 Run the root pass uphill with a stringer bead, as shown in *Figure 10*. Repeat for the opposite side of the pipe.

Step 4 Brush the root pass to clean the weld.

Step 5 Run the remaining passes uphill at the appropriate work angles. Overlap the passes, and clean the weld after each pass.

6.3.0 Inclined Multiple (6G) Position

Practice the inclined multiple (6G) (45°) position groove pipe welds using a shielding gas and a solid filler rod with a diameter specified by your instructor. Start at the bottom of the pipe and weld uphill toward the top.

The torch should be at approximately a 10° to 15° push angle for the root, fill, and cover passes. (See *Figure 12*.) When running the fill and cover passes, use stringer beads and both the walking-the-cup and freehand torch handling techniques. Ensure tie-in at the toes of the weld bead. Pay particular attention at the termination of the weld to fill the crater.

NOTE

If required for training purposes, a restricting ring may be added to the 6G position coupon to form a 6GR position coupon.

Follow these steps to practice groove pipe welds in the 6G position:

Step 1 Tack weld together the practice pipe weld coupon, as explained earlier.

Step 2 Clamp or tack weld the pipe weld coupon into position with the pipe axis inclined 45° to the horizontal plane.

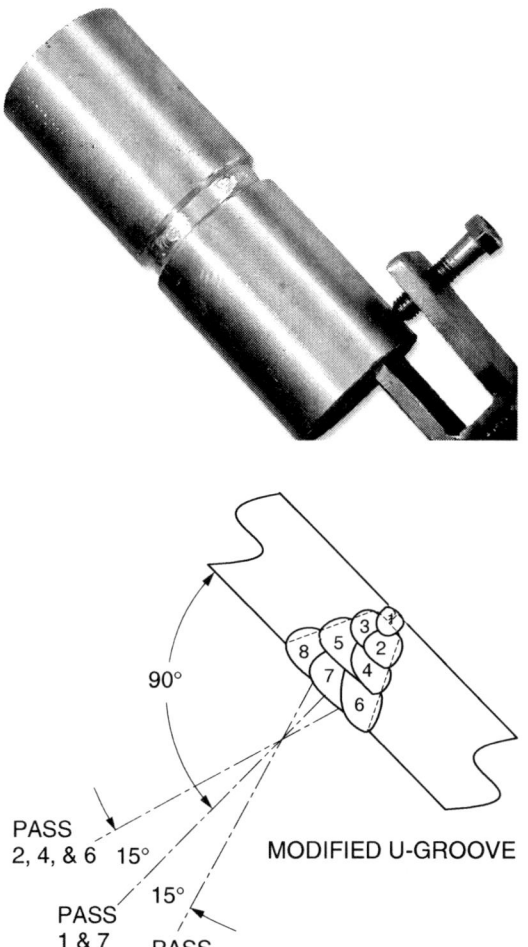

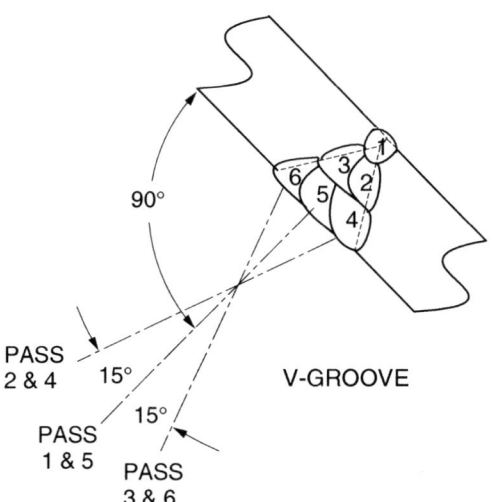

PASS
2, 4, & 6 15°

PASS
1 & 7 15°

PASS
3, 5, & 8

MODIFIED U-GROOVE

PASS
2 & 4 15°

PASS
1 & 5 15°

PASS
3 & 6

V-GROOVE

STRINGER BEAD SEQUENCES

307F12.EPS

Figure 12 ◆ Multiple-pass 6G bead sequences and work angles.

Step 3 Run the root pass uphill with a stringer bead as shown in *Figure 10*. Repeat for the opposite side of the pipe.

Step 4 Brush the root pass to clean the weld.

Step 5 Run the remaining passes uphill at the appropriate work angles. Overlap the passes, and clean the weld after each pass.

Summary

The ability to make V- or U-groove welds on aluminum pipe in all positions is one of the more difficult skills you must develop as a welder. The groove weld is the most common weld joint used for joining medium- and thick-walled pipe. You must be able to set up the equipment, perform the welding, and recognize acceptable welds. Groove pipe welds can be made in the 1G-ROTATED, 2G, 5G, and 6G positions. Practice the 2G, 5G, and 6G position welds until you can consistently produce acceptable welds.

Review Questions

1. In GTAW of aluminum pipe, the preferred welding table surface is _____.
   a. low-alloy steel
   b. high-carbon steel
   c. aluminum
   d. stainless steel

2. The preferred GTAW joint for aluminum pipe without backing is the _____ weld.
   a. double U-groove
   b. double V-groove
   c. open-root V-groove
   d. modified U-groove

3. As the GTAW torch is raised, arc length and voltage _____.
   a. increase, and bead width increases as well
   b. decrease, but bead width increases
   c. decrease, and bead width decreases as well
   d. increase, but bead width decreases

4. _____ angles of _____ are normally used for GTAW of aluminum.
   a. Push; 10° to 15°
   b. Push; 15° to 20°
   c. Pull; 10° to 15°
   d. Pull; 15° to 20°

5. The two basic GTAW handling techniques are _____.
   a. freehand and on-the-wire
   b. freehand and walking-the-cup
   c. side-to-side and freehand
   d. on-the-wire and side-to-side

6. Groove welds on aluminum pipe are typically used for joining _____.
   a. thin-walled pipe for critical systems
   b. medium- and thick-walled pipe for non-critical systems
   c. thin-walled pipe for non-critical systems
   d. medium- and thick-walled pipe for critical systems

7. GTAW of aluminum pipe usually requires _____ as a shielding gas.
   a. alternating current with argon
   b. direct current with helium
   c. alternating current with nitrogen
   d. direct current with argon

8. The modified U-groove is preferred for non-backed joints because it _____.
   a. prevents complete root penetration
   b. decreases the heat conducted away from the weld joint
   c. allows for a larger gap in the root opening
   d. allows for a larger weld pool

9. To increase penetration during GTAW of aluminum pipe, use a(n) _____ mix as shielding gas.
   a. argon/nitrogen
   b. argon/carbon dioxide
   c. helium/carbon dioxide
   d. argon/helium

10. High thermal conductivity in the base metal means that _____ must be used in GTAW of aluminum.
   a. higher currents
   b. lower currents
   c. higher voltage
   d. lower voltage

Performance Accreditation Tasks

The Performance Accreditation Tasks (PATCs) correspond to and support learning objectives in the *AWS Guide for the Training and Qualification of Welding Personnel*.

PATCs provide specific acceptable criteria for performance and help to ensure a true competency-based welding program for students.

The following tasks are designed to evaluate your ability to run groove welds with GTAW equipment in three standard test positions using aluminum filler rod of the appropriate diameter and argon shielding gas. Perform each task when you are instructed to do so by your instructor. As you complete each task, show it to your instructor for evaluation. Do not proceed to the next task until told to do so by your instructor. For AWS 2G and 5G certifications, refer to *AWS EG3.0-96* for bend test requirements. For AWS 6G certifications, refer to *AWS EG4.0-96* for bend test requirements.

GROOVE PIPE WELD IN THE 2G POSITION

Using aluminum filler rod of the appropriate diameter, argon shielding gas, and stringer beads, make a groove weld on aluminum pipe in the 2G position with either type of joint.

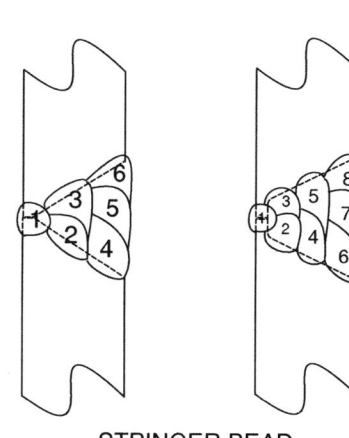

STRINGER BEAD
SEQUENCES

307A01.EPS

Criteria for Acceptance

- Uniform appearance on the bead face _____

- Craters and restarts filled to the full cross section of the weld _____

- Uniform weld width ±¹⁄₁₆" _____

- Acceptable weld profile in accordance with the *ASME Boiler and Pressure Vessel Code, Section IX* _____

- Smooth transition with complete fusion at the toes of the weld _____

- No porosity _____

- No excessive undercut _____

- No cracks _____

- No overlap _____

- No incomplete fusion _____

GROOVE PIPE WELD
IN THE 5G POSITION

Using aluminum filler rod of the appropriate diameter, argon shielding gas, and stringer beads, make a groove weld on aluminum pipe in the 5G position with either type of joint.

GTAW 75°

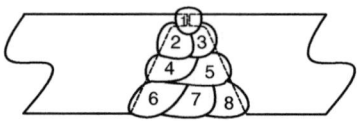

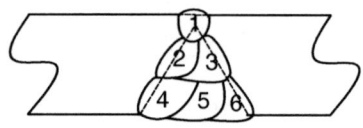

STRINGER BEAD SEQUENCES

307A02.EPS

Criteria for Acceptance

* Uniform appearance on the bead face   _____

* Craters and restarts filled to the full cross section of the weld   _____

* Uniform weld width ±¹⁄₁₆"   _____

* Acceptable weld profile in accordance with the
  ASME Boiler and Pressure Vessel Code, Section IX   _____

* Smooth transition with complete fusion at the toes of the weld   _____

* No porosity   _____

* No excessive undercut   _____

* No cracks   _____

* No overlap   _____

* No incomplete fusion   _____

GROOVE PIPE WELD IN THE 6G (OR 6GR) POSITION

Using aluminum filler rod of the appropriate diameter, argon shielding gas, and stringer beads, make a groove weld on aluminum pipe in the 6G (or 6GR) position with either type of joint, as shown.

NOTE: IF REQUIRED FOR QUALIFICATION PURPOSES, A RESTRICTING RING MAY BE ADDED TO THE 6G POSITION COUPON TO FORM A 6GR POSITION COUPON.

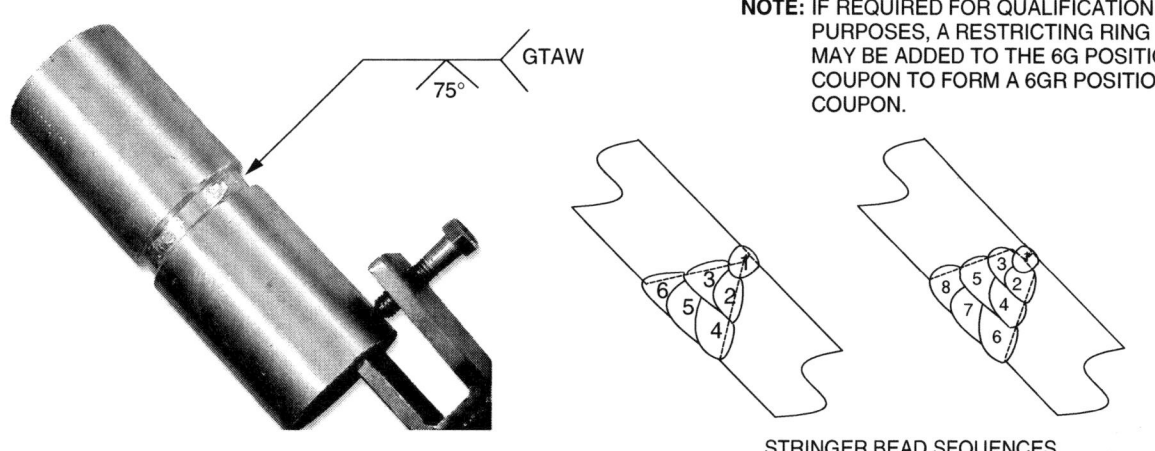

STRINGER BEAD SEQUENCES

307A03.EPS

Criteria for Acceptance

* Uniform appearance on the bead face _____

* Craters and restarts filled to the full cross section of the weld _____

* Uniform weld width ±1/16" _____

* Acceptable weld profile and guided bend test in accordance with the *ASME Boiler and Pressure Vessel Code, Section IX* _____

* Smooth transition with complete fusion at the toes of the weld _____

* No porosity _____

* No excessive undercut _____

* No cracks _____

* No overlap _____

* No incomplete fusion

Additional Resources

This module is intended to be a thorough resource for task training. The following reference works are suggested for further study. These are optional materials for continued education rather than for task training.

ASME – Boiler and Pressure Vessel Code: Section 9, Welding and Brazing Qualifications, Current Edition. New York, NY: ASME International.

AWS D10.7M/D10.7, Guide for the Gas Shielded Arc Welding of Aluminum and Aluminum Alloy Pipe. Miami, FL: American Welding Society.

MIG Welding Handbook. Florence, SC: L-TEC Welding & Cutting Systems.

Welding of Pipelines and Related Facilities, API Standard 1104, Latest Edition. New York, NY: ASME International.

Welding Pressure Pipe Lines and Piping Systems. Cleveland, OH: The Lincoln Electric Company.

Welding Skills. R.T. Miller. Homewood, IL: American Technical Publishers.

Acknowledgments

Terry Lowe   307F03, 307F09, 307F11, 307F12, 307A01, 307A02, 307A03

Miller Electric Mfg. Co.   307SA01

CONTREN® LEARNING SERIES — USER UPDATES

The NCCER makes every effort to keep these textbooks up-to-date and free of technical errors. We appreciate your help in this process. If you have an idea for improving this textbook, or if you find an error, a typographical mistake, or an inaccuracy in NCCER's Contren® textbooks, please write us, using this form or a photocopy. Be sure to include the exact module number, page number, a detailed description, and the correction, if applicable. Your input will be brought to the attention of the Technical Review Committee. Thank you for your assistance.

Instructors – If you found that additional materials were necessary in order to teach this module effectively, please let us know so that we may include them in the Equipment/Materials list in the Instructor's Guide.

Write: Curriculum Revision and Development Department
National Center for Construction Education and Research
P.O. Box 141104, Gainesville, FL 32614-1104

Fax: 352-334-0932

E-mail: curriculum@nccer.org

Craft _____ Module Name _____

Copyright Date _____ Module Number _____ Page Number(s) _____

Description _____

_____

_____

_____

_____

(Optional) Correction _____

_____

_____

(Optional) Your Name and Address _____

_____

_____

Gas Metal Arc Welding (GMAW) – Aluminum Plate and Pipe

COURSE MAP

This course map shows all of the modules in the third level of the Welding curriculum. The suggested training order begins at the bottom and proceeds up. Skill levels increase as you advance on the course map. The local Training Program Sponsor may adjust the training order.

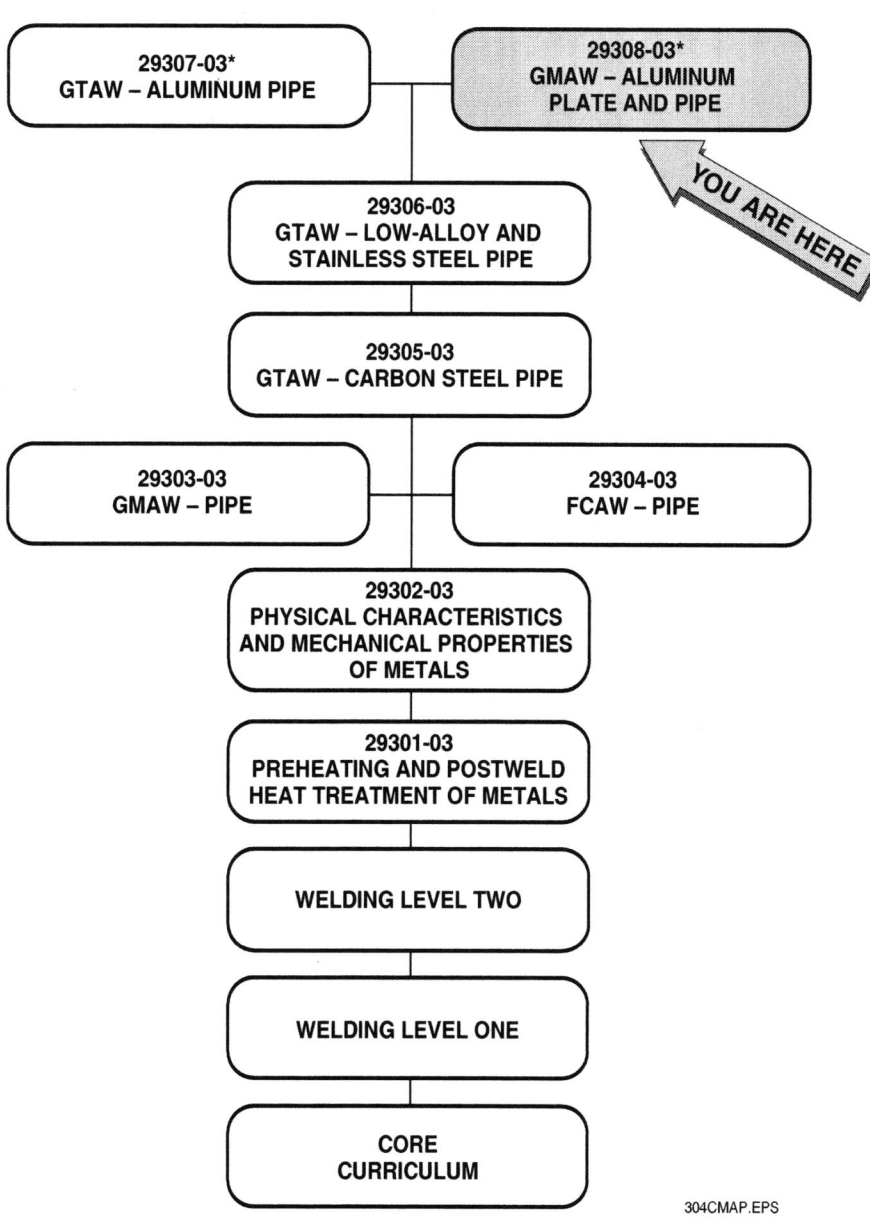

WELDING LEVEL THREE

29307-03*
GTAW – ALUMINUM PIPE

29308-03*
GMAW – ALUMINUM
PLATE AND PIPE

YOU ARE HERE

29306-03
GTAW – LOW-ALLOY AND
STAINLESS STEEL PIPE

29305-03
GTAW – CARBON STEEL PIPE

29303-03
GMAW – PIPE

29304-03
FCAW – PIPE

29302-03
PHYSICAL CHARACTERISTICS
AND MECHANICAL PROPERTIES
OF METALS

29301-03
PREHEATING AND POSTWELD
HEAT TREATMENT OF METALS

WELDING LEVEL TWO

WELDING LEVEL ONE

CORE
CURRICULUM

304CMAP.EPS

*Please note that Modules 29307-03 and 29308-03 are electives
for those progressing through the *Welding Level Three* program.

Figures

Gas Metal Arc Welding (GMAW) – Aluminum Plate and Pipe

Objectives

When you have completed this module, you will be able to do the following:

1. Explain GMAW, and set up equipment to weld aluminum.
2. Build a pad with stringer beads and weave beads, using aluminum wire and shielding gas.
3. Perform multiple-pass fillet welds on aluminum plate in the following positions, using aluminum wire and shielding gas:

   - 1F (flat)
   - 2F (horizontal)
   - 3F (vertical)
   - 4F (overhead)

4. Perform V-groove welds on aluminum plate in the following positions, using aluminum wire and shielding gas:

   - 1G (flat)
   - 2G (horizontal)
   - 3G (vertical)
   - 4G (overhead)

5. Perform V-groove welds on aluminum pipe in the following positions, using aluminum wire and shielding gas:

   - 1G-ROTATED (flat)
   - 2G (horizontal)
   - 5G (multiple)
   - 6G (inclined multiple)

Prerequisites

Before you begin this module, it is recommended that you successfully complete the following: Core Curriculum; Welding Levels One and Two; Welding Level Three, Modules 29301-03 through 29306-03.

Required Trainee Materials

1. Pencil and paper
2. Appropriate personal protective equipment

1.0.0 ◆ INTRODUCTION

GMAW is an arc welding process that uses a continuous, consumable solid wire electrode for the filler metal and shielding gas to protect the weld zone. (See *Figure 1*.) The GMAW process is commonly used to make welds on carbon, low-alloy, and stainless steel, as well as aluminum and other metals. *Figure 2* shows typical GMAW equipment.

GMAW is fast and effective for producing high-quality welds. Because this type of welding can be continuous, discontinuities and restarts are

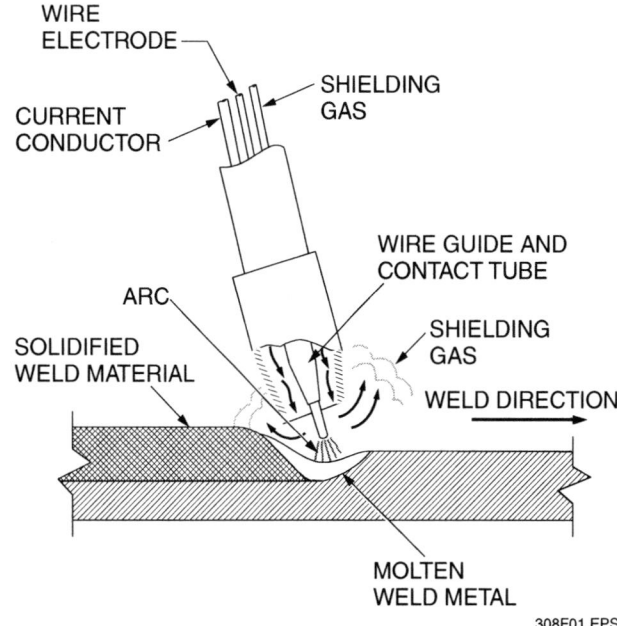

Figure 1 ◆ GMAW process.

POWER
SOURCE

GMAW
PUSH-PULL
WIRE FEEDER

PUSH-PULL
GUN

308F02.EPS

Figure 2 ◆ Gas shielded GMAW equipment.

reduced. With some materials, such as aluminum, it is common field practice to use GTAW for the root pass and then GMAW to complete the fill and cover passes. For the purposes of this training module, the dimensions used are representative of various codes and standards. Always refer to the applicable code, standard, or site welding procedure specification (WPS).

2.0.0 ◆ SAFETY SUMMARY

The following is a summary of safety procedures and practices you must observe while cutting or welding. Keep in mind that this is just a summary. Complete safety coverage is provided in the Level One module *Welding Safety*. If you have not completed that module, do so before continuing. Above all, be sure to wear appropriate protective clothing and equipment when welding or cutting.

2.1.0 Protective Clothing and Equipment

- Always use safety goggles with a full face shield or a helmet. The goggles, face shield, or helmet lens must have the proper light-reducing tint for the type of welding or cutting to be performed. Never directly or indirectly view an electric arc without using a properly tinted lens.

- Wear proper protective leather and/or flame retardant clothing and welding gloves to protect from flying sparks, molten metal, and heat.

- Wear 8" or taller high-top safety shoes or boots. Make sure that the tongue and lace area of the footwear is covered by a pant leg. If the tongue and lace area is exposed or the footwear must be protected from burn marks, wear leather spats under the pants or chaps and over the front top of the footwear.

- Wear a solid material (non-mesh) hat with a bill pointing to the rear or, if much overhead cutting or welding is required, a full leather hood with a welding faceplate and the correct tinted lens. If a hard hat is required, use one that allows the attachment of both rear deflector material and a face shield.

- If a full leather hood is not worn, wear a face shield and snug fitting welding goggles over safety glasses for gas welding or cutting. Either the face shield or the lenses of the welding goggles must be an approved shade 5 or 6 filter. For electric arc welding or cutting, wear safety goggles and a welding hood with the correct tinted lens (shade 10 to 14).

- If a full leather hood is not worn, wear earplugs to protect your ear canals from sparks.

2.2.0 Fire/Explosion Prevention

- Never carry matches or gas-filled lighters in your pockets. Sparks can cause the matches to ignite or the lighter to explode, causing serious injury.

- Never perform any type of heating, cutting, or welding until a hot-work permit is obtained and an approved fire watch is established. Most work-site fires caused by these types of operations are started by cutting torches.

- Never use oxygen to blow off clothing. The oxygen can remain trapped in the fabric for a time. If a spark hits the clothing during this period, the clothing can burn rapidly and violently out of control.

- Make sure that any flammable material in the work area is moved or is shielded by a fire-resistant covering. Approved fire extinguishers must be available before attempting any heating, welding, or cutting operations.

- Never release a large amount of oxygen nor use oxygen as compressed air. Its presence around flammable materials or sparks can cause rapid and uncontrolled combustion. Keep oxygen away from oil, grease, and other petroleum products.

- Never release a large amount of fuel gas, especially acetylene. Methane and propane tend to concentrate in and along low areas and can ignite at a considerable distance from the release point. Acetylene is lighter than air but is even more dangerous than methane. When mixed with air or oxygen, it will explode at much lower concentrations than any other fuel.
- To prevent fires, maintain a neat and clean work area, and make sure that any metal scrap or slag is cold before disposal.
- Before cutting or welding containers such as tanks or barrels, check to see if they have contained any explosive, hazardous, or flammable materials, including petroleum products, citrus products, or chemicals that decompose into toxic fumes when heated. As a standard practice, always clean and then fill any tanks or barrels with water, or purge them with a flow of inert gas to displace any oxygen.

2.3.0 Work Area Ventilation

- Make sure to follow confined space procedures before conducting any welding or cutting in the confined space.
- Never use oxygen to ventilate confined spaces.
- Always perform cutting or welding operations in a well-ventilated area. Such operations involving zinc or cadmium materials or coatings result in toxic fumes. For long-term cutting or welding of these materials, always wear an approved, full-face, supplied-air respirator (SAR) that uses breathing air supplied from outside of the work area. For occasional, very short-term exposure, you may use a high-efficiency particulate arresting (HEPA)-rated or metal-fume filter on a standard respirator.
- Make sure confined spaces are ventilated properly for cutting or welding operations.

3.0.0 ◆ WELDING PREPARATION

Before welding can begin, the area has to be readied, the welding equipment must be set up, and the metal to be welded must be prepared. The following sections explain how to set up the equipment for welding.

To practice welding, you need a welding table, bench, or stand. The welding surface can be steel, but an aluminum surface is preferred. Provisions must be available for placing weld coupons out of position.

To set up the area for welding, follow these steps:

Step 1 Check to ensure the area is properly ventilated. Make use of doors, windows, and fans.

Step 2 Check the area for fire hazards. Remove any flammable materials before proceeding.

Step 3 Locate the nearest fire extinguisher. Do not proceed unless the extinguisher is charged and you know how to use it.

Step 4 Position a welding table near the welding machine.

Step 5 Set up flash shields around the welding area.

3.1.0 Practice Welding Coupons

If possible, cut welding coupons from ⅜"-thick aluminum plate. If this size is not readily available, you can use aluminum plate ¼" to ¾" thick.

Because aluminum tends to be very susceptible to contamination, careful coupon preparation is important. Use an approved solvent, such as acetone, to remove oil and grease from the coupons. Use a grinder with a stainless steel buffer wheel, as necessary, to remove all other contaminants and oxidation. Use only light pressure.

 **WARNING!**
To prevent injury, always comply with the MSDS guidelines for the cleaning solvent(s) being used.

 *Conserve Aluminum Practice Coupons*

Aluminum for practice welding is expensive and difficult to obtain. Make every effort to conserve and not waste available material. Completely use welding coupons until all surfaces have been welded. Weld on both sides of the joint and then cut the welding coupon apart and reuse the pieces. To practice running beads, use any material that cannot be cut into welding coupons.

3.1.1 Plate Coupons

The coupons must be shaped to allow the following welds:

- *Stringer or weave beads (both running and overlapping)* – The coupons can be any easily handled size or shape.
- *Fillet welds* – Cut the metal into 4" × 5" rectangles for the base and 3" × 5" rectangles for the web. (See *Figure 3*.)
- *V-groove welds with backing* – Referring to both joint H of *Figure 4* and to *Figure 5*, cut the metal into 3" × 5" rectangles with one or both of the 5" lengths beveled at 30°. Grind a 0" to ³⁄₁₆" root face on the bevel as directed by your instructor. Cut backing strips at least 1½" wide and 6" long from the same base metal as the beveled pieces.

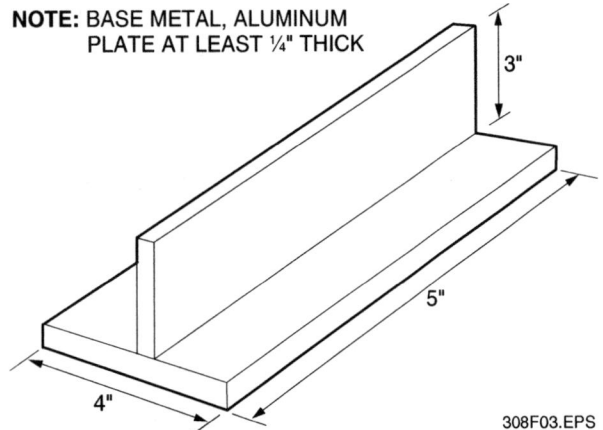

NOTE: BASE METAL, ALUMINUM PLATE AT LEAST ¼" THICK

3"

5"

4"

308F03.EPS

Figure 3 ◆ Fillet weld coupons.

NOTE

This module uses joint H shown in *Figure 4* for practice purposes; however, the other configurations can also be used at the direction of the instructor.

Follow these steps to tack weld each V-groove weld coupon with backing.

Step 1 Check the bevel angle. It should be approximately 30°.

Step 2 Center the beveled strips on the backing strip with a ³⁄₈" root opening (*Figure 6*), and tack weld them in place. (If desired, use a piece of metal or rod with the required thickness as a temporary spacer.) Place the tack welds on the back side of the joint in the lap formed by the backing strip and the beveled plate. Use three to four ½" tack welds on each beveled plate. Be sure the backing strip is tight against the beveled plates.

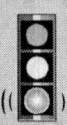

NOTE

Before tacking the strip in place, make sure that the surface of the backing strip under the root opening has been cleaned of all oxide coating.

Backing Strip

When the backing strip is ½" to 1" longer at each end, as shown in *Figure 6*, it allows you to start and stop the bead outside the weld groove.

An alternate joint preparation can be used when welding in the horizontal position: it has one plate beveled at approximately 45° and the other plate at approximately 15°. The 45°-beveled plate is positioned above the 15° plate with a ³⁄₈" root opening, as shown in *Figure 7*. (Note that degrees shown are approximate.)

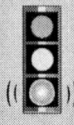

NOTE

Check with your instructor on whether to use the standard V-groove preparation or the alternate preparation for horizontal weld coupons.

3.1.2 Pipe Coupons

Pipe weld coupons should be cut from 3" to 12" diameter Schedule 40 aluminum pipe, and each welded joint will require two coupons of the same size. *Figure 8* shows typical AWS aluminum pipe specifications.

To prepare pipe coupons for V-groove weld joints with backing, follow these steps:

Step 1 Use an approved solvent and a clean rag to remove any grease or oil within a minimum of 1" from the weld joint. Clean the inside and outside of the aluminum pipe with a grinder or stainless steel brush.

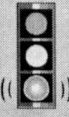

NOTE

If you use a grinding wheel, make sure it is a special high-speed grinding wheel approved to use on aluminum. Ensure that it has not been used on any other metal or used to remove hydrocarbon contaminants.

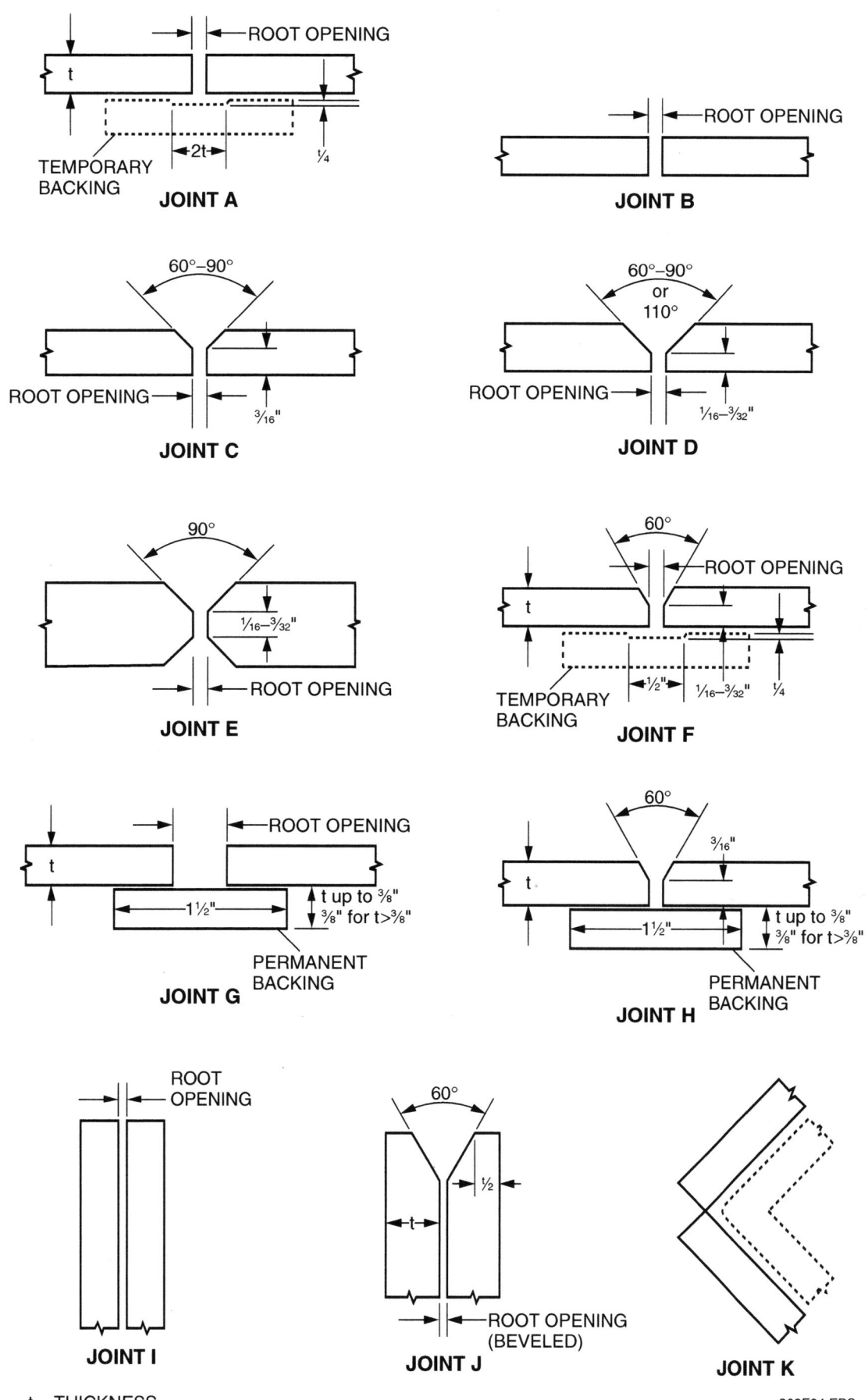

Figure 4 ◆ Joint preparations and geometrics.

308F04.EPS

NOTE: BASE METAL, ALUMINUM
PLATE AT LEAST ¼" THICK

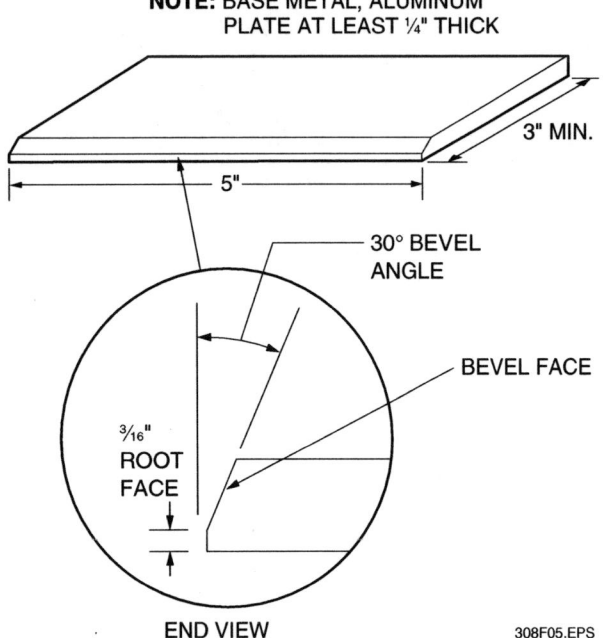

30° BEVEL
ANGLE

BEVEL FACE

³⁄₁₆"
ROOT
FACE

END VIEW

308F05.EPS

Figure 5 ◆ Weld coupon dimensions.

BACKING STRIP 1½" MIN. WIDTH
EXTENDED ½" MIN. FROM PLATES

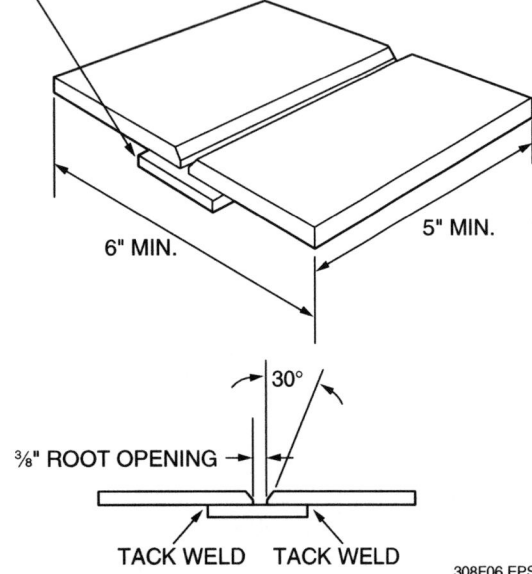

6" MIN.

5" MIN.

30°

⅜" ROOT OPENING

TACK WELD   TACK WELD

308F06.EPS

Figure 6 ◆ V-groove with metal backing weld coupon.

NOTE: BASE METAL, ALUMINUM
PLATE AT LEAST ¼" THICK

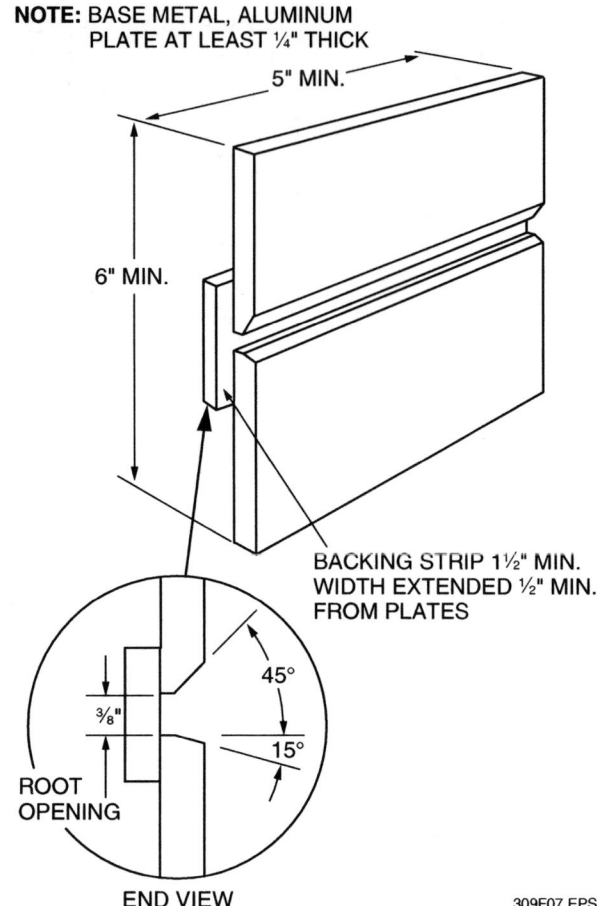

5" MIN.

6" MIN.

BACKING STRIP 1½" MIN.
WIDTH EXTENDED ½" MIN.
FROM PLATES

45°

⅜"

15°

ROOT
OPENING

END VIEW

309F07.EPS

Figure 7 ◆ Alternate horizontal weld coupon.

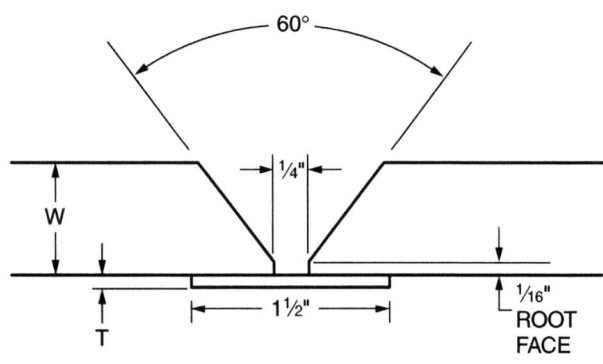

60°

¼"

W

1½"

T

1⁄16"
ROOT
FACE

(W) WALL THICKNESS (IN.)	(T) BACKING THICKNESS (IN.)
0.13	0.07
0.14	0.07
0.15	0.09
0.20	0.09
0.22	0.09
0.23	0.09
0.24	0.13
0.26	0.13
0.28	0.20
0.32	0.20
0.37	0.20
0.41	0.20

309F08.EPS

Figure 8 ◆ AWS aluminum pipe groove specifications for use with backing.

WARNING!

To prevent injury, always comply with the MSDS guidelines for the cleaning solvent(s) being used.

Step 2 Bevel the end of the pipe as shown in *Figure 8* by any acceptable beveling method, such as by using a mechanical cutter or by thermal cutting or grinding.

Step 3 Cut off a section of the beveled pipe end (4" minimum).

NOTE

For 1G-ROTATED welding, you may have to cut longer coupons to fit the rollers, if used.

Step 4 Check the bevel. There should be no dross, and the groove angle should be as shown in *Figure 8* with no notches more than ¹⁄₁₆" deep.

Step 5 Grind or file a root face on the bevel as specified by your instructor or as shown in *Figure 8*.

For this module, the practice pipe coupons will be tacked with a root opening and a permanent backing ring without any backing gas.

Step 1 Using a permanent backing ring compatible with the size and alloy of the aluminum pipe, insert the ring inside one of the beveled edges of a pipe coupon section. Make an allowance for the ¹⁄₄" root opening, and center the backing ring at the joint edge. Clamp the ring to the pipe at several places.

Step 2 Tack weld the ring to the pipe at four equally spaced positions around the pipe. The tack welds should be narrow and should not exceed 1" in length. Clean the tack welds.

Step 3 Place a beveled edge of the other pipe coupon section over the backing ring. Adjust the section on the ring to allow for the proper root opening. Then, make the first of four tacks parallel to one of the existing tack welds on the ring.

NOTE

Heavy-wall or large-diameter pipe may require tacks of greater length or number.

Step 4 After the first tack weld, check the root opening on the opposite side. Adjust the gap if necessary, and make the second tack weld on the opposite side from the first tack made in Step 3 and parallel to the existing tack weld made in Step 2.

Step 5 Check the root opening again and weld the third tack midway between the first two tacks and parallel to the existing tack weld from Step 2.

Step 6 Weld the fourth tack opposite the third tack and midway between the first and second tacks. There should now be four sets of tack welds evenly spaced (90°) around the pipe coupon. This is illustrated by *Figure 9*.

Step 7 Clean and feather the tack welds with the edge of a grinding wheel. Feathering the ends of the tack welds with a grinder helps fuse the tack welds into the root pass.

3.2.0 The Welding Machine

Identify a proper welding machine for GMAW and follow these steps to set it up for use.

Step 1 Verify that the welding machine can be used for GMAW.

Step 2 Check to be sure that the welding machine is properly grounded through the primary current receptacle.

Machined Grooves and Root Faces

Machining the grooves and root faces of aluminum pipe is the cleanest and most uniform method of preparing pipe for GMAW. This is especially true for aluminum where you must remove all contaminants from thermal cutting to eliminate weld discontinuities.

Welding Large-Diameter Pipe

The practice coupons on large-diameter pipe must be greater than 6" in length. They need more metal to help absorb the increased heat used to make these welds.

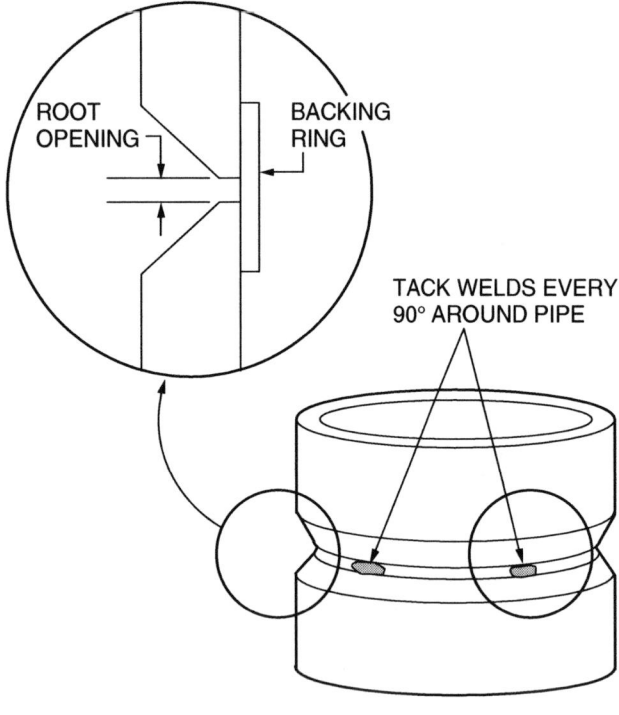

308F09.EPS

Figure 9 ◆ Tacked open-root V-groove weld coupon.

Step 3 Verify the location of the primary disconnect.

Step 4 Configure the welding machine for GMAW parameters as directed by your instructor (*Figure 10*). Configure the gun polarity (DCEP) and equip the gun with the correct nozzle for the application and the correct liner material and contact tube for the diameter of the wire being used.

Step 5 In accordance with the manufacturer's instructions, configure and load the wire feeder and gun with the proper diameter solid electrode wire as directed by your instructor.

Step 6 Connect the proper shielding gas (argon or argon/helium) for the application as described in the previous level and as specified by the wire electrode manufacturer, WPS, site quality standards, or your instructor.

Step 7 Connect the clamp of the workpiece lead to the workpiece.

Step 8 Turn on the welding machine, and purge the gun as directed by the gun manufacturer's instructions.

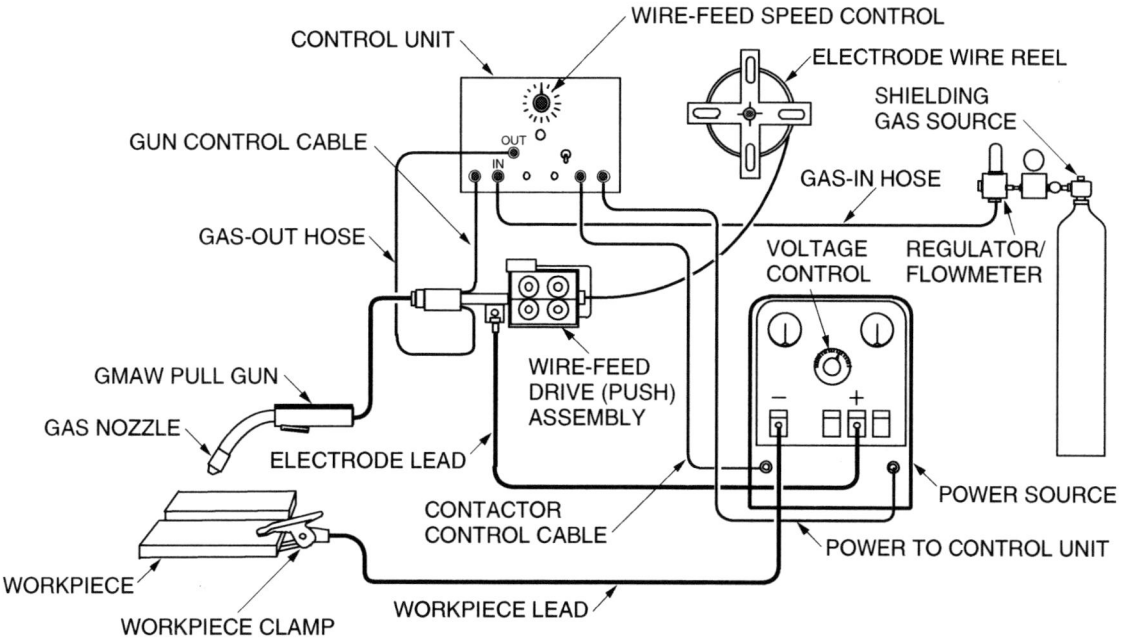

NOTE: THE POLARITY OF THE GUN AND WORKPIECE LEADS IS DETERMINED BY THE TYPE OF FILLER METAL AND APPLICATION.

308F10.EPS

Figure 10 ◆ Configuration diagram of typical GMAW machine.

Step 9 Set the initial welding voltage and wire feed speed as recommended by the manufacturer for the type and size electrode wire being used.

3.2.1 Voltage

Arc length is determined by the voltage, which is set at the power source. Arc length is the distance from the wire electrode tip to the base metal or to the molten pool at the base metal. (See *Figure 11*.) If voltage is set too high, the arc will be too long and can cause the wire to melt and fuse to the contact tube. Excessively high voltage also causes porosity and excessive spatter. Voltage must be increased or decreased as wire feed speed is increased or decreased. Set the voltage to maintain consistent spray transfer.

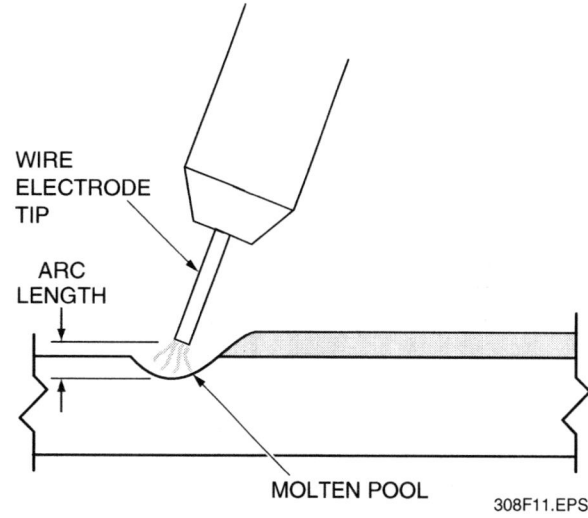

308F11.EPS

Figure 11 ◆ Arc length.

Wire Electrode Manufacturer's Recommendations

Always obtain and follow the manufacturer's recommendations for using shielding gas and for setting the initial welding voltage and wire feed speed parameters. These balanced parameters are critical and are based on the welding position, size, and composition of the solid wire.

Arc Voltage

A minimum arc voltage is needed to maintain spray transfer. However, penetration is not directly related to voltage. Penetration will increase with voltage for a time but will actually decrease if the voltage is increased above its optimum.

3.2.2 Amperage

With a standard constant-voltage power source, the electrode feed speed controls the welding amperage after the initial recommended setting. The welding power source provides the amperage needed to melt the wire electrode while maintaining the selected welding voltage. Within limits, when the wire electrode feed speed is increased, the welding amperage and deposition rate also increase. This results in higher welding heat, deeper penetration, and higher beads. When the wire electrode feed speed is decreased, the welding amperage automatically decreases. With lower welding amperage and less heat, the deposition rate drops and the weld beads are smaller with less penetration.

Note that some constant-voltage power sources used for GMAW/FCAW provide varying degrees of current modification, such as slope or induction adjustments. Power sources with a slope adjustment allow you to vary the amount of amperage change in relation to voltage range of the unit. Standard constant-voltage GMAW/FCAW units have a current slope fixed by the manufacturer for general welding applications and conditions. Pulse transfer power sources allow you to adjust the peak current for the pulse and the background current between pulses to match specific welding applications and/or wire electrode requirements. In other cases, where a voltage sensing wire feeder is used with a constant-current power source, the current is adjustable, and the wire feeder varies the wire feed speed to maintain proper arc voltage at the wire electrode tip.

3.2.3 Weld Travel Speed

Weld travel speed is the speed at which the electrode tip passes across the base metal in the direction of the weld. It is measured in inches per minute (ipm). Travel speed has a great effect on penetration and bead size: slower travel speeds build higher beads with deeper penetration while faster travel speeds build smaller beads with less penetration. Ideally, the welding parameters should be adjusted so that the electrode tip is positioned at the leading edge of the weld puddle during travel.

3.2.4 Gun Position

The gun position in relation to the direction of the weld is defined for aluminum as shown in *Figure 12*. A push angle is used to allow base metal cleaning.

3.2.5 Electrode Extension, Stickout, and Standoff Distance

Electrode extension is the length of the wire that extends beyond the tip of the welding gun's contact tube. Increasing extension is useful for bridging gaps and compensating for mismatch, but it can cause overlap or lack of fusion and a ropy bead appearance. When the extension is decreased, the power source is forced to increase its current output to burn off the wire. Too little extension can cause the wire to weld to the contact tube and can develop porosity in the weld. For high-conductivity metal wires, the preheating effect of wire resistance is minimal, and wire speed and voltage settings have a more direct effect on the amount of weld penetration and deposit rate.

GMAW electrode extension spray transfer varies between ½" to 1". *Figure 13* shows the typical electrode extension for the spray transfer GMAW gun configuration and defines other gun nomenclature. Stickout is the distance from the gas nozzle or insulating nozzle to the end of the electrode. Standoff distance is the distance from the gas nozzle or insulating nozzle to the workpiece. Contact tube extension or setback is usually dependent on the transfer mode for the GMAW application. Spray transfer or pulsed spray transfer is normally used for aluminum.

3.2.6 Gas Nozzle Cleaning

As the welding machine is used, weld spatter accumulates on the gas nozzle and the contact tube. If the gas nozzle is not occasionally cleaned, it will restrict the shielding gas flow, which will cause porosity in the weld.

Travel Speed and Wire Feed Speed

Inexperienced welders are tempted to turn down the wire feed speed if they experience difficulty controlling the weld puddle. Wire electrodes must be run at certain balanced parameters that cannot be changed individually. Voltage, current (if variable), wire feed speed, and travel speed are adjusted and balanced together to control the weld puddle.

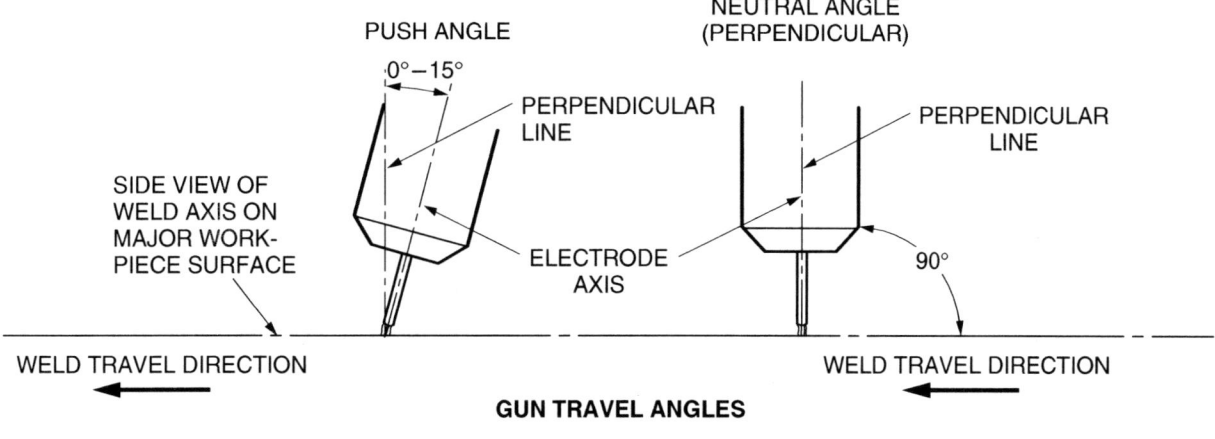

Figure 12 ◆ Gun work and travel angles.

Clean the gas nozzle with a reamer, round file, or the tang of a standard file. After cleaning, the nozzle can be sprayed with or dipped in a special antispatter compound. The antispatter compound helps to prevent the spatter from sticking to the nozzle.

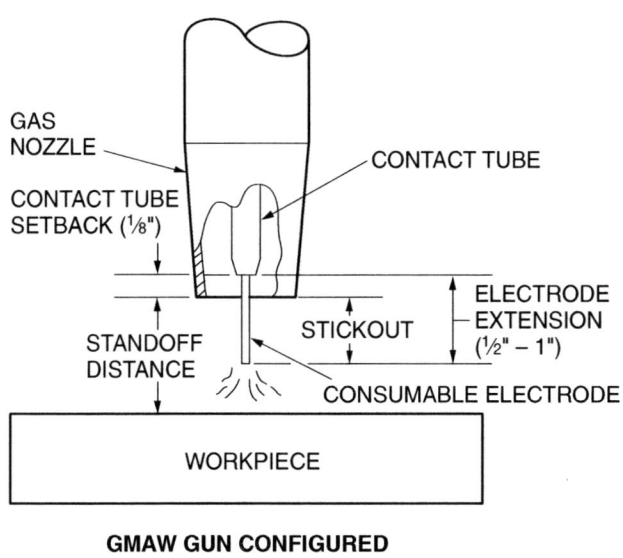

GMAW GUN CONFIGURED FOR SPRAY TRANSFER

308F13.EPS

Figure 13 ◆ GMAW gun configuration for aluminum welding.

4.0.0 ◆ WELDING BEADS

There are two basic bead types: stringer beads and weave beads. Aluminum tends to get contaminated more easily than steels, so you must take additional care to clean the base metals just before starting either type of bead. Always refer to the WPS for cleaning instructions and for the proper chemicals to remove oil and grease.

4.1.0 Bead Types

Stringer beads (*Figure 14*) are made with little or no side-to-side motion of the gun. Practice running stringer beads in the flat position. Experiment with different push angles and stickouts.

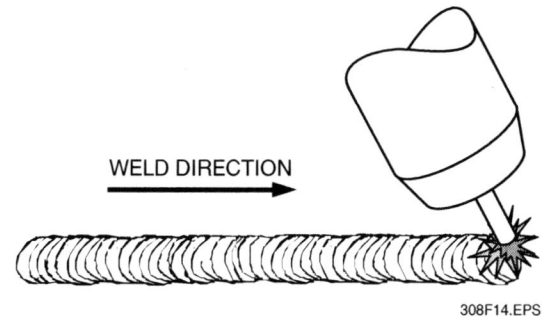

308F14.EPS

Figure 14 ◆ Stringer bead.

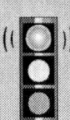

Hot Tip

Antispatter Compound

Use only antispatter material specifically designed for GMAW gas nozzles. Water-based compounds may cause porosity in aluminum. Always refer to material safety data sheets (MSDS) for specific safety considerations.

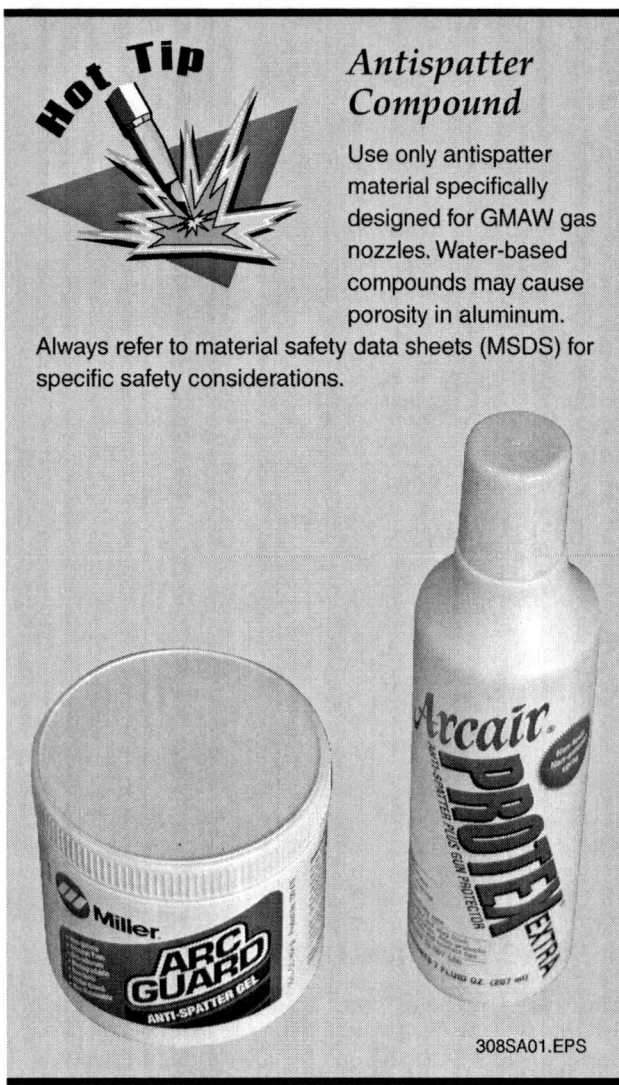

308SA01.EPS

Follow these steps to make stringer beads:

Step 1 Clean the base metal before welding.

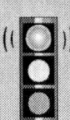

WARNING!

To prevent personal injury, always comply with the MSDS guidelines for the cleaning solvent(s) being used.

Step 2 Hold the gun at the desired angle, with the electrode tip directly over the point where the weld is to begin, and pull the gun trigger.

Step 3 Hold the arc in place until the weld puddle begins to form.

Step 4 Slowly advance the arc while you maintain the gun angle.

CAUTION

Stay in the leading edge of the puddle to prevent lack of fusion.

Step 5 Finish the bead by pausing until the crater is filled and then releasing the trigger.

Step 6 Inspect the bead for the following:
- Straightness
- Uniform appearance
- Smooth, flat transition with complete fusion at the toes of the weld
- Filled crater
- No porosity
- No undercut
- No cracks

Step 7 Continue practicing stringer beads until you can make acceptable welds every time.

Weave beads (*Figure 15*) are made with wide side-to-side motions, called oscillations, of the electrode. The width of a weave bead is determined by the amount of side-to-side motion.

When making a weave bead, you must use care at the toes to ensure proper tie-in to the base metal by slowing down or pausing slightly at the edges. Pausing at the edges will also flatten out the weld and give it the proper profile.

CAUTION

Do not exceed the width of weave beads that is often specified in the welding code or WPS used at your site.

Practice running weave beads in the flat position. Experiment with different weave motions, push angles, and stickouts.

308F15.EPS

Figure 15 ◆ Weave bead.

Follow these steps to make weave beads:

Step 1 Clean the base metal before welding.

Step 2 Hold the gun at the desired angle, with the electrode tip directly over the point where the weld is to begin, and pull the gun trigger.

Step 3 Hold the arc in place until the weld puddle begins to form.

Step 4 Slowly advance the arc in a weaving motion (*Figure 16*) while you maintain the gun angle.

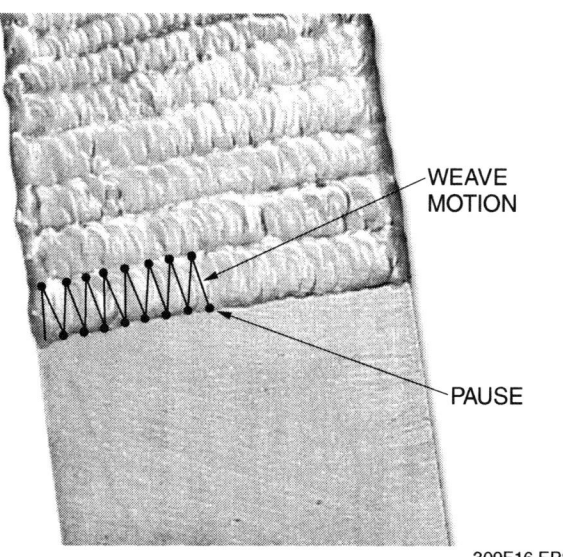

309F16.EPS

Figure 16 ◆ Weave motion.

Step 5 Finish the bead by pausing until the crater is filled and then releasing the trigger.

Step 6 Inspect the bead for the following:
- Straightness
- Uniform appearance
- Smooth, flat transition with complete fusion at the toes of the weld
- Filled crater
- No porosity
- No excessive undercut
- No cracks

Step 7 Continue practicing weave beads until you can make acceptable welds every time.

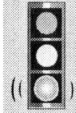

NOTE

Clean the beads between each pass.

4.2.0 Restarts

A restart is the junction where a new weld connects to and continues the bead of a previous weld. A restart must be made properly so that it blends smoothly with the previous weld and does not stand out or create a weld defect. The technique for making a GMAW restart is the same for both stringer and weave beads.

Follow these steps to make a restart:

Step 1 Clean the base metal before welding.

Step 2 Hold the gun at the proper angle while restarting the arc directly over the center of the crater. (Welding codes do not allow arc strikes outside the area that is to be welded.)

Step 3 Move the electrode tip in a small circular motion over the crater to fill the crater with a molten puddle.

Step 4 Move to the leading edge of the puddle, and, as soon as the puddle fills the crater, continue the stringer or weave bead pattern.

Step 5 Inspect the restart.

NOTE

A properly made restart blends into the bead and is hard to detect. If the restart has undercut, not enough time was spent in the crater to fill it. If undercut is on one side or the other, use more of a side-to-side motion as you move back into the crater. If the restart has a lump, it was overfilled; too much time was spent in the crater before the forward motion was resumed.

Step 6 Continue to practice restarts until they are correct.

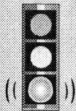

NOTE

Use the same techniques for making restarts whenever performing GMAW.

4.3.0 Terminations

A weld termination normally leaves a crater. Most welding codes require that when making a termination, the crater should be filled to the full cross section of the weld. This can be difficult because most terminations are at the edge of a plate where welding heat tends to build up, which makes it harder to fill the crater.

Follow these steps, illustrated in *Figure 17*, to terminate a weld:

Step 1 Start to bring the gun up to a 0° travel angle and slow the forward travel as you approach the end of the weld.

Step 2 Stop forward movement and back up about ⅛" from the end of the plate.

Step 3 Release the trigger when the crater is filled.

CAUTION

Do not remove the gun from the weld until the puddle has solidified. The shielding gas postflow that continues after the welding stops protects the molten aluminum. If you remove the gun and shielding gas before the crater has solidified, crater porosity or cracks can occur.

Step 4 Inspect the termination.

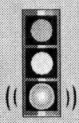

NOTE

The crater should be filled to the full cross section of the weld.

4.4.0 Overlaps

Overlapping beads are made by depositing connective weld beads parallel to one another. The parallel beads overlap to form a flat surface. This is also called padding. Overlapping beads are used to build up a surface and to make multiple-pass welds. Both stringer and weave beads can be overlapped. Properly overlapped beads, when viewed from the end, form a relatively flat surface. *Figure 18* illustrates overlapping beads.

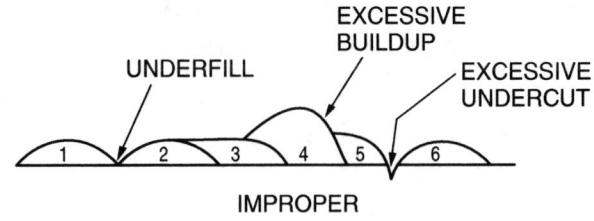

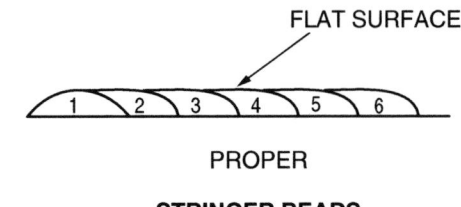

STRINGER BEADS

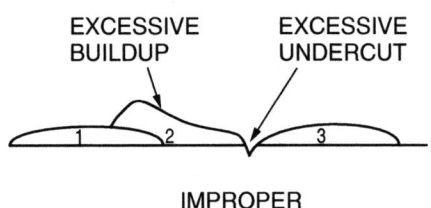

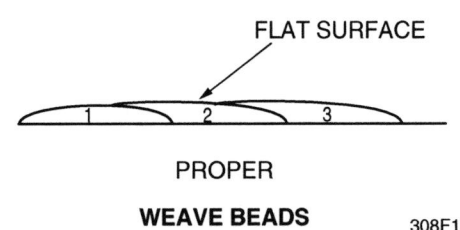

WEAVE BEADS   308F18.EPS

Figure 18 ◆ Proper and improper overlapping beads.

Using aluminum wire electrodes and the appropriate shielding gas, follow these steps to weld overlapping stringer or weave beads:

Step 1 Mark out a square on a piece of aluminum.

Step 2 Clean and deoxidize the base metal.

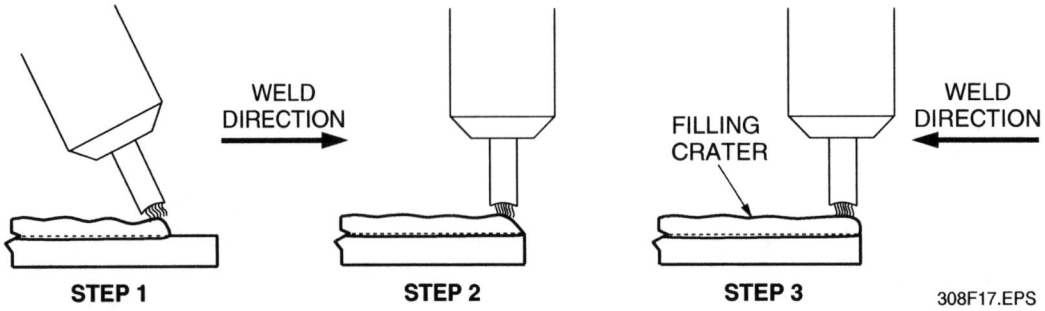

Figure 17 ◆ Terminating a weld.

Step 3 Weld a stringer or weave bead along one edge.

Step 4 Clean the bead with a stainless steel brush.

Step 5 Position the gun at a work angle of 10° to 15° toward the side of the previous bead to get proper tie-in, and then pull the trigger.

Step 6 Continue running overlapping stringer or weave beads until the square is covered.

Step 7 Continue building layers of stringer beads, one on top of the other, until the technique is perfected.

5.0.0 ◆ FILLET WELDS

The most common fillet welds are made in lap and T-joints. The weld position is determined by the axis of the weld. The positions for fillet welding are 1F (flat), 2F (horizontal), 3F (vertical), and 4F (overhead). (See *Figure 19*.) In the 1F and 2F positions, the weld axis can be inclined up to 15°. Any weld axis inclination for the other positions varies with the rotational position of the weld face as specified in AWS Standards.

Fillet welds can be concave or convex, depending on the WPS or site quality standards. Welding codes require a fillet weld to have a uniform concave or convex face, although a slightly non-uniform face is acceptable. The convexity of a fillet weld or individual surface bead must not exceed that permitted by the applicable code or standard. A fillet weld with profile defects is unacceptable and must be repaired. *Figure 20* shows various fillet weld profiles.

5.1.0 Flat (1F) Position

Practice 1F-position (flat) fillet welds by making multiple-pass (six-pass) convex fillet welds in a T-joint using an appropriate aluminum filler metal as directed by your instructor. When making flat fillet welds, pay close attention to the gun angle and travel speed. For the first bead, the gun angle is vertical (45° to both plate surfaces). Adjust the angle for all subsequent beads. (See *Figure 21*.)

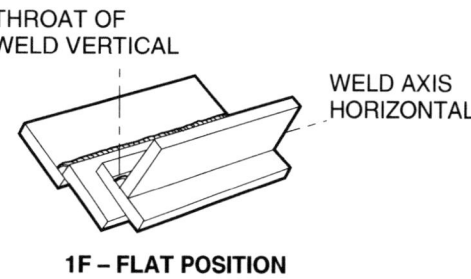

1F – FLAT POSITION

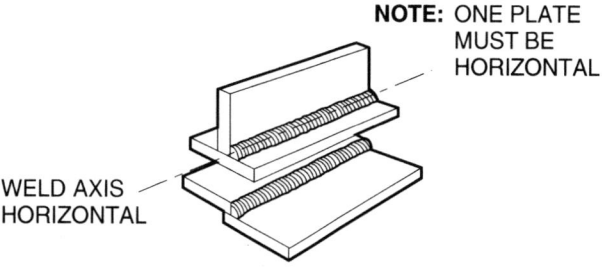

2F – HORIZONTAL POSITION

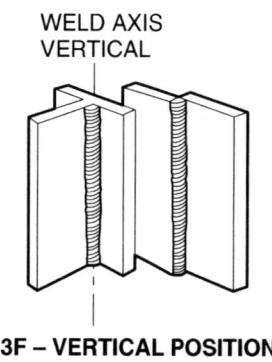

3F – VERTICAL POSITION

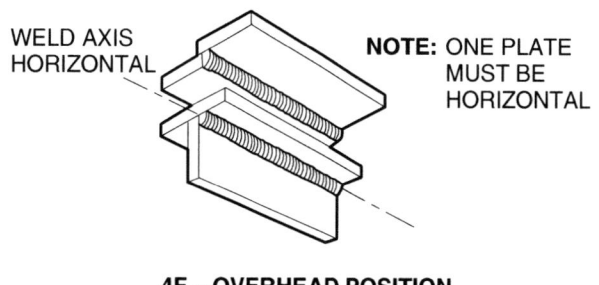

4F – OVERHEAD POSITION

308F19.EPS

Figure 19 ◆ Fillet welding positions.

Alternate Crater Filling or Elimination Techniques

One method to eliminate the crater when you terminate a GMAW weld on aluminum is to use run-off tabs. Where tabs cannot be used or are not practical, such as with pipe welds, you can reduce the weld pool before you extinguish the arc by increasing the travel speed just before you release the gun trigger. This results in a small crater that may be free of cracks. Another technique to fill the crater without adding heat to increase its size is to make several quick arc starts into the crater as it cools.

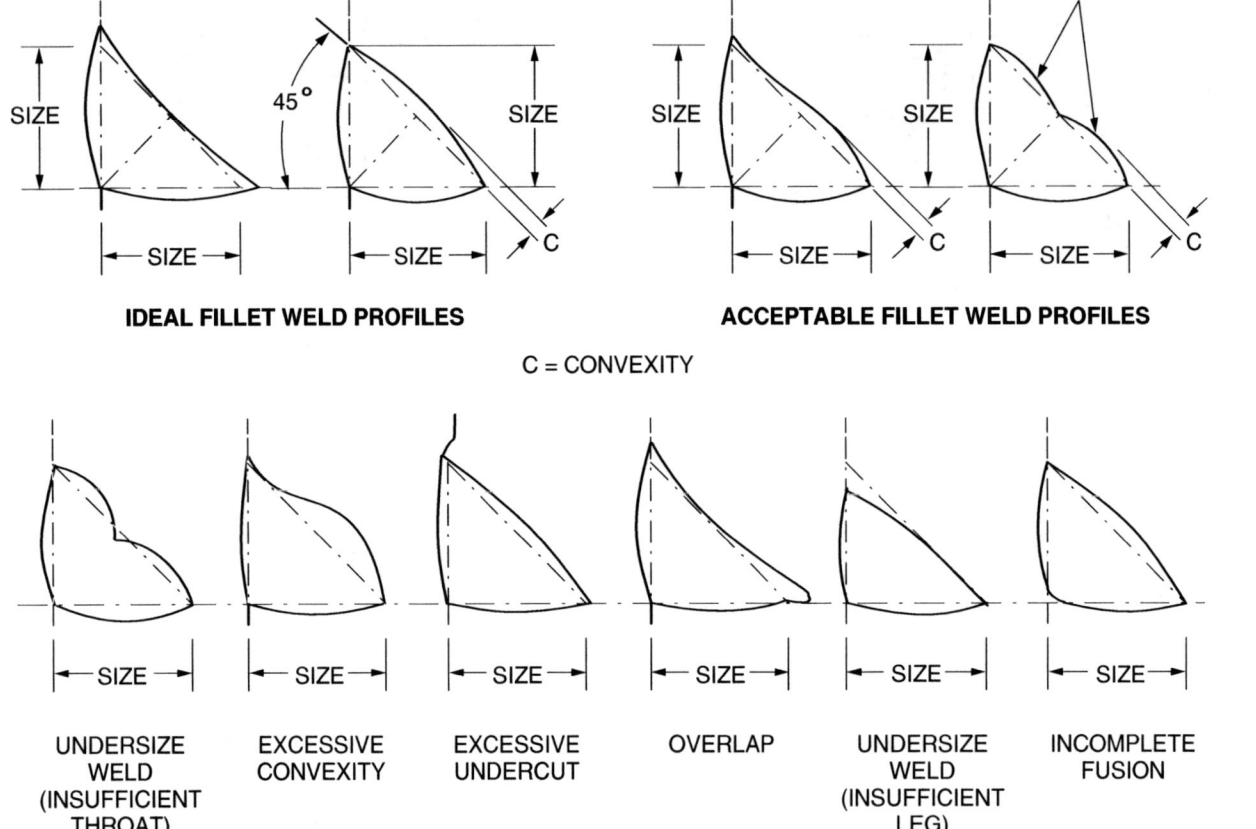

IDEAL FILLET WELD PROFILES

ACCEPTABLE FILLET WELD PROFILES

C = CONVEXITY

UNDERSIZE WELD (INSUFFICIENT THROAT)

EXCESSIVE CONVEXITY

EXCESSIVE UNDERCUT

OVERLAP

UNDERSIZE WELD (INSUFFICIENT LEG)

INCOMPLETE FUSION

UNACCEPTABLE FILLET WELD PROFILES

308F20.EPS

Figure 20 ◆ Acceptable and unacceptable fillet weld profiles.

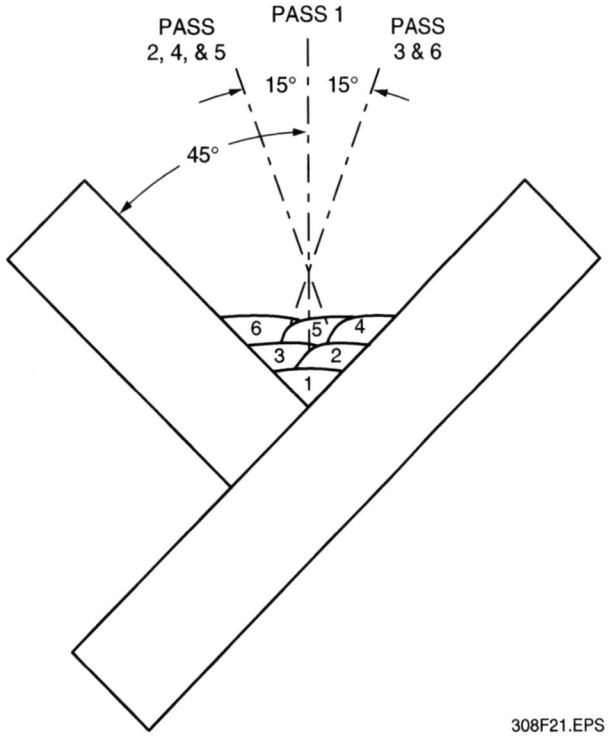

308F21.EPS

Figure 21 ◆ Multiple-pass 1F bead sequence and work angles.

Clean each completed bead with a stainless steel wire brush before starting the next bead.

Follow these steps to make a flat fillet weld:

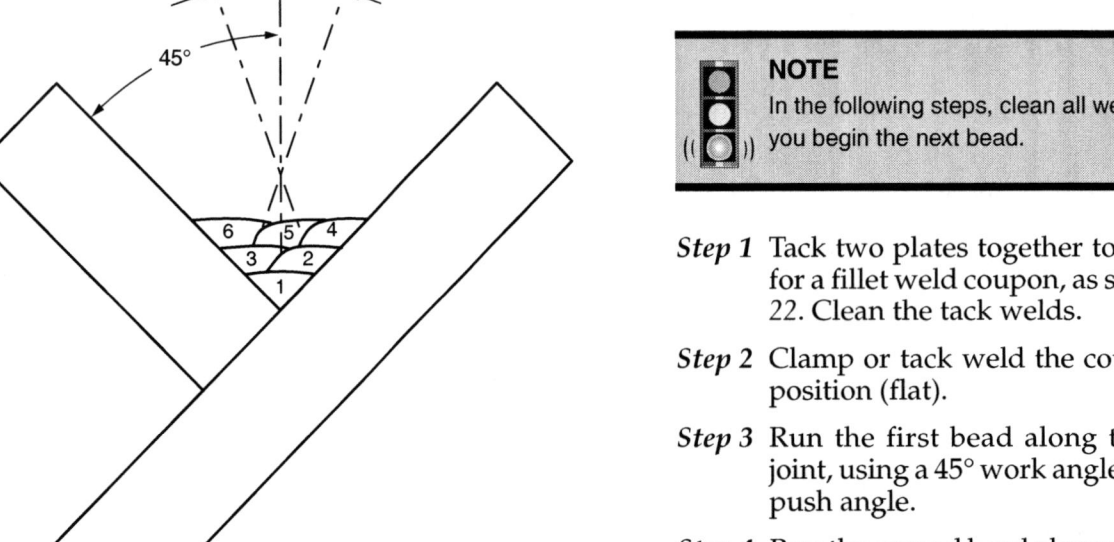

> **NOTE**
>
> In the following steps, clean all weld beads before you begin the next bead.

Step 1 Tack two plates together to form a T-joint for a fillet weld coupon, as shown in *Figure 22*. Clean the tack welds.

Step 2 Clamp or tack weld the coupon in the 1F position (flat).

Step 3 Run the first bead along the root of the joint, using a 45° work angle and a 5° to 10° push angle.

Step 4 Run the second bead along a toe of the first weld, overlapping about 75% of the first bead. Alter the work angle as shown in *Figure 21*, and use a 5° to 10° push angle with a slight oscillation.

Tacking and Aligning Workpieces

When tacking workpieces together, to position the workpieces and minimize distortion when the final welds are made, both sides of the workpieces are usually tacked with welds that are about ½" long. After the first tack weld, use a hammer or another tool to align the workpieces side to side and end to end; then tack the opposite side. Tack the far ends of the workpieces in the same manner. Intermediate tack welds can be made every 5" to 6", as necessary, to minimize lengthwise distortions.

T-Joint Heat Dissipation

In T-joints, the welding heat dissipates more rapidly in the thicker or non-butting member. On various bead passes, the arc may have to be concentrated slightly more on the thicker or the nonbutting member to compensate for the heat loss.

NOTE: BASE METAL, ALUMINUM PLATE AT LEAST ¼" THICK

308F22.EPS

Figure 22 ◆ Fillet weld coupon.

Step 5 Run the third bead along the other toe of the first weld, filling the groove created when the second bead was run. Use the work angle shown in *Figure 21* and a 5° to 10° push angle with a slight oscillation.

Step 6 Run the fourth bead along the outside toe of the second weld, overlapping about half the second bead. Use the work angle shown in *Figure 21* and a 5° to 10° push angle with a slight oscillation.

Step 7 Run the fifth bead along the inside toe of the fourth weld, overlapping about half the fourth bead. Use the work angle shown in *Figure 21* and a 5° to 10° push angle with a slight oscillation.

Step 8 Run the sixth bead along the toe of the fifth weld, filling the groove created when the fifth bead was run. Use the work angle shown in *Figure 21* and a 5° to 10° push angle with a slight oscillation.

Step 9 Inspect the weld. The weld is acceptable if it has the following:

- Uniform appearance on the bead face
- Craters and restarts filled to the full cross section of the weld
- Uniform weld size ±¹⁄₁₆"
- Acceptable weld profile in accordance with the applicable code or standard
- Smooth transition with complete fusion at the toes of the weld
- No porosity
- No excessive undercut
- No overlap
- No inclusions
- No cracks

5.2.0 Horizontal (2F) Position

Practice horizontal (2F) fillet welding by placing multiple-pass fillet welds in a T-joint, using an aluminum filler metal as directed by your instructor. When making horizontal fillet welds, pay close attention to the gun angles. For the first bead, the electrode work angle is 45°. The work angle is adjusted for all other welds.

Follow these steps to make a horizontal fillet weld:

Step 1 Tack two plates together to form a T-joint for the fillet weld coupon. Clean the tack welds.

Step 2 Clamp or tack weld the coupon in the horizontal position.

Step 3 Run the first bead along the root of the joint using a work angle of approximately 45° with a 5° to 10° push angle (*Figure 23*).

Step 4 Clean the weld.

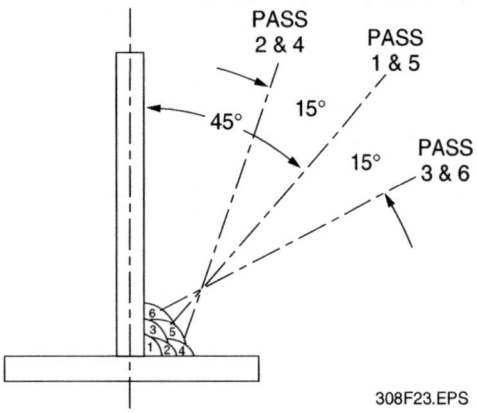

Figure 23 ◆ Multiple-pass 2F bead sequence and work angles.

Step 5 Run the remaining passes using a 5° to 10° push angle and a slight oscillation at the appropriate work angles (*Figure 23*). Overlap each previous pass. Clean the weld after each pass.

Step 6 Have the instructor inspect the weld. The weld is acceptable if it has the following:

- Uniform appearance on the bead face
- Craters and restarts filled to the full cross section of the weld
- Uniform weld size ±¹⁄₁₆"
- Acceptable weld profile in accordance with the applicable code or standard
- Smooth transition with complete fusion at the toes of the weld
- No porosity
- No excessive undercut
- No overlap
- No inclusions
- No cracks

5.3.0 Vertical (3F) Position

Practice vertical (3F) fillet welding by placing multiple-pass fillet welds in a T-joint, using an aluminum filler metal as directed by your instructor. Normally, vertical welds are accomplished by welding uphill from the bottom to the top and using a gun push angle (up-angle). Because of the uphill welding and push angle, this type of welding is sometimes called vertical-up fillet welding. Either stringer or weave beads can be used for vertical welding. The site WPS or site quality standard will specify which technique to use when on the job.

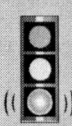

NOTE

Check with your instructor to see if you should practice stringer beads, weave beads, or both.

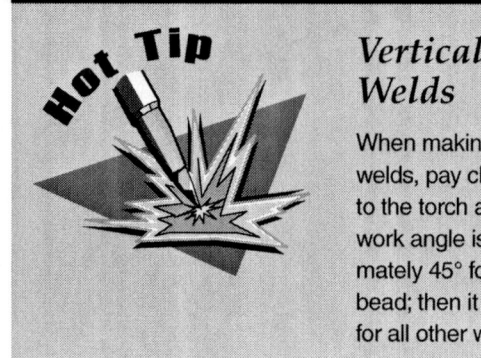

Vertical Fillet Welds

When making vertical fillet welds, pay close attention to the torch angles. The work angle is approximately 45° for the first bead; then it is adjusted for all other welds.

5.3.1 Weave Beads

Follow these steps to make an uphill fillet weld:

Step 1 Tack two plates together to form a T-joint for the fillet weld coupon.

Step 2 Clamp or tack weld the coupon in the vertical position.

Step 3 Starting at the bottom, run the first bead along the root of the joint, using a work angle of approximately 45° with a 10° to 15° push angle. Pause in the weld puddle to fill the crater.

Step 4 Clean the weld.

Step 5 Run the remaining passes using a 10° to 15° push angle and a side-to-side weave technique with a 45° work angle (*Figure 24*). Use a slow motion across the face of the weld, pausing at each toe to penetrate and fill the crater. Clean the weld after each pass.

Step 6 Have the instructor inspect the weld. The weld is acceptable if it has the following:

- Uniform appearance on the bead face
- Craters and restarts filled to the full cross section of the weld
- Uniform weld size ±¹⁄₁₆"
- Acceptable weld profile in accordance with the applicable code or standard
- Smooth transition with complete fusion at the toes of the weld
- No porosity
- No excessive undercut
- No overlap
- No inclusions
- No cracks

5.3.2 Stringer Beads

Repeat vertical fillet (3F) welding using stringer beads. Use a slight oscillation motion, pausing

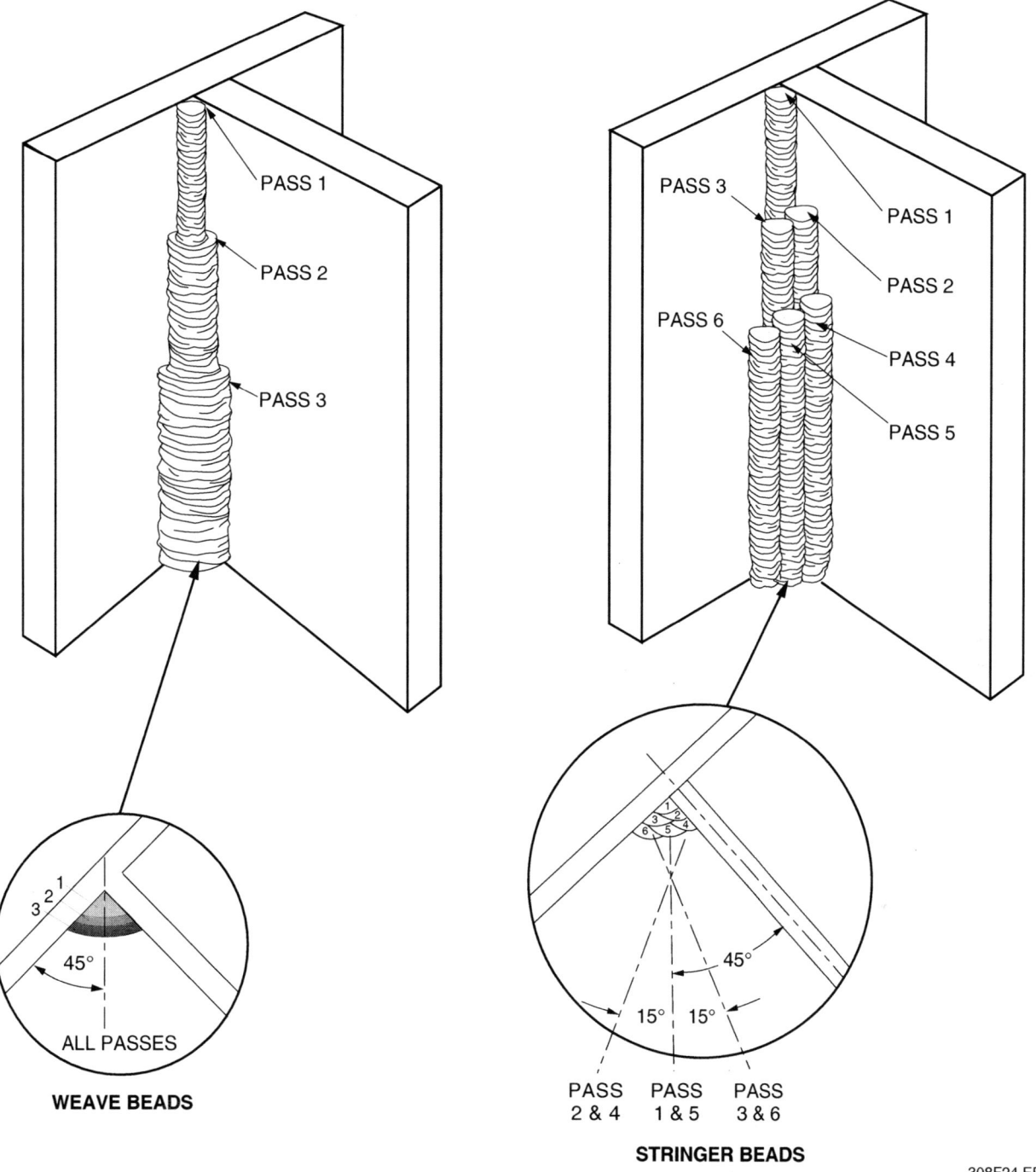

Figure 24 ◆ Multiple-pass 3F bead sequences and work angles for stringer and weave beads.

slightly at each toe to prevent undercut. For stringer beads, use a 10° to 15° push angle and the required work angles shown in *Figure 24*.

5.4.0 Overhead (4F) Position

Practice overhead (4F) fillet welding by welding multiple-pass fillet welds in a T-joint, using an aluminum filler metal as directed by your instructor. When making overhead fillet welds, pay close attention to the gun angles. The work angle is approximately 45° for the first bead, and then it is adjusted for all other welds.

To make an overhead fillet weld:

Step 1 Tack two plates together to form a T-joint for the fillet weld coupon.

Step 2 Clamp or tack weld the coupon so it is in the overhead position.

Step 3 Run the first bead along the root of the joint, using a work angle of approximately 45° with a 5° to 10° push angle.

Step 4 Clean the weld.

Step 5 Run the remaining passes using a 5° to 10° push angle and the work angles shown in *Figure 25*. Overlap each previous pass. Clean the weld after each pass.

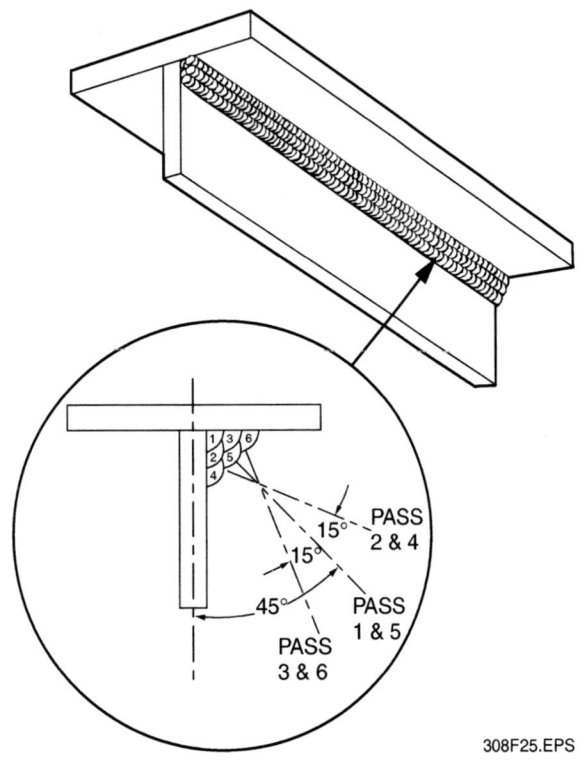

308F25.EPS

Figure 25 ◆ Multiple-pass 4F bead sequence and work angles.

6.0.0 ◆ V-GROOVE PLATE WELDS

The V-groove weld is a common groove weld normally made on plate and pipe. In this module, the backing method of welding an aluminum V-groove joint with GMAW is used because it is easier to master. Practicing the V-groove welds on plate will prepare you to make the more difficult pipe welds.

V-groove welds with backing can be made in all positions. The weld position is determined by the axis of the weld. Groove weld positions are flat (1G), horizontal (2G), vertical (3G), and overhead (4G), as shown in *Figure 26*.

After you complete the root pass, clean and inspect it. Use the appropriate cleaning method(s) to clean the root pass of the oxidation that tends to form along the joint. Inspect the root pass for the following:

- Uniformly smooth face
- Complete fusion
- No excessive buildup
- No excessive undercut
- No porosity

Practice making GMAW V-groove welds with backing in the 1G, 2G, 3G, and 4G positions, using appropriately sized aluminum filler wire for the passes. Pay particular attention to fill the crater at the weld termination. Clean each completed bead with a stainless steel wire brush before you start the next bead.

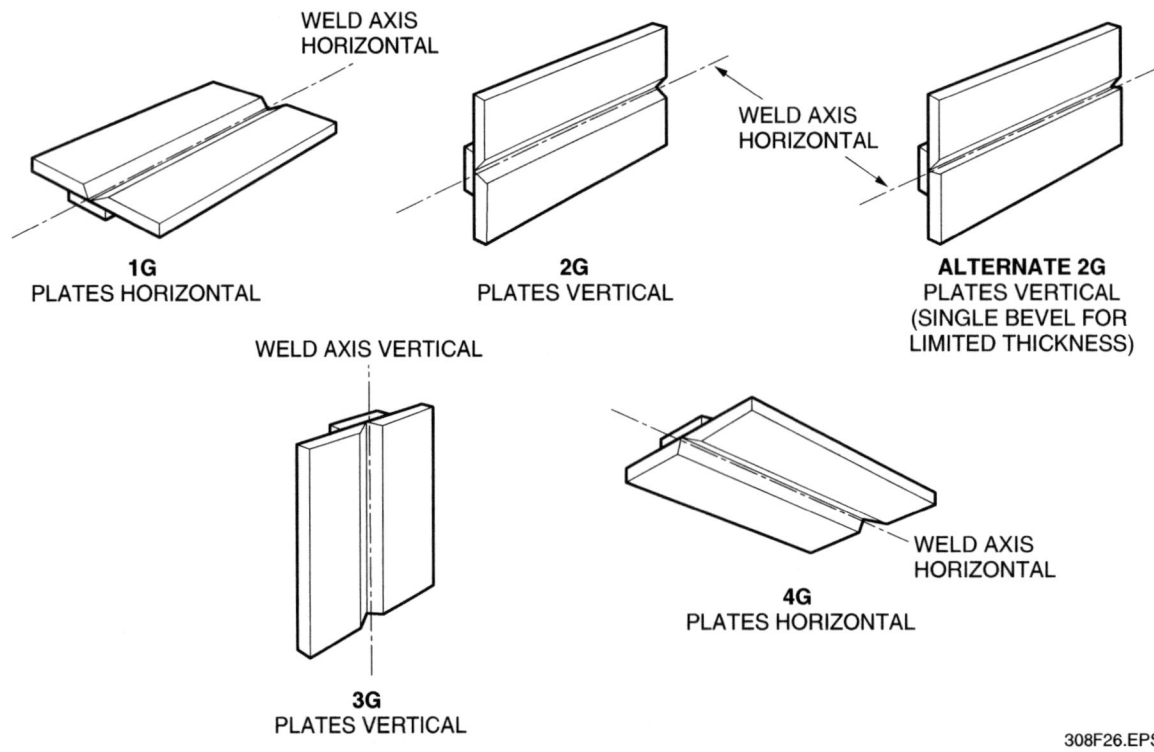

308F26.EPS

Figure 26 ◆ V-groove weld positions.

V-groove welds with backing should be made with slight reinforcement not exceeding ⅛" and a gradual transition to the base metal at each toe. The root pass should have complete fusion to the backing. Groove welds must not have excessive reinforcement, underfill, excessive undercut, or overlap. If a groove weld has any of these defects (shown in *Figure 27*), it must be repaired.

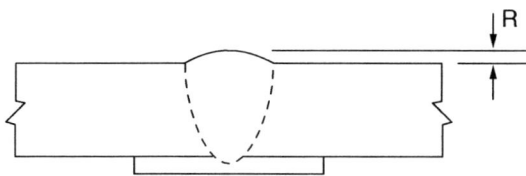

R = FACE REINFORCEMENT NOT TO
EXCEED ⅛" OR AS SPECIFIED BY CODE

**PROFILE OF ACCEPTABLE
V-GROOVE WELD WITH BACKING**

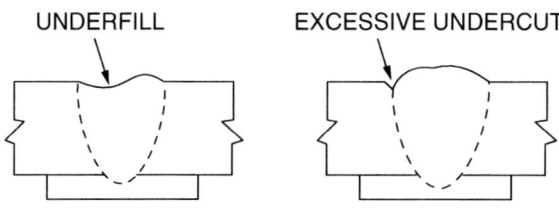

UNDERFILL   EXCESSIVE UNDERCUT

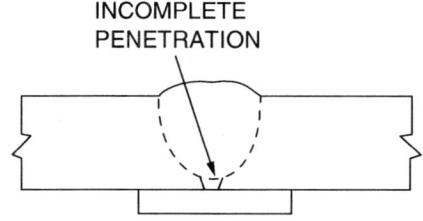

INCOMPLETE
PENETRATION

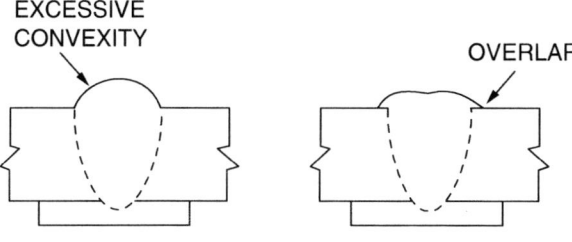

EXCESSIVE
CONVEXITY

OVERLAP

**PROFILES OF UNACCEPTABLE
V-GROOVE WELDS WITH BACKING**

308F27.EPS

Figure 27 ◆ Acceptable and unacceptable V-groove weld profiles.

6.1.0 Flat (1G) Position

Follow these steps to make V-groove welds with backing in the flat (1G) position (*Figure 28*):

Step 1 Tack weld together the practice coupon, following the example given in the section on preparing welding coupons.

Step 2 Position the weld coupon in the flat position on the welding table.

Step 3 Use a 10° to 15° push angle and a 0° work angle to run the root pass, using an appropriate aluminum filler metal.

Step 4 Clean the root pass.

Step 5 Run the remaining passes, using a 5° to 10° push angle and the bead sequence and work angles shown in *Figure 28*. Clean the weld between each pass.

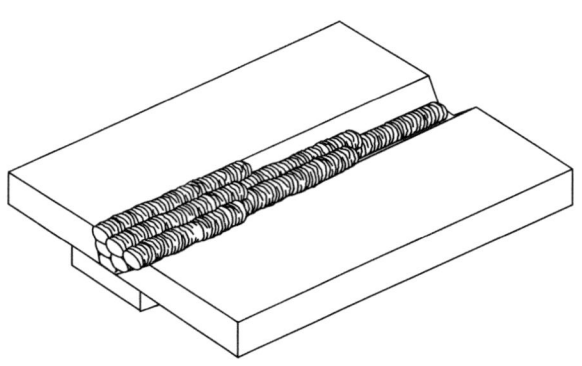

NOTE: THE ACTUAL NUMBER OF WELD BEADS WILL VARY DEPENDING ON THE PLATE THICKNESS

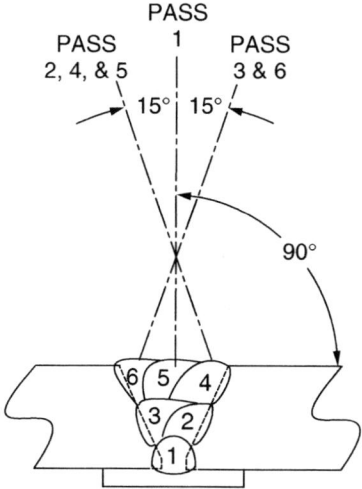

308F28.EPS

Figure 28 ◆ Multiple-pass 1G bead sequence and work angles.

Specific V-Groove Requirements

Refer to your site's WPS for specific requirements on groove welds. The information in this module is provided only as a general guideline. The WPS or site quality standards must be followed for all welds. Check with your supervisor if you are unsure of the specifications for your application.

6.2.0 Horizontal (2G) Position

Follow these steps to practice welding V-groove welds with backing in the horizontal (2G) position (*Figure 29*):

Step 1 Tack weld together the practice coupon, as explained earlier. Use the standard or alternate weld coupon as directed by your instructor.

Step 2 Clamp or tack weld the coupon in the horizontal position.

Step 3 Use a 10° to 15° push angle to run the root pass, using an appropriate aluminum filler metal. Use the work angle for pass 1 that corresponds to the alternate or standard joint shown in *Figure 29*.

Step 4 Clean the weld.

Step 5 Run the remaining passes, using a 5° to 10° push angle and the bead sequences and work angles shown in *Figure 29*. Clean the weld between each pass.

6.3.0 Vertical (3G) Position

Follow these steps to practice V-groove welds with backing in the vertical (3G) position (*Figure 30*):

NOTE

Stringer or weave beads may be used for vertical (3G) position welds, as specified by your instructor, WPS, or site quality standard.

Step 1 Tack weld together the practice coupon, as explained earlier.

Step 2 Clamp or tack weld the coupon in the vertical position.

Step 3 Run the root pass uphill using an appropriate aluminum filler metal and a stringer bead. Use a push angle of 10° to 15° and a 0° work angle for pass 1, as shown in *Figure 30*.

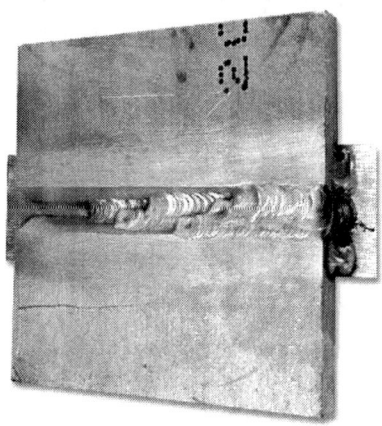

NOTE: THE ACTUAL NUMBER OF WELD BEADS WILL VARY DEPENDING ON THE PLATE THICKNESS

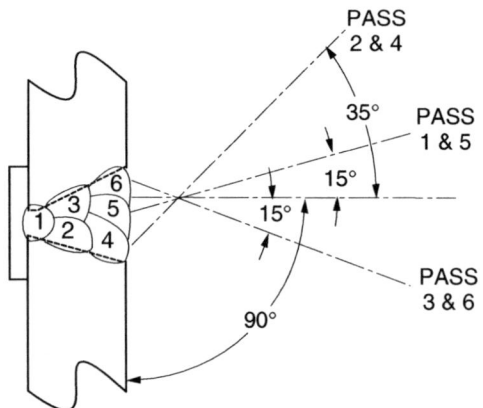

ALTERNATE JOINT REPRESENTATION

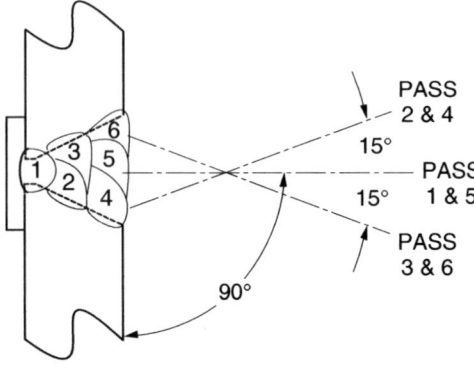

STANDARD JOINT REPRESENTATION

308F29.EPS

Figure 29 ◆ Multiple-pass 2G bead sequences and work angles.

NOTE: THE ACTUAL NUMBER OF WELD
BEADS WILL VARY DEPENDING ON
THE METAL THICKNESS.

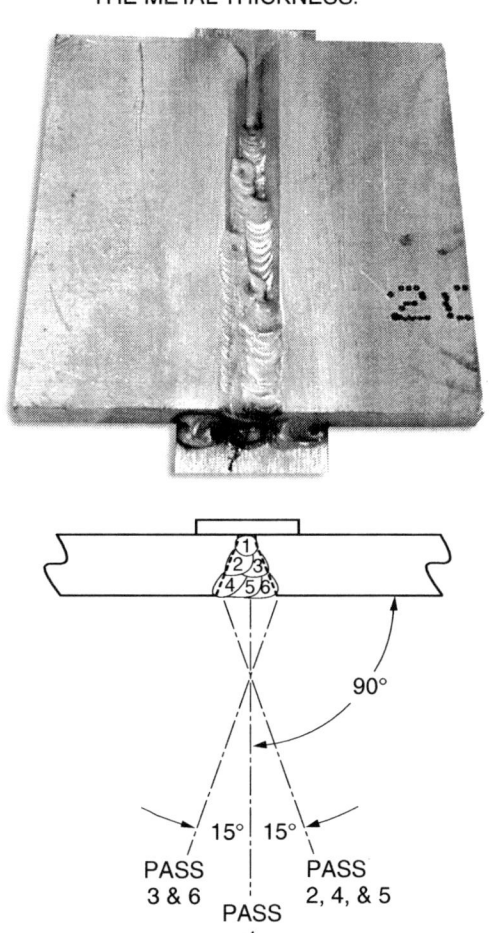

STRINGER BEAD SEQUENCE

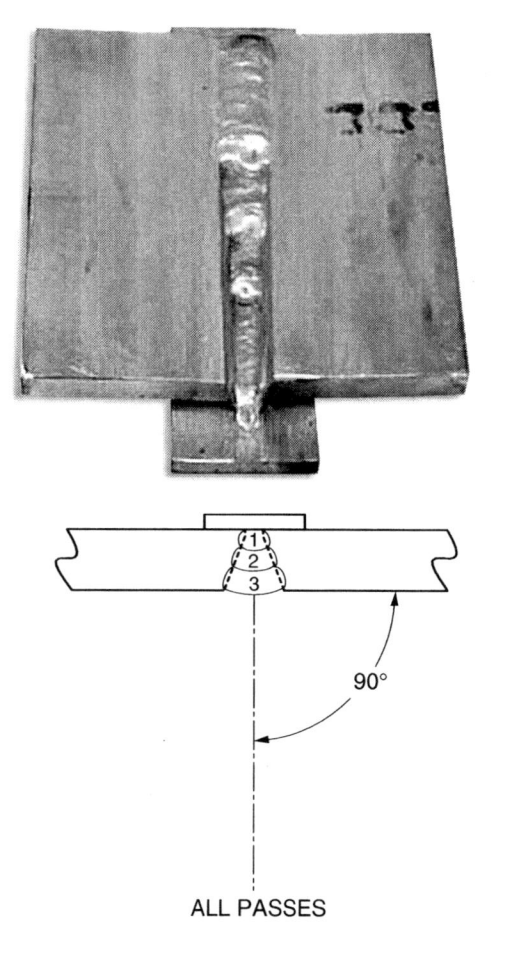

WEAVE BEAD SEQUENCE   308F30.EPS

Figure 30 ◆ Multiple-pass 3G bead sequences and work angles.

Step 4 Clean the weld.

Step 5 Run the remaining passes uphill to complete the weld. Use a 5° to 10° push angle and the bead sequences and work angles shown in *Figure 30* for stringer or weave beads. Clean the weld between each pass.

6.4.0 Overhead (4G) Position

Follow these steps to practice V-groove welds with backing in the overhead (4G) position (*Figure 31*):

Step 1 Tack weld together the practice coupon, as explained earlier.

Step 2 Clamp or tack weld the coupon in the overhead position.

Step 3 Run the root pass using an appropriate aluminum filler metal. Use a 10° to 15° push angle and a 0° work angle for pass 1, as shown in *Figure 31*.

Cooling Practice Coupons

Cooling coupons with water can cause weld cracks and affect the mechanical properties of the base metal. Practice coupons can be cooled with water, but never cool test coupons with water.

Step 4 Clean the weld.

Step 5 Run the remaining passes, using a 5° to 10° push angle and the bead sequence and work angles shown in *Figure 31*. Clean the weld between each pass.

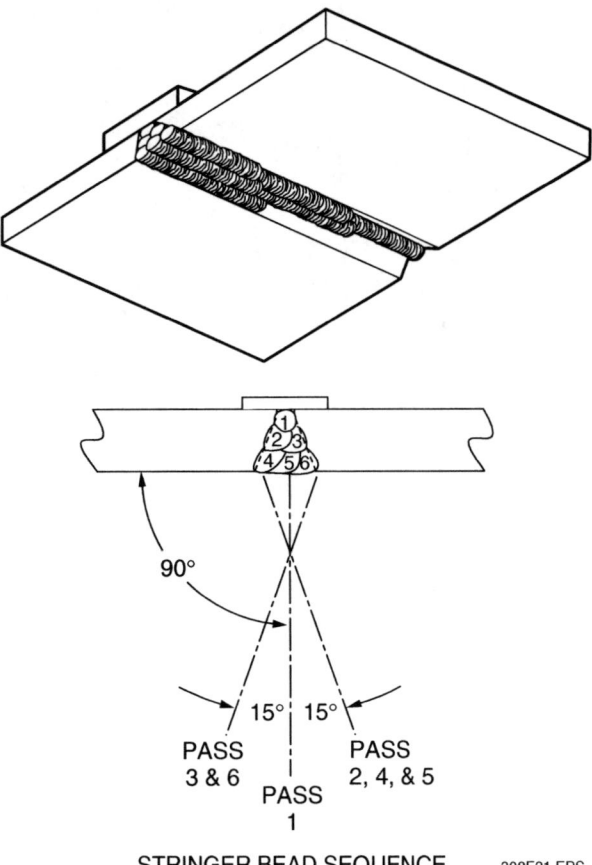

STRINGER BEAD SEQUENCE   308F31.EPS

Figure 31 ◆ Multiple-pass 4G bead sequence and work angles.

7.0.0 ◆ V-GROOVE PIPE WELDS

The V-groove weld with backing is the most common groove weld normally made on aluminum pipe. These welds are typically used for joining pipe used in non-critical piping systems. Critical piping systems, including pipe, fittings, and welded joints, contain or carry material with the potential to cause long- or short-term catastrophic danger and damage to personnel and/or the environment, should the components fail to contain or carry the material as designed. Non-critical piping is low-pressure piping used for heating and air conditioning, simple water systems, and other service installations. Less-stringent code requirements are used to evaluate non-critical piping welds.

> **CAUTION**
>
> When welding pipe, do not make arc strikes outside the weld groove on the surface of the pipe. An arc strike can cause a hardened spot that can crack as the pipe expands and contracts. An arc strike on the pipe surface is considered a defect and will require repair or rework.

Groove welds may be made in all positions on pipe. The weld position is determined by the axis of the pipe. Four standard weld test positions are used with pipe and are shown in *Figure 32*.

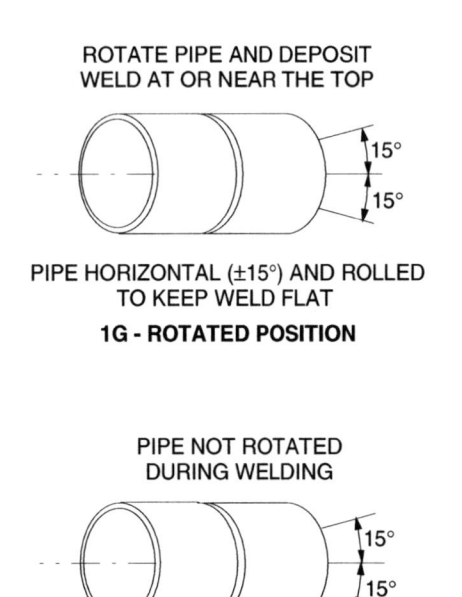

PIPE HORIZONTAL (±15°) AND ROLLED
TO KEEP WELD FLAT
1G - ROTATED POSITION

PIPE NOT ROTATED
DURING WELDING
PIPE HORIZONTAL (±15°)
5G POSITION

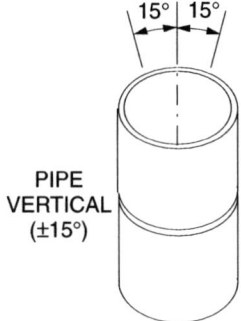

PIPE
VERTICAL
(±15°)

PIPE NOT ROTATED DURING WELDING
2G POSITION

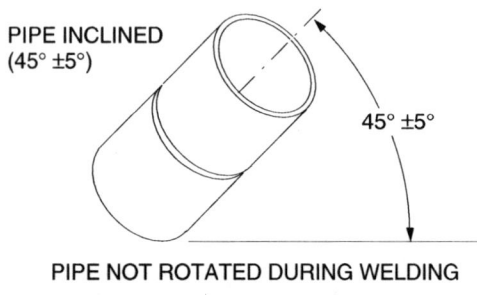

PIPE INCLINED
(45° ±5°)

PIPE NOT ROTATED DURING WELDING
6G POSITION   308F32.EPS

Figure 32 ◆ Four basic pipe groove weld test positions.

To destructively test 5G or 6G welds, the test specimens are cut from the four regions illustrated in *Figure 33*: midway between the 12-o'clock and 3-o'clock, 3-o'clock and 6-o'clock, 6-o'clock and 9-o'clock, and 9-o'clock and 12-o'clock positions. (The 12-o'clock position is the top of the coupon groove; 6-o'clock is the bottom.)

This section of the module explains how to perform the following V-groove pipe weld positions with backing:

- Flat (1G-ROTATED)
- Horizontal (2G)
- Multiple (5G)
- Inclined multiple (6G)

Acceptable and unacceptable V-groove welds with backing are the same for pipe as for plate. These were covered in an earlier section and were shown in *Figure 27*.

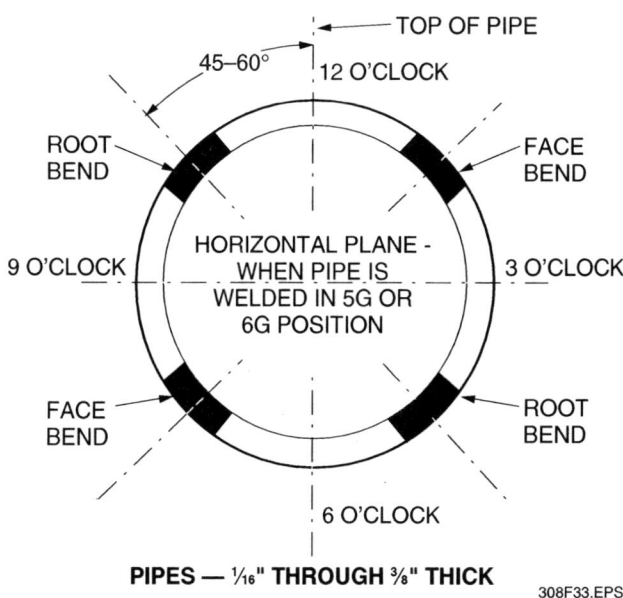

PIPES — ¹⁄₁₆" THROUGH ³⁄₈" THICK

308F33.EPS

Figure 33 ◆ Test specimen regions of 5G or 6G position pipe.

7.1.0 Flat (1G-ROTATED) Position

Practice the 1G-ROTATED position V-groove pipe welds with backing, using a shielding gas and aluminum filler wire with a diameter specified by your instructor. For the root pass, keep the gun angle at 90° (0° work angle) to the pipe axis with a 10° to 15° push angle. Use a slight side-to-side oscillation to run the root pass and to control penetration, paying particular attention to tie into the tack welds. Clean the face of the root pass with a brush.

When running the fill and cover passes, use stringer or weave beads with a slight side-to-side motion to ensure tie-in at the toes of the weld bead. Try to keep the wire at the leading edge of the weld puddle to ensure proper penetration.

Take particular care to fill the crater at the weld termination. Run all passes at or near the top of the pipe as the pipe is rotated.

Follow these steps to practice V-groove pipe welds with backing in the 1G-ROTATED position:

Step 1 Tack weld together the practice pipe weld coupon, as explained earlier.

Step 2 Horizontally position the pipe weld coupon on two sets of rollers at a comfortable welding height. *Figure 34* shows roller supports commonly found in pipe welding shops.

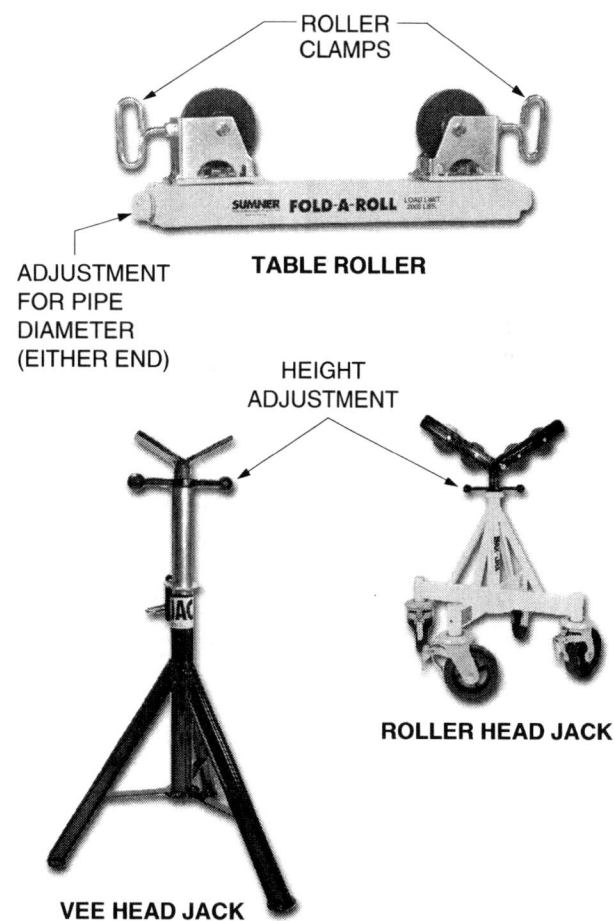

FLOOR STAND ROLLER

308F34.EPS

Figure 34 ◆ Pipe roller supports.

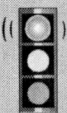
Step 3 Make sure the workpiece clamp is attached directly to the pipe coupon. This will prevent the welding current from passing through the roller bearings or arcing between the rollers and the pipe coupon.

Test Position

Welding pipe in the 6G position is a common welding test. A ring may be added to test for restricted accessibility (6GR).

Fill Pass

Follow these guidelines when running the fill pass:

- Fill craters to preclude cracking.
- Stagger the location of the start and stop spots for each weld.

Powered Pipe Roller

A pipe roller that is operated by an electric motor activated by a foot switch is shown here. These devices are primarily used for large, heavy pipe sections.

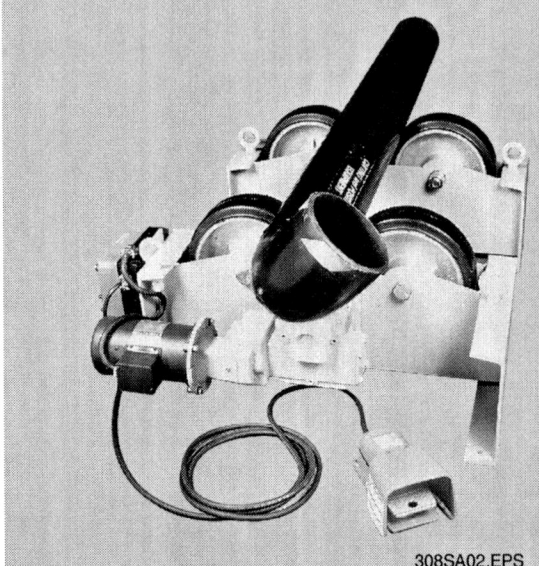

308SA02.EPS

Step 4 Run the root pass using a slight side-to-side oscillation. Position a tack weld at the 11 o'clock position, start the weld bead on the tack weld, and advance toward the 12 o'clock position.

Step 5 Roll the pipe as necessary to keep the weld in the flat position.

Step 6 Brush the root pass to clean the weld.

Step 7 Use the same rolling procedure to make the remaining passes. Use stringer or weave beads as applicable. Pay particular attention at the tie-ins to ensure proper fusion and prevent excess buildup. Overlap the passes (*Figure 35*) and clean the weld after each pass.

7.2.0 Horizontal (2G) Position

Practice 2G position V-groove pipe welds with backing using a shielding gas and an aluminum filler wire with a diameter specified by your instructor. The gun should be at a 10° to 15° push angle and at 90° (0° work angle) to the surface of the pipe for the root pass. To prevent the weld puddle from sagging, the gun work angle can be dropped slightly but not more than 10°. For the root pass, use a slight side-to-side oscillation to control penetration, as shown in *Figure 36*. Pay particular attention to tie into the tack welds.

When running the fill/cover passes (see *Figure 37*), use stringer beads and a slight side-to-side motion to ensure tie-in at the toes of the weld bead. Try to keep the wire electrode at the leading edge of the weld puddle. Take particular care to fill the crater at the weld termination.

Follow these steps to practice V-groove pipe welds with backing in the 2G position:

Step 1 Tack weld together the practice pipe weld coupon, as explained earlier.

Step 2 Clamp or tack weld the pipe coupon with the axis of the pipe vertical.

Step 3 Starting on a tack weld, run the root pass as shown in *Figure 36*. Clean the crater between any necessary restarts. Continue this procedure until the root pass is completed.

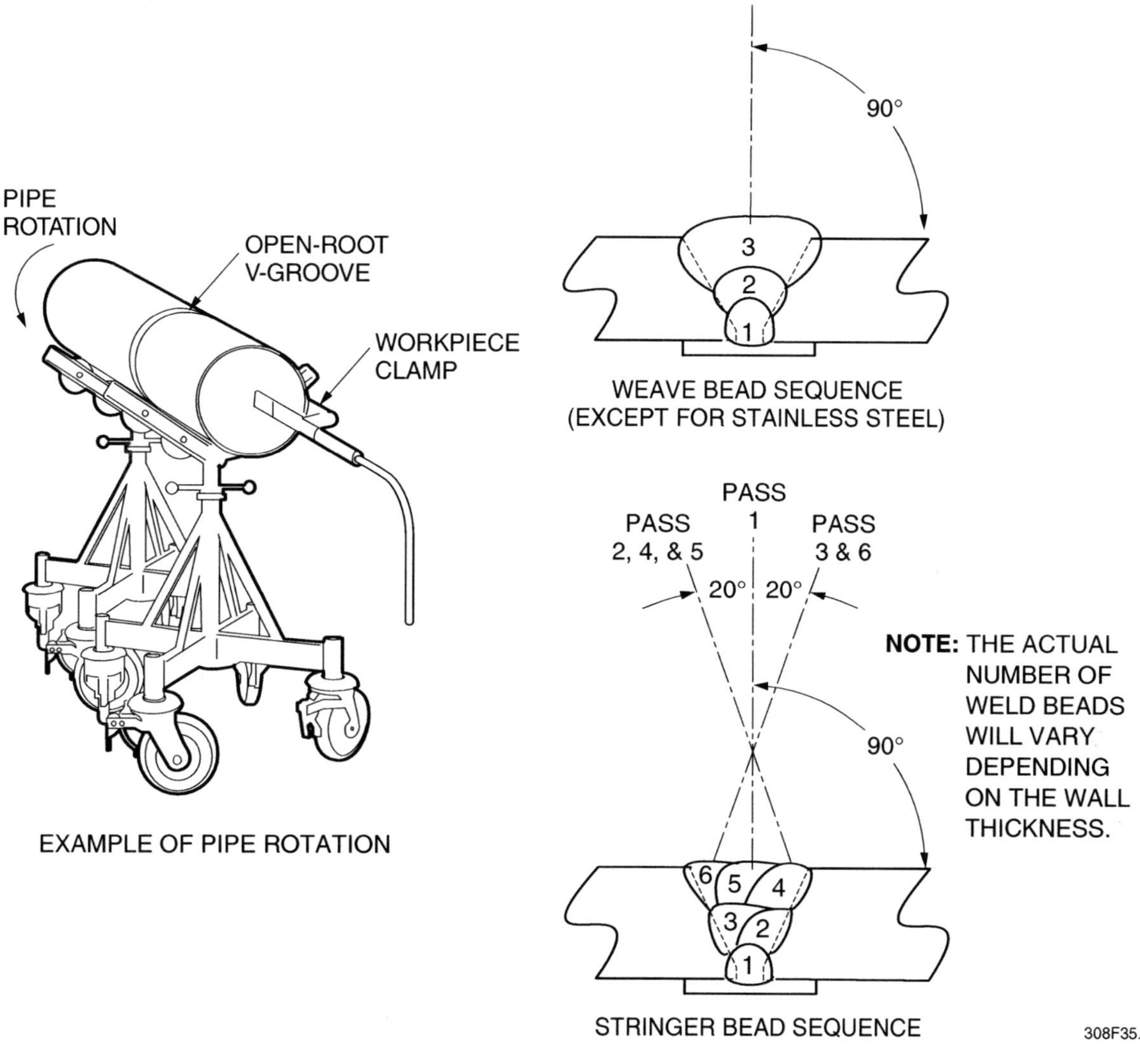

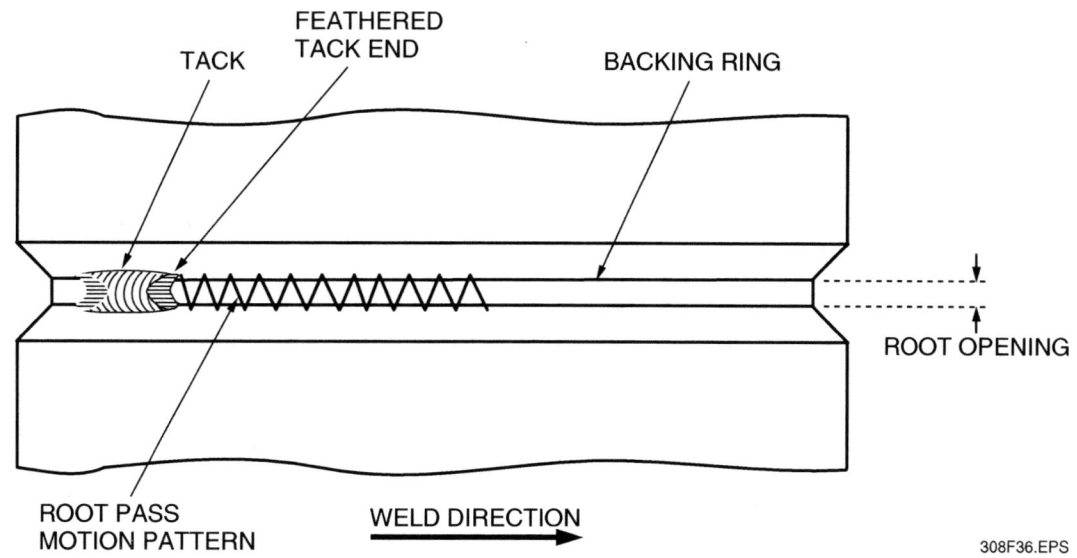

PIPE ROTATION

OPEN-ROOT V-GROOVE

WORKPIECE CLAMP

EXAMPLE OF PIPE ROTATION

90°

3
2
1

WEAVE BEAD SEQUENCE
(EXCEPT FOR STAINLESS STEEL)

PASS 1

PASS 2, 4, & 5

PASS 3 & 6

20° 20°

NOTE: THE ACTUAL NUMBER OF WELD BEADS WILL VARY DEPENDING ON THE WALL THICKNESS.

90°

6 5 4
3 2
1

STRINGER BEAD SEQUENCE

308F35.EPS

Figure 35 ◆ Multiple-pass 1G-ROTATED bead sequences and work angles.

TACK

FEATHERED TACK END

BACKING RING

ROOT OPENING

ROOT PASS MOTION PATTERN

WELD DIRECTION

308F36.EPS

Figure 36 ◆ Root pass motion pattern.

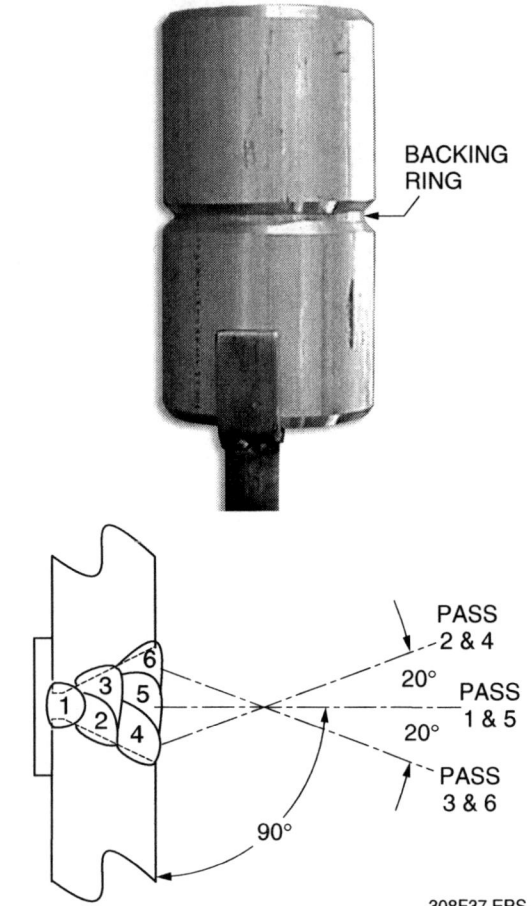

Figure 37 ◆ Multiple-pass 2G bead sequence and work angles.

Step 4 Brush the root pass to clean the weld.

Step 5 Run the remaining passes at the appropriate work angles. Overlap the passes, and clean the weld after each pass.

7.3.0 Multiple (5G) Position

Practice the multiple (5G) position V-groove pipe welds with backing using a shielding gas and an aluminum filler wire with a diameter specified by your instructor. The gun should be at approximately a 10° to 15° push angle for the root, fill, and cover passes, depending on whether your instructor directs you to run downhill or uphill passes. In some cases, the root pass is made downhill and the fill and cover passes are made uphill. However, downhill passes are not recommended for aluminum welds. Refer to *Figure 38* for the travel positions necessary for the passes that are made on both sides of the pipe.

When running the fill and cover passes (see *Figure 39*), use weave beads or stringer beads with a slight side-to-side motion to ensure tie-in at the toes of the weld bead. Try to keep the wire elec-

trode at the leading edge of the weld puddle. Pay particular attention to fill the crater at the weld termination.

Follow these steps to practice V-groove pipe welds with backing in the 5G position:

Step 1 Tack weld together the practice pipe weld coupon, as explained earlier.

Step 2 Clamp or tack weld the pipe weld coupon into position with the pipe axis horizontal.

Step 3 Run the root pass with a stringer bead, modifying the travel angles as necessary (*Figure 38*) for an uphill pass.

Step 4 Brush the root pass to clean the weld.

Step 5 Run the remaining passes as stringer or weave beads in an uphill direction at the appropriate work angles. Overlap the passes and clean the weld after each pass.

7.4.0 Inclined Multiple (6G) Position

Practice the inclined multiple (6G) (45°) position V-groove pipe welds with backing using a shielding gas and an aluminum filler wire with a diameter specified by your instructor. The gun should be at approximately a 10° to 15° push angle for the root, fill, and cover passes, depending on whether your instructor directs you to run downhill or uphill passes. In some cases, the root pass is made downhill and the fill and cover passes are made uphill. However, downhill passes are not recommended for aluminum welds. Refer to *Figure 38* for the travel positions necessary for the passes that are made on both sides of the pipe.

When running the fill and cover passes (see *Figure 40*), use weave beads or stringer beads and a slight side-to-side motion to ensure tie-in at the toes of the weld bead. Try to keep the wire electrode at the leading edge of the weld puddle. Pay particular attention to fill the crater at the weld termination.

Follow these steps to practice V-groove pipe welds with backing in the 6G position:

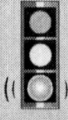 **NOTE**

If required for training purposes, a restricting ring may be added to the 6G position coupon to form a 6GR position coupon.

Step 1 Tack weld together the practice pipe weld coupon, as explained earlier.

Step 2 Clamp or tack weld the pipe weld coupon into position with the pipe axis inclined 45° to the horizontal plane.

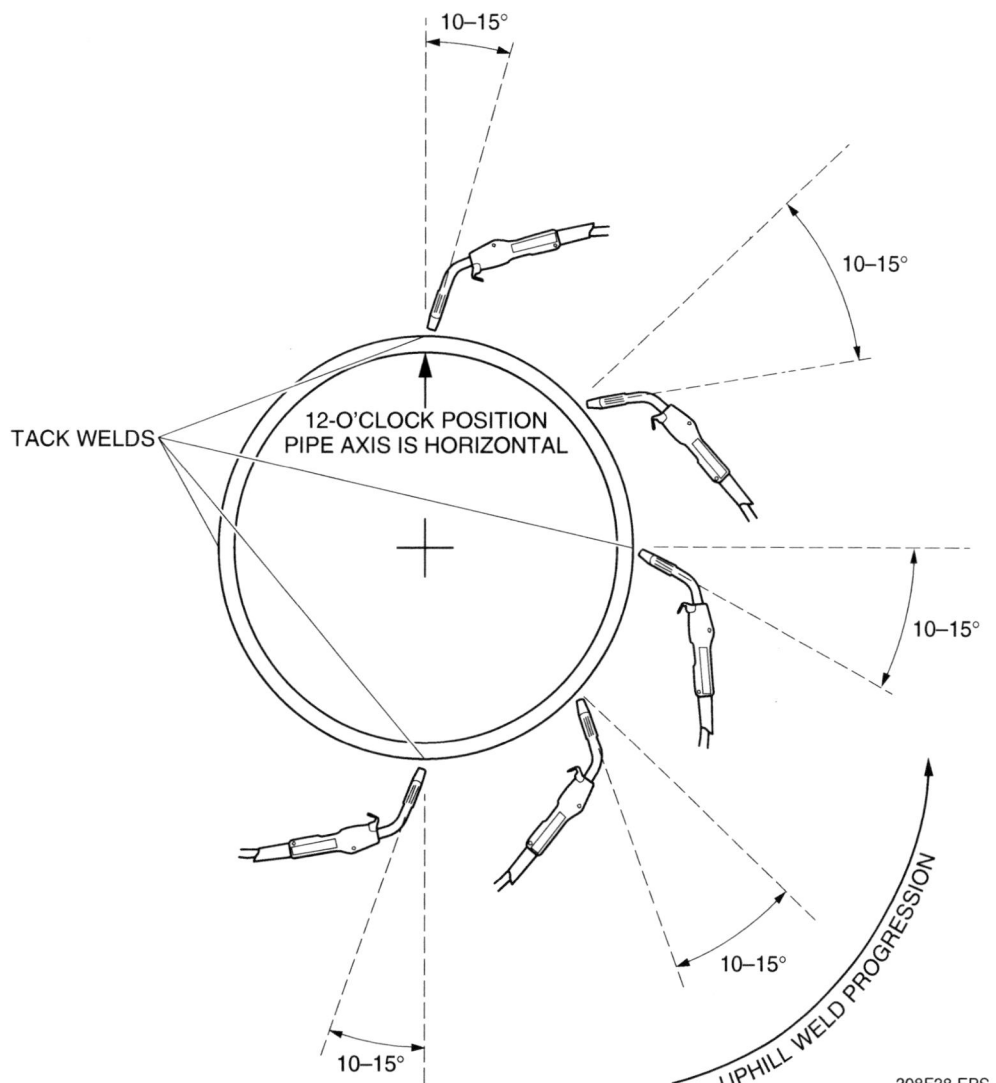

Figure 38 ◆ Multiple-pass weld progression for 5G position.

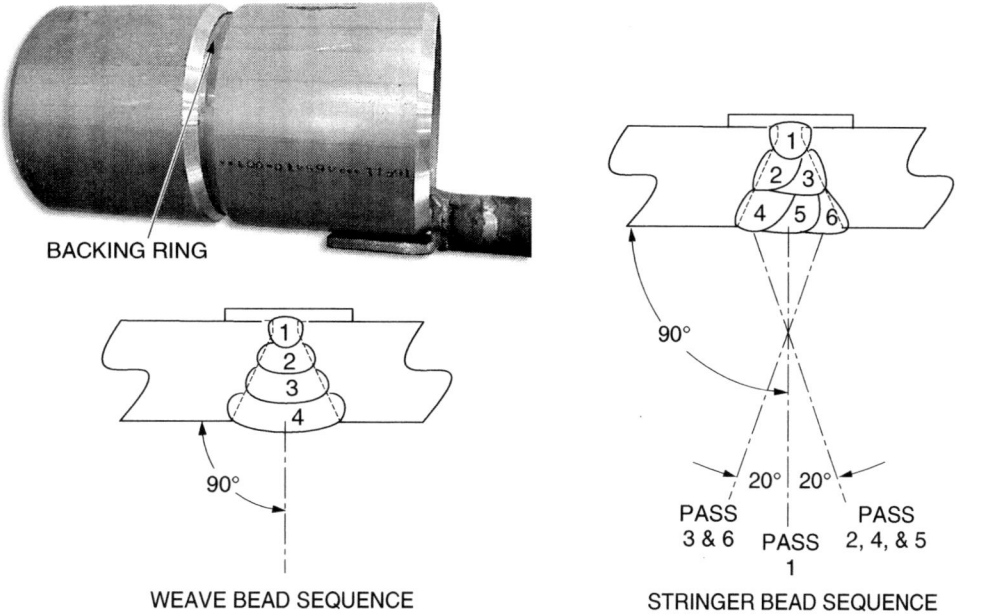

Figure 39 ◆ Multiple-pass 5G bead sequences and work angles.

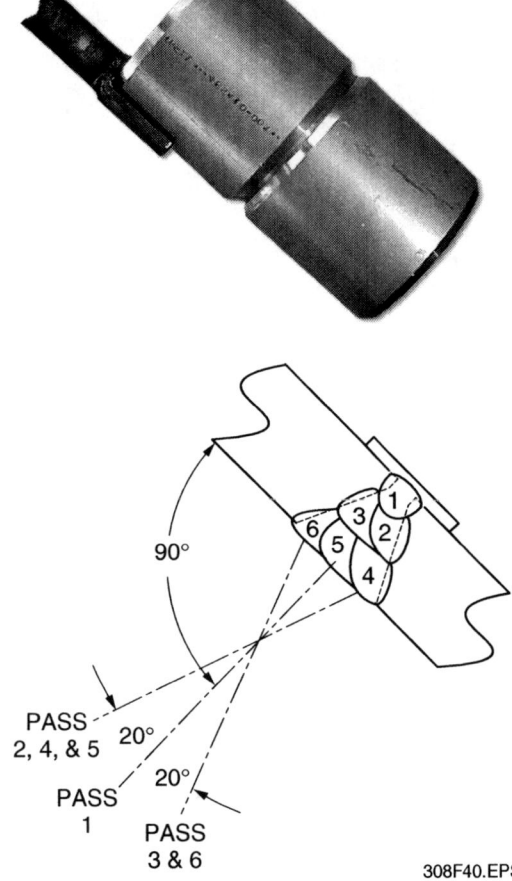

Figure 40 ◆ Multiple-pass 6G bead sequence and work angles.

The V-groove weld is the most common weld joint used for joining pipe used in non-critical piping systems. For aluminum pipe welding, you must be able to set up the equipment, prepare the coupons for welding, and recognize acceptable welds. V-groove pipe welds can be made in the 1G-ROTATED, 2G, 5G, and 6G positions. Practice these and the aluminum plate welds until you can consistently produce acceptable welds as defined in the criteria for acceptance.

Review Questions

1. The GMAW process for aluminum uses _____ for the filler metal.
   a. continuous wire electrode
   b. short lengths of electrode
   c. nickel-coated aluminum rods
   d. tin-coated aluminum rods

2. When welding aluminum, GTAW is often used for the root pass, and _____ is used for the fill and cover passes.
   a. SMAW
   b. FCAW-G
   c. GMAW
   d. FCAW-S

3. To remove oil and grease from aluminum coupons, use a(n) _____.
   a. approved solvent
   b. grinder
   c. stainless steel buffer
   d. welding torch

4. Before welding aluminum pipe, clean the inside and outside of the pipe with a grinder or _____ brush.
   a. aluminum
   b. stainless steel
   c. copper
   d. nonmetallic

5. The GMAW application normally used for aluminum is spray transfer or _____ transfer.
   a. globular
   b. deposit
   c. short-circuiting
   d. pulsed spray

Step 3 Run the root pass with a stringer bead, modifying the travel angles as necessary (*Figure 38*) for an uphill pass.

Step 4 Brush the root pass to clean the weld.

Step 5 Run the remaining passes in an uphill direction at the appropriate work angles. Overlap the passes and clean the weld after each pass.

Summary

Setting up GMAW equipment, preparing the welding work area, running stringer and weave beads, and making acceptable fillet and V-groove welds with backing on aluminum plate in all positions are among the more difficult skills you must develop as a welder. Making V-groove welds with backing on aluminum pipe in all positions is also very difficult.

6. Contact tube extension or setback is usually dependent on the _____.
   a. amperage
   b. distance from the gun nozzle to the work-piece
   c. transfer mode for the GMAW application
   d. voltage

7. If weld spatter that accumulates in the nozzle is not removed, it will restrict the shielding gas flow and cause _____ in the weld.
   a. cold fusion
   b. porosity
   c. cracking
   d. worm holes

8. To clean weld spatter from a gas nozzle, you can use _____.
   a. a grinder
   b. chemical cleaners
   c. a round file
   d. a stainless steel brush

9. When making a weave bead, take care to _____ the base metal.
   a. ensure proper tie-in to
   b. touch the electrode to
   c. avoid oscillations on
   d. avoid pauses on the edges of

10. When making a root pass on a fillet weld, keep the gun at a _____ work angle to the plate surface.
    a. 15°
    b. 45°
    c. 55°
    d. 90°

11. Reinforcement that exceeds _____ is considered excessive _____.
    a. ¹⁄₁₆"; and must be repaired
    b. ⅛"; and must be repaired
    c. ¹⁄₁₆"; but cannot be repaired
    d. ⅛"; but cannot be repaired

12. The most common groove weld normally made on aluminum pipe is the _____.
    a. single-groove weld
    b. bevel-groove weld
    c. open-root V-groove weld
    d. V-groove weld with backing

13. When welding pipe, do not make arc strikes outside the weld groove because it can cause a _____.
    a. thick, rough area that must be filed
    b. keyhole that must be filled
    c. hardened spot that can crack
    d. thin, weak spot that might expand

14. When welding pipe in the 1G-ROTATED position, be sure to keep the filler metal in the _____ of the weld puddle to ensure proper penetration.
    a. overlapping side edge
    b. trailing edge
    c. leading edge
    d. center

15. When welding aluminum, avoid using _____ passes.
    a. overhead
    b. uphill
    c. downhill
    d. horizontal

Performance Accreditation Tasks

The Performance Accreditation Tasks (PATCs) correspond to and support learning objectives in the *AWS Guide for the Training and Qualification of Welding Personnel.*

PATCs provide specific acceptable criteria for performance and help to ensure a true competency-based welding program for students.

The following tasks are designed to evaluate your ability to run V-groove welds with GMAW equipment in four standard test positions using aluminum filler wire of the appropriate diameter and shielding gas. Perform each task when you are instructed to do so by your instructor. As you complete each task, show it to your instructor for evaluation. Do not proceed to the next task until told to do so by your instructor. For AWS 2G and 5G certifications, refer to *AWS EG3.0-96* for bend test requirements. For AWS 6G certifications, refer to *AWS EG 4.0-96* for bend test requirements.

V-GROOVE WELDS WITH BACKING ON PLATE IN THE HORIZONTAL POSITION

Using aluminum filler wire of the appropriate diameter, proper shielding gas, and stringer beads, make a V-groove weld with backing on plate in the horizontal (2G) position, as shown.

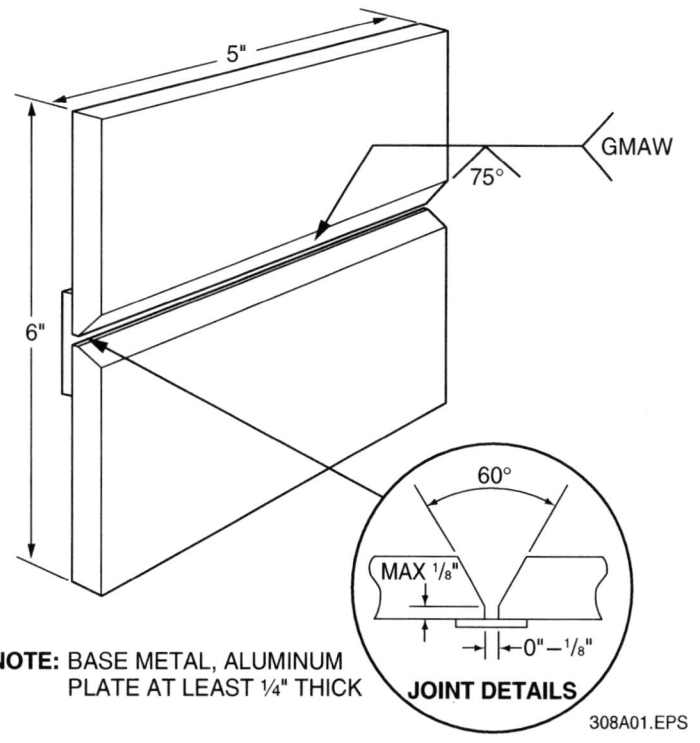

5"

GMAW

75°

6"

60°

MAX 1/8"

0" – 1/8"

NOTE: BASE METAL, ALUMINUM
PLATE AT LEAST 1/4" THICK

JOINT DETAILS

308A01.EPS

Criteria for Acceptance

• Uniform rippled appearance on the bead face _____

• Craters and restarts filled to the full cross section of the weld _____

• Uniform weld size ±1/16" _____

• Acceptable weld profile in accordance with the *ASME Boiler and Pressure Vessel Code* _____

• Smooth transition with complete fusion at the toes of the weld _____

• No porosity _____

• No overlap _____

• No excessive undercut _____

• No inclusions _____

• No cracks _____

• No incomplete fusion _____

V-GROOVE WELDS WITH BACKING ON PLATE IN THE VERTICAL POSITION

Using aluminum filler wire of the appropriate diameter, proper shielding gas, and stringer beads, make a V-groove weld with backing on plate in the vertical (3G) position, as shown.

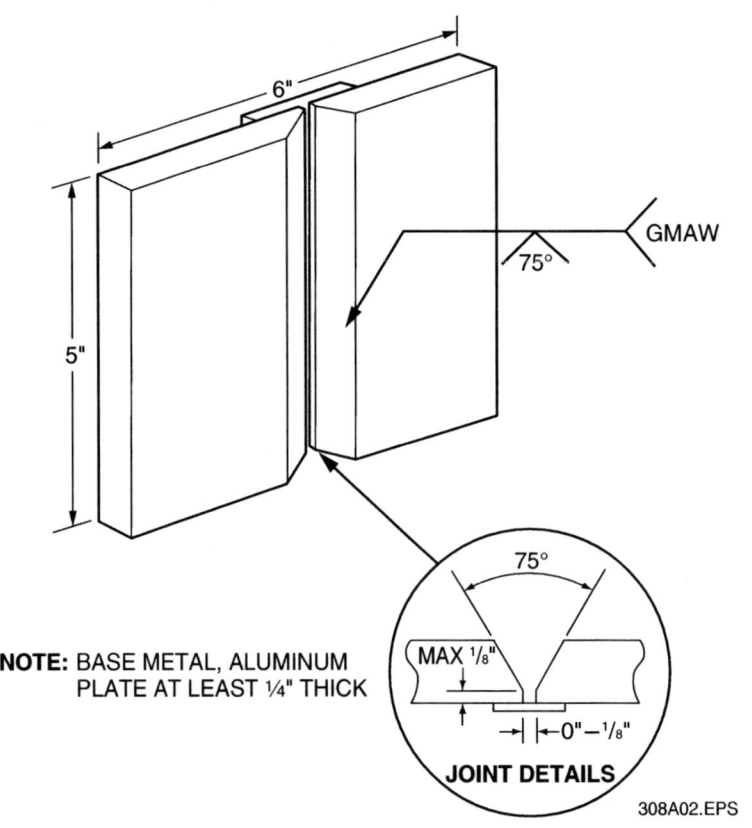

NOTE: BASE METAL, ALUMINUM PLATE AT LEAST ¼" THICK

JOINT DETAILS

308A02.EPS

Criteria for Acceptance

- Uniform rippled appearance on the bead face　　　　　　_____

- Craters and restarts filled to the full cross section of the weld　　　_____

- Uniform weld size ±¹⁄₁₆"　　　　　　　　　　　　　　　_____

- Acceptable weld profile in accordance with the
  ASME Boiler and Pressure Vessel Code　　　　　　　　　_____

- Smooth transition with complete fusion at the toes of the weld　　_____

- No porosity　　　　　　　　　　　　　　　　　　　　　_____

- No overlap　　　　　　　　　　　　　　　　　　　　　_____

- No excessive undercut　　　　　　　　　　　　　　　　_____

- No cracks　　　　　　　　　　　　　　　　　　　　　　_____

- No incomplete fusion　　　　　　　　　　　　　　　　　_____

V-GROOVE WELDS WITH BACKING ON PLATE IN THE OVERHEAD POSITION

Using aluminum filler wire of the appropriate diameter, proper shielding gas, and stringer beads, make a V-groove weld with backing on plate in the overhead (4G) position, as shown.

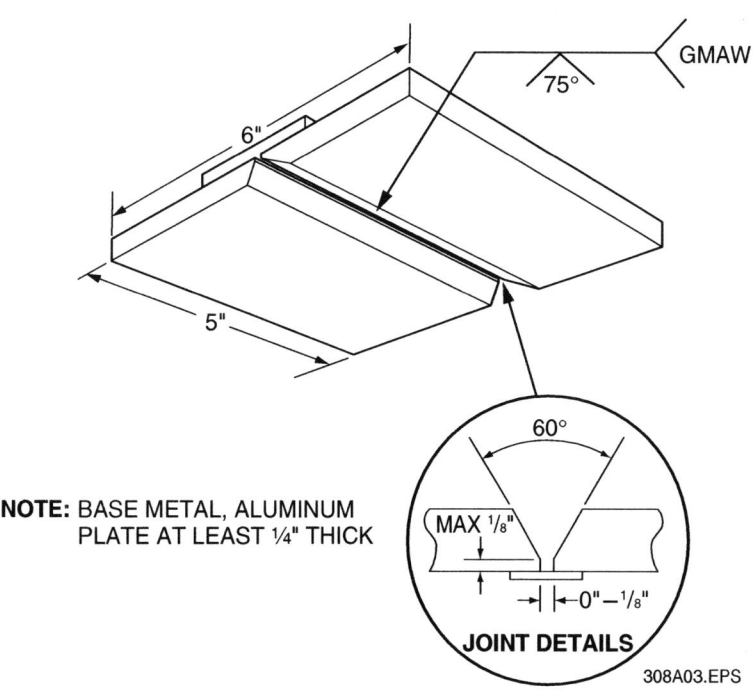

NOTE: BASE METAL, ALUMINUM PLATE AT LEAST ¼" THICK

JOINT DETAILS

308A03.EPS

Criteria for Acceptance

- Uniform rippled appearance on the bead face   _____
- Craters and restarts filled to the full cross section of the weld   _____
- Uniform weld size ±¹⁄₁₆"   _____
- Acceptable weld profile in accordance with the
  ASME Boiler and Pressure Vessel Code   _____
- Smooth transition with complete fusion at the toes of the weld   _____
- No porosity   _____
- No overlap   _____
- No excessive undercut   _____
- No cracks   _____
- No incomplete fusion

V-GROOVE PIPE WELD WITH BACKING IN THE 6G (OR 6GR) POSITION

Using aluminum filler wire of the appropriate diameter, proper shielding gas, and stringer beads, make a V-groove weld on aluminum pipe in the 6G (or 6GR) position, as shown.

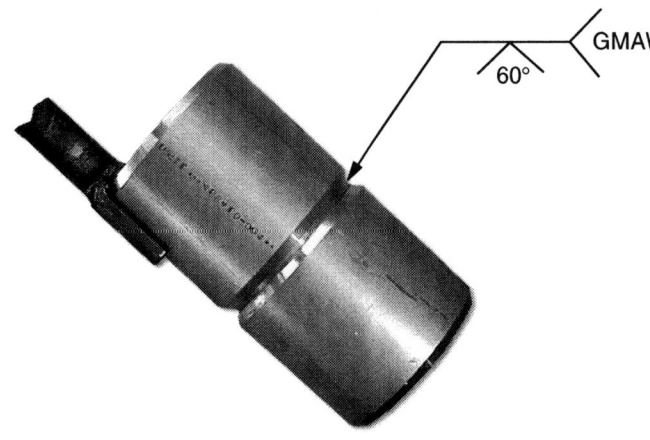

NOTE: IF REQUIRED FOR QUALIFICATION PURPOSES, A RESTRICTING RING MAY BE ADDED TO THE 6G POSITION COUPON TO FORM A 6 GR POSITION COUPON.

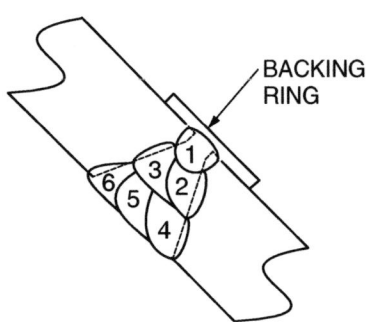

308A04.EPS

Criteria for Acceptance

- Uniform appearance on the bead face   _____

- Craters and restarts filled to the full cross section of the weld   _____

- Uniform weld width ±¹⁄₁₆"   _____

- Acceptable weld profile and guided bend test in accordance with the
  ASME Boiler and Pressure Vessel Code, Section IX   _____

- Smooth transition with complete fusion at the toes of the weld   _____

- No porosity   _____

- No excessive undercut   _____

- No cracks   _____

- No overlap   _____

- No incomplete fusion   _____

Additional Resources

This module is intended to be a thorough resource for task training. The following reference works are suggested for further study. These are optional materials for continued education rather than for task training.

MIG Welding Handbook (P/N 791F18 F-3690-E). Florence, SC: L-TEC Welding and Cutting Systems.

The Procedure Handbook of Arc Welding. Cleveland, OH: The Lincoln Electric Company.

Welding Aluminum: Theory and Practice. Washington, DC: The Aluminum Association.

Welding Pressure Pipe Lines and Piping Systems. Cleveland, OH: The Lincoln Electric Company.

Welding Skills. R.T. Miller. Homewood, IL: American Technical Publishers, Inc.

Welding Technology. Giachino, Weeks, and Johnson. Homewood, IL: American Technical Publishers, Inc.

Figure Credits

The NCCER makes every effort to keep these textbooks up-to-date and free of technical errors. We appreciate your help in this process. If you have an idea for improving this textbook, or if you find an error, a typographical mistake, or an inaccuracy in NCCER's Contren® textbooks, please write us, using this form or a photocopy. Be sure to include the exact module number, page number, a detailed description, and the correction, if applicable. Your input will be brought to the attention of the Technical Review Committee. Thank you for your assistance.

Instructors – If you found that additional materials were necessary in order to teach this module effectively, please let us know so that we may include them in the Equipment/Materials list in the Instructor's Guide.

Write: Curriculum Revision and Development Department
National Center for Construction Education and Research
P.O. Box 141104, Gainesville, FL 32614-1104

Fax: 352-334-0932

E-mail: curriculum@nccer.org

Craft _____ Module Name _____

Copyright Date _____ Module Number _____ Page Number(s) _____

Description _____

_____

_____

_____

_____

(Optional) Correction _____

_____

_____

(Optional) Your Name and Address _____

_____

_____

_____

Index

Index

Welds, *continued*
  FCAW, 4.11–4.18
  GMAW, 3.10–3.18, 8.20–8.30
  GTAW, 5.12–5.20, 6.14–6.19, 7.7–7.13
  overview, 3.10–3.13
  on plate, 8.20–8.30
Wind resistance, 4.1
Wire. *See* Electrodes, extension; Wire feeder
Wire feeder, 3.8, 4.8, 8.9, 8.10
Work angle
  for GMAW of aluminum, 3.8, 8.11, 8.18, 8.29
  for GTAW
    of aluminum, 7.7, 7.8, 7.11
    of carbon steel, 5.10, 5.18
    of low-alloy and stainless steel, 6.9, 6.17
Work area
  preparation for welding, 3.3, 4.4, 5.2–5.3, 6.2–6.3,
    7.2–7.3, 8.3
  ventilation. *See* Ventilation
WPS. *See* Welding procedure specifications
Wrought iron, 2.1, 2.45

X-ray fluorescence (XRF), 2.23

Y-adapter, 5.9, 6.11
Yield strength, 2.27

Zinc
  in nonferrous alloys, 2.13, 2.15, 2.18, 2.20
  properties, 2.24, 2.25
Zirconium, 2.15, 2.16